Encyclopaedia of Mathematical Sciences

Volume 72

Editor-in-Chief: R. V. Gamkrelidze

Springer
Berlin
Heidelberg
New York
Barcelona
Budapest
Hong Kong
London
Milan
Paris
Tokyo

V. P. Havin · N. K. Nikol'skij (Eds.)

Commutative Harmonic Analysis III

Generalized Functions. Applications

With 34 Figures

Springer

Title of the Russian edition:
Itogi nauki i tekhniki, Sovremennye problemy matematiki,
Fundamental'nye napravleniya, Vol. 72. Kommutativnyi garmonicheskij analiz 3,
Publisher VINITI, Moscow 1991

Cataloging-in-Publication Data applied for

Die Deutsche Bibliothek - CIP- Einheitsaufnahme
Commutative harmonic analysis / V.P. Khavin; N.K. Nikol'skij (ed.). - Berlin; Heidelberg; New York;
London; Paris; Tokyo; Hong Kong; Barcelona; Budapest: Springer.
Einheitssacht.: Kommutativnyi garmoničeskij analiz (engl.) Literaturangaben
NE: Chavin, Viktor P. (Hrsg); EST
3. Generalized functions, applications. - 1995
(Encyclopaedia of mathematical sciences; Vol. 72)
ISBN 3-540-57034-9 (Berlin ...)
ISBN 0-387-57034-9 (New York ...)
NE: GT

Mathematics Subject Classification (1991): 42-02, 44-02, 78A97, 78-02

ISBN 3-540-57034-9 Springer-Verlag Berlin Heidelberg New York
ISBN 0-387-57034-9 Springer-Verlag New York Berlin Heidelberg

Printed in Germany
Typesetting: Camera-ready copy from the translator using a Springer $\TeX$ macro package
SPIN: 10031231 41/3140 - 5 4 3 2 1 0 - Printed on acid-free paper

List of Editors, Authors and Translators

Editor-in-Chief

R. V. Gamkrelidze, Russian Academy of Sciences, Steklov Mathematical Institute, ul. Vavilova 42, 117966 Moscow, Institute for Scientific Information (VINITI), ul. Usievicha 20a, 125219 Moscow, Russia, e-mail: gam@ips.eps.msk.su

Consulting Editors

V. P. Havin, St. Petersburg State University, Department of Mathematics, Bibliotechnaya pl. 2, 198904 St. Petersburg, Staryj Peterhof, Russia, e-mail: havin@math.lgu.spb.su and Department of Mathematics and Statistics, Mc Gill University, 805 Sherbrooke Street West, Montreal QC, Canada H3A 2K6 e-mail: havin@math.mcgill.ca

N. K. Nikol'skij, Steklov Mathematical Institute, ul. Fontanka 27, 191011 St. Petersburg, Russia and Département de Mathématiques, Université de Bordeaux I, 351, Cours de la Liberation, 33405 Talence, Cedex, France, e-mail: nikolski@ceremab.u-bordeaux.fr

Authors

V. S. Buslaev, Department of Physics, St. Petersburg State University, Bibliotechnaya pl. 2, 198904 St. Petersburg, Staryj Peterhof, Russia, e-mail: buslaev@imi.spb.su

V. P. Havin, St. Petersburg State University, Department of Mathematics, Bibliotechnaya pl. 2, 198904 St. Petersburg, Staryj Peterhof, Russia, e-mail: havin@math.lgu.spb.su

B. Jöricke, Arbeitsgruppe „Algebraische Geometrie und Zahlentheorie, Max-Planck-Gesellschaft zur Förderung der Wissenschaften e.V., Humboldt-Universität Berlin, Jägerstraße 10-11, 10117 Berlin, Germany, e-mail: joericke@zahlen.ag-berlin.mpg.de

V. P. Palamodov, Mathematical College, Independent Moscow University, Fotievoy 18, Moscow, Russia

Translator

R. Cooke, Department of Mathematics and Statistics, University of Vermont, Burlinton, VT 05405, USA, e-mail: cooke@emba.uvm.edu

Contents

I. Distributions and Harmonic Analysis
V. P. Palamodov
1

II. Optical and Acoustic Fourier Processors
V. S. Buslaev
129

III. The Uncertainty Principle in Harmonic Analysis
V. P. Havin and B. Jöricke
177

Index
261

I. Distributions and Harmonic Analysis

V. P. Palamodov

Translated from the Russian
by Roger Cooke

Contents

Introduction 3

Chapter 1. The Elementary Theory 5

§1. Test Functions and Distributions on the Line 5
§2. Localization and the Structure of Generalized Functions 7
§3. The Finite Parts of Divergent Integrals 11

Chapter 2. The General Theory 15

§1. Generalized Functions and Distributions on a Manifold 15
§2. De Rham Currents 21
§3. Inverse and Direct Images 23
§4. Partial Smoothness of Generalized Functions 29
§5. The Wave Front 34

Chapter 3. The Fourier Transform 38

§1. Elementary Results 38
§2. The Convolution of Distributions 42
§3. The Fourier Transform of Distributions of Unrestricted Growth .. 46
§4. The Fourier Transform of Ultradistributions and Smooth Functions 52
§5. The Radon Transform 57

Chapter 4. Special Problems 60

§1. The Inverse Image of Generalized Functions in the Presence of Critical Points 60
§2. The Division Problem 67
§3. Translation-Invariant Spaces of Generalized Functions 71
§4. Causal Generalized Functions 78
§5. Multiplication of Generalized Functions 81

Chapter 5. Contact Structures and Distributions 83

Introduction 83

§1. Contact Structures 84
§2. Distributions Connected with a Wave Manifold 87
§3. Versal Wave Manifolds 96
§4. The Geometry of Wave Fronts and Singularities of Distributions . 101
§5. Versal Integrals 114

Commentary on the References 120

References 121

Introduction

The theory of generalized functions is a general method that makes it possible to consider and compute divergent integrals, sum divergent series, differentiate discontinuous functions, perform the operation of integration to any complex power and carry out other such operations that are impossible in classical analysis. Such operations are widely used in mathematical physics and the theory of differential equations, where the ideas of generalized functions first arose, in other areas of analysis and beyond.

The point of departure for this theory is to regard a function not as a mapping of point sets, but as a linear functional defined on smooth densities. This route leads to the loss of the concept of the value of function at a point, and also the possibility of multiplying functions, but it makes it possible to perform differentiation an unlimited number of times. The space of generalized functions of finite order is the minimal extension of the space of continuous functions in which coordinate differentiations are defined everywhere. In this sense the theory of generalized functions is a development of all of classical analysis, in particular harmonic analysis, and is to some extent the perfection of it. The more general theories of ultradistributions or generalized functions of infinite order make it possible to consider infinite series of generalized derivatives of continuous functions. However the possibility of a well-defined summation of such series turns out to be connected with the rate of growth of these continuous functions at infinity – a fact that has no analog in classical analysis.

In the half-century during which the theory of generalized functions has existed its concepts and language have become irreplaceable in the majority of divisions of analysis and in many applied areas. The achievements of the theory were preceded by a long prehistory, in which Cauchy, Riemann, Kirchhoff, Hadamard, Bochner, de Rham, Dirac, and many other mathematicians and physicists made contributions (cf. the historical survey of Lützen (1982) and also the commentary to (Sobolev 1988)). The basic concepts of the theory of generalized functions appeared in their modern form in the work of Sobolev (1936) as a new method of studying a classical problem of analysis. The achievement of L. Schwartz (1955) was to develop and enrich the methods of this theory, thereby greatly extending the possibility for applying it. Of particular importance was the unification of the theory of generalized functions and harmonic analysis on $\mathbb{R}^n$, and subsequently also harmonic analysis on manifolds (Fourier integrals).

At present the number of papers connected with the theory of generalized functions is too large to be encompassed in a single survey. The principle of selection of material for the present survey was to encompass first of all the internal questions of the theory connected with the general problems of analysis. The result of this selection, of course, cannot help but reflect the author's own tastes.

To achieve an acceptable exposition of heterogeneous, sometimes contradictory material, the author was compelled to conduct a certain unification of terminology and notation. In particular, contrary to the traditional confusion the distinction between generalized functions and distributions is maintained, which makes it possible to develop the theory for any smooth manifold, in particular, to define direct and inverse images. The essential difference between generalized functions and distributions is similar to that between cohomology and homology: one may interchange these last two if a canonical class is fixed.

The first chapter of the survey is designed as an elementary introduction to the subject. The second contains a survey of certain little-known parts of the theory, mainly geometrical questions. The third and fourth chapters are devoted to harmonic analysis and its applications, which are one of the basic areas of the theory of generalized functions. This area has a rich literature and is therefore represented in our survey in relatively compressed form. In the fifth chapter the author has attempted to give a coherent exposition of the theory of special generalized functions that arise in hyperbolic problems of mathematical physics and in problems of integral geometry. The subject is nonclassical special functions whose construction is based on the ideas of contact geometry and theory of deformations.

Chapter 1
The Elementary Theory

§1. Test Functions and Distributions on the Line

1.1. The Space of Test Functions $\mathcal{D} = \mathcal{D}(\mathbb{R}^1)$. The space $\mathcal{D} = \mathcal{D}(\mathbb{R}^1)$ consists of the infinitely differentiable complex-valued functions of compact support on the line. The function φ is said to have compact support if there exists a number R such that $\varphi(x) = 0$ if $|x| > R$. The set $\mathcal{D}$ is a vector space over the field $\mathbb{C}$ of complex numbers. A sequence $\varphi_k \in \mathcal{D}$, $k = 1, 2, \ldots$, is said to *converge* to a function $\varphi \in \mathcal{D}$ if the following two conditions hold:

1) for every i, $\varphi_k^{(i)} \Rightarrow \varphi^{(i)}$, where $\Rightarrow$ denotes uniform convergence and $\varphi^{(i)} = \dfrac{d^i\varphi}{dx^i}$;
2) there exists a number R such that $\varphi_k(x) = 0$ for $|x| > R$ for all k. Convergent sequences preserve the linear operations.

A *distribution* on the line is any continuous linear functional $u : \mathcal{D} \to \mathbb{C}$. Continuity of such a linear functional means that $u(\varphi_k) \to u(\varphi)$ if $\varphi_k \to \varphi$. The distributions form a vector space denoted $\mathcal{D}'$. Weak convergence $u_k \to u$ is defined in $\mathcal{D}'$ to mean that $u_k(\varphi) \to u(\varphi)$ for each test function φ.

Consider the space $\mathcal{K} = \mathcal{K}(\mathbb{R})$ of smooth densities with compact support on the line. If we choose a coordinate function x on $\mathbb{R}$, then any such density ρ can be written in the form $\rho = \varphi\, dx$, where $\varphi \in \mathcal{D}$ and vice versa. Therefore there is an isomorphism $\mathcal{K} \cong \mathcal{D}$, which we use to define a convergence in $\mathcal{K}$. Any continuous linear functional $v : \mathcal{K} \to \mathbb{C}$ is called a *generalized function.* Hence we can write $u = v\,|dx|$ for any distribution u with some generalized function v and vice versa.

1.2. Ordinary Functions as Generalized Functions. An ordinary function is defined as any measurable locally integrable function on the line. Local integrability means that the restriction of this function to any compact set $K \subset \mathbb{R}$ is an integrable function with respect to Lebesgue measure. To such a function f one can assign the functional $[f] : \mathcal{K} \to \mathbb{C}$ acting according to the rule

$$[f](\rho) \equiv \int_{-\infty}^{\infty} f(x)\rho(x)\, dx\,. \tag{1.1.1}$$

This functional is obviously linear and continuous, i.e., it is a generalized function. Thus a linear mapping $L \to \mathcal{D}'$ is defined, where L is the space of ordinary functions on the line. When this is done, two functions f and g that differ only on a set of measure zero give the same functionals $[f] = [g]$. Conversely, if $[f] = [g]$, then $f = g$ almost everywhere. The image of L in $\mathcal{D}'$ is dense with respect to weak convergence. Indeed every generalized function is the weak limit of a sequence of infinitely differentiable functions (cf. Sect. 1.3). The space $\mathcal{D}'$ is complete with respect to weak convergence.

1.3. Operations on Generalized Functions. All the operations we shall consider satisfy the following compatibility principle: on the image of the space L (or a dense portion of it) they agree with analogous operations from ordinary analysis. We now list the fundamental operations:

I) addition of generalized functions and multiplication of generalized functions by a complex number;

II) multiplication of a generalized function by an infinitely differentiable function h:

$$hu(\varphi) \equiv u(h\varphi) .$$

This formula defines a generalized function hu, since the operator $\varphi \mapsto h\varphi$ in $\mathcal{D}$ is continuous, i.e., it maps convergent sequences into convergent sequences.

III) Differentiation

$$u'(\varphi) \equiv -u(\varphi') , \quad u^{(i)}(\varphi) \equiv (-1)^i u(\varphi^{(i)}) .$$

Thus the action of any linear differential operator with infinitely differentiable coefficients is defined in the space $\mathcal{D}'$.

1.4. Generalized Functions Defined on an Open Set $U \subset \mathbb{R}$. The space $\mathcal{D}(U)$ of test functions is by definition the subspace of $\mathcal{D}$ consisting of functions φ satisfying the condition that $\varphi = 0$ outside some compact set $K \subset U$. Convergent sequences $\{\varphi_k\} \subset \mathcal{D}(U)$ are distinguished by two conditions, the first of which coincides with 1) of Sect. 1.1 and the second of which can be written as follows:

2) There exists a compact set $K \subset U$ such that $\varphi_k(x) = 0$ for $x \in \mathbb{R}^1 \setminus K$ and any k. Continuous linear functionals $\mathcal{D}(U) \to \mathbb{C}$ are called *distributions* on U and functionals on $\mathcal{K}(U)$ are called *generalized functions* on U.

IV) The operation of restriction of generalized functions defined on U to an open subset V. By definition this operation is defined as the restriction of the functional $\mathcal{D}(U) \to \mathbb{C}$ to the subspace $\mathcal{D}(V)$.

V) Change of variables. Let $a : U \to V$ be a diffeomorphism of class $\mathbb{C}^\infty$ between the open subsets $U, V \subset \mathbb{R}^1$ and v a generalized function on V. The action of the diffeomorphism on v is defined by the formula

$$a^*(v)(\varphi) = v(\psi) , \qquad (1.1.2)$$

where $\varphi(x) = \psi(a(x))|a'(x)|$. This last relation can be written as follows: $\psi(y)|dy| = \varphi(x)|dx|$, where $y = a(x)$. Thus under changes of variables the elements of the test space transform as densities, i.e., expressions of the form $\varphi(x)|dx|$.

1.5. The Support of a Generalized Function. The *support* of a continuous function f is defined as the closure of the set of points where it is different from zero. In the case $f \in L$ the support of f, denoted $\operatorname{supp} f$, is the smallest closed set outside which $f = 0$ almost everywhere. A generalized function u is said to be zero on an open subset V if its restriction to V is the zero linear functional. Consider now the union W of all open subsets $V \subset U$ on which the generalized function u equals zero. On the set W this generalized function is also zero; the complement $U \setminus W$ is called the *support* of the generalized function u and denoted $\operatorname{supp} u$. All three definitions of the support are consistent with one another. The operations I)–IV) do not increase the support, and $\operatorname{supp} a^*(v) = a^{-1}(\operatorname{supp} v)$ under a change of variable, as in V).

1.6. An Example. The delta-function at the point $\xi \in R$ is the functional defined by the formula $\delta_\xi(\varphi) \equiv \varphi(\xi)$. It is clear that $\delta_\xi \in \mathcal{K}'$, and that $\operatorname{supp} \delta_\xi = \{\xi\}$. We have $\delta_\xi = Y'(x - \xi)$, where Y is the Heaviside function, i.e., the ordinary function equal to 1 for $x \geq 0$ and to 0 for $x < 0$. We write the formula for change of variables in the delta-function as

$$a^*(\delta_\eta) = \frac{1}{|a'(\xi)|}\delta_\xi\,, \quad \eta = a(\xi)\,.$$

§2. Localization and the Structure of Generalized Functions

2.1. Localization

Theorem 1. *Let $\Omega \subset \mathbb{R}^1$ be any open set and $\{\Omega_a\}$ an open covering of it. Every generalized function on Ω can be written in the form*

$$u = \sum_\alpha u_\alpha\,,$$

where $u_\alpha \in \mathcal{K}'(\Omega_\alpha)$, $\operatorname{supp} u_\alpha \subset \Omega_\alpha$, and the sum is locally finite.

Local finiteness means that any compact set $K \subset \Omega$ intersects only a finite number of the sets $\operatorname{supp} u_\alpha$. Such a representation can be obtained by multiplying u by a suitable partition of unity. It is sometimes called the localization principle for generalized functions and to some extent compensates for the fact that generalized functions do not have values at individual points. The converse assertion is also true: *every locally finite sum of generalized functions $\sum u_\alpha$ defined in Ω is also a generalized function.* In general the space $\mathcal{K}'(\Omega)$ can be characterized as the smallest extension of the space of ordinary functions in Ω in which the operations I) and III) are admissible and this converse of Theorem 1 holds. If we omit the latter condition we obtain the subspace of generalized functions of finite order (cf. Sect. 2.2), which is isomorphic to $\mathcal{D}'_F(\Omega)$.

2.2. The Local Structure

Theorem 2 (cf. (Schwartz 1950/51)). *For any open set $\Omega \subset \mathbb{R}$ every generalized function $u \in \mathcal{K}'(\Omega)$ can be written in the form*

$$u = \sum_{i=0}^{\infty} \frac{d^i}{dx^i}[f_i] , \tag{1.2.1}$$

where f_i are locally integrable functions in Ω, and for any compact set $K \subset \Omega$ the functions f_i vanish on K starting from some subscript $i(K)$.

As a result we obtain an equivalent definition of a generalized function—it is a functional on $\mathcal{K}(\Omega)$ having the form (1.2.1). The functions f_i are not uniquely determined; any of them can be replaced by an expression $\frac{dF}{dx}$, where F is a primitive of it. Therefore we can assume that all the functions f_i in (1.2.1) are continuous. Every functional u having a representation (1.2.1) in which the sum is finite is called a generalized function of finite order. Its order is the minimal number $q = q(u)$ for which a representation (1.2.1) is possible with $f_i \equiv 0$ for $i > q$. According to Theorem 2 every generalized function in Ω, when restricted to any subset $\omega \Subset \Omega$, has finite order on ω.

2.3. Generalized Functions of Compact Support. A generalized function or a distribution is said to be *of compact support* if its support is compact.

Let Ω be an open set on the line, and $\mathcal{E}(\omega)$ the space of infinitely differentiable functions on ω. It is endowed with a notion of convergence: $\psi_k \to \psi$ if for any order i $\psi_k^{(i)} \to \psi^{(i)}$ uniformly on each compact set $K \subset \omega$. Let u be a distribution of compact support and ω any open neighborhood of $\operatorname{supp} u$. The functional u can be canonically extended to a continuous functional $\mathcal{E}(\omega) \to \mathbb{C}$. To do this we choose any function $h \in \mathcal{D}(\omega)$ equal to 1 in a neighborhood of $\operatorname{supp} u$ and for any function $\psi \in \mathcal{E}(\omega)$ we set

$$\check{u}(\psi) \equiv u(h\psi) .$$

It is clear that $\check{u}$ is a continuous functional on $\mathcal{E}(\omega)$, and this functional is independent of the choice of the "truncating" function h. The following proposition holds: *every continuous functional on $\mathcal{E}(\omega)$ can be obtained from some distribution of compact support u,* $\operatorname{supp} u \subset \omega$ *using the construction of the extension just described.*

Theorem 3. *Every generalized function $u \in \mathcal{K}'(\Omega)$ vanishes on the test functions φ that vanish on* $\operatorname{supp} u$ *together with all their derivatives. If u is of compact support, it suffices that the derivatives of φ up to a sufficiently high order vanish on* $\operatorname{supp} u$.

The following theorem is a consequence of Theorem 3.

Theorem 4 (cf. (Schwartz 1950/51)). *If* $\operatorname{supp} u$ *is a single point a, then*

$$u = \sum_{i=0}^{m} c_i \delta_a^{(i)} , \tag{1.2.2}$$

and the coefficients c_i are uniquely determined.

If the support of u is a finite or countable set $\{a_j,\, j = 1,2,\ldots\}$, having no limit points in Ω, then by the localization principle u can be written in the form $\sum u_j$, where $\operatorname{supp} u_j = \{a_j\}$, and consequently each generalized function u_j has the form (1.2.2). But if the sequence a_j has a limit point $a_0 \in \Omega$, no such representation is possible.

Example. Let $a_j \to a_0$, and let $\sum |c_j|$ be a convergent numerical series such that the series $\sum (a_j - a_0)^{-1} c_j$ diverges. The functional

$$u(\varphi) = \sum c_j \frac{\varphi(a_j) - \varphi(a_0)}{a_j - a_0}$$

is a generalized function, and its support is contained in the closure of the sequence $\{a_j,\, j = 1,2,\ldots\}$. However it cannot be written as a sum of generalized functions $u_0 + \sum u_j$, where $\operatorname{supp} u_j \subset \{a_j\}$, $j = 0,1,2,\ldots$.

2.4. Expansion of a Generalized Function in a Fourier Series. A generalized function $u \in \mathcal{K}'(\mathbb{R})$ is said to be *periodic* with period t if it is invariant under translation by t, i.e., $u(\varphi(x+t)) = u(\varphi(x))$ for any test function φ.

Theorem 5. *Every 1-periodic generalized function u has a unique expansion into a weakly convergent Fourier series*

$$u = \sum_{-\infty}^{\infty} c_n e^{2\pi i n x} = \sum_{0}^{\infty} a_n \cos(2\pi n x) + b_n \sin(2\pi n x) , \tag{1.2.3}$$

and the coefficients satisfy inequalities

$$|c_n|^2 + |c_{-n}|^2 = \frac{1}{2}(|a_n|^2 + |b_n|^2) \le c(n+1)^q, \quad n = 0,1,2,\ldots .$$

Conversely, if the coefficients of the series (1.2.3) satisfy these inequalities, then the series converges weakly to a 1-periodic generalized function.

In particular

$$\sum_{k=-\infty}^{\infty} \delta_k = \sum_{n=-\infty}^{\infty} e^{2\pi i n x} = 1 + 2\cos 2\pi x + 2\cos 4\pi x + \cdots + 2\cos 2\pi n x + \cdots ,$$

which leads to the Poisson summation formula when applied to test functions (cf. Chapt. 3, §2).

The generalized functions on the circle S are continuous linear functionals on the space $\mathcal{K}(S)$ of infinitely differentiable functions on S in which convergence is defined by condition 1) of Sect. 1.1. Consider the linear mapping $\sigma : \mathcal{K}(\mathbb{R}) \to \mathcal{K}(S)$ acting according to the formula

$$\varphi(x) \mapsto \psi(x) = \sum_{k=-\infty}^{\infty} \varphi(x+k) .$$

It is obviously continuous.

Theorem 6. *The image of the adjoint mapping $\sigma' : \mathcal{K}'(S) \to \mathcal{K}'(\mathbb{R})$ coincides with the space of* 1*-periodic generalized functions on the line.*

To prove this it is necessary to construct $v \in \mathcal{K}'(S)$ from a given 1-periodic generalized function u. Consider a certain function $h \in \mathcal{D}(\mathbb{R})$ such that $\sum h(x+k) \equiv 1$. Every element $\psi \in \mathcal{D}(S)$ can be regarded as a 1-periodic density on the line, and $h\psi \in \mathcal{K}(\mathbb{R})$. We set $v(\psi) \equiv u(h\psi)$. It follows from this construction in particular that the coefficients (1.2.3) can be found from the formulas

$$c_n = u(h(x)e^{-2\pi inx}), \quad n = \ldots -2, -1, 0, 1, 2, \ldots .$$

2.6. The Convolution of Generalized Functions with Test Densities. The convolution of two ordinary integrable functions on the line is the integral

$$(f * g)(x) = \int_{-\infty}^{\infty} f(x-y)g(y)\,dy . \tag{1.2.4}$$

This integral is defined for almost all x and the right-hand side belongs to the space $L_1(\mathbb{R})$ of functions that are integrable with respect to the Lebesgue measure $|dx|$. Convolution is a bilinear associative and commutative operation in $L_1(\mathbb{R})$. The convolution is also defined if f is any locally integrable function and g is an integrable function of compact support (or conversely). In accordance with the compatibility principle (Sect. 1.3) we define the convolution of any generalized function u and test density φ on the line by the formula

$$(u * \varphi)(\xi) = u(t_\xi^*(\varphi)) , \tag{1.2.5}$$

where t_ξ is the change of variables $t_\xi(x) = \xi - x$, $t_\xi^*(\varphi) = \varphi(\xi - x)$. By (1.1.2) the convolution can also be written as follows: $(u * \varphi)(\xi) = t_\xi^*(u)(\varphi)$. The function $u * \varphi$ is always infinitely differentiable and

$$\frac{d}{d\xi}(u * \varphi) = u' * \varphi = u * \varphi' .$$

Convolution is a bilinear operation and the relation $\varphi_k \to \varphi$ in $\mathcal{K}$ implies that $u * \varphi_k \to u * \varphi$ in $\mathcal{E}(\mathbb{R})$.

A convolution operation $\mathcal{D}' \times \mathcal{D} \to \mathcal{E}$ is defined similarly.

2.7. Approximate Identities. An approximate identity is usually taken to mean a sequence of ordinary functions on the line that converges weakly to δ_0. Let $\{\varphi_k\}$ be any approximate identity consisting of test functions satisfying condition 2) of Sect. 1.1. For any generalized function u on the line the convolution $u * \rho_k$ converges weakly to u, where $\rho_k = \varphi_k|dx| \in \mathcal{K}$. The sequence

φ_k can be chosen in the form $\varphi_k(x) = k\varphi_1(kx)$, where $\varphi_1 \in \mathcal{D}$ is any function for which $\int \varphi\, dx = 1$.

The following are examples of approximate identities:

I) $f_\varepsilon(x) = \dfrac{1}{\pi}\dfrac{\varepsilon}{x^2+\varepsilon}$, $\varepsilon \to 0$;

II) $f_n(x) = \dfrac{1}{2\sqrt{\pi n}} \exp\Big(-\dfrac{x^2}{4n}\Big)$, $n \to \infty$;

III) $f_n(x) = \dfrac{1}{\pi}\dfrac{\sin nx}{x}$, $n \to \infty$.

§3. The Finite Parts of Divergent Integrals

3.1. Regularization. Let Ω be an open set on the line $a \in \Omega$, and f a locally integrable function on $\Omega \setminus \{a\}$. From formula (1.1.1) it can be associated with the generalized function $[f]$ in $\Omega \setminus \{a\}$. In which cases does there exist an extension of the generalized function $[f]$ to a functional $u \in \mathcal{K}'(\Omega)$? If such a functional exists, it is called the *finite part* or the *regularization* of a divergent integral of the form (1.1.1). If u_1 and u_2 are any regularizations, the difference $u_1 - u_2$ is a generalized function in Ω with support at the point a and consequently by the theorem of L. Schwartz (Theorem 4 of Sect. 2.3), equals a linear combination of derivatives of the delta-function at the point a.

Theorem 1. *If for some ε and q the function f has bounded q-fold primitives F^- and F^+ on the intervals $[a-\varepsilon, a)$ and $(a, a+\varepsilon]$, then $[f]$ can be extended to a generalized function in Ω. This condition is also necessary in order for regularization to be possible.*

The sufficiency can be verified immediately: on the test densities φ whose supports belong to $[a-\varepsilon, a+\varepsilon]$ we set

$$u(\varphi) = (-1)^q \left[\int_{a-\varepsilon}^{a} F^- \varphi^{(q)}\, dx + \int_a^{a+\varepsilon} F^+ \varphi^{(q)}\, dx\right] .$$

In particular if the polynomial estimate $f = O(|x-a|^{-q})$ holds, then the hypothesis of the theorem holds, and the integral has a finite part.

Similarly one can speak of the finite part of an integral (1.1.1) in which f has a nonintegrable singularity on a discrete subset $A \subset \Omega$.

3.2. Hadamard Kernels. Consider the family of functions $x_+^\lambda = |x|^\lambda Y(x)$ depending on a complex parameter λ. For Re$\lambda > -1$ these functions are locally integrable on the entire line and therefore define a family of generalized functions

$$x_+^\lambda(\varphi) \equiv \int_0^\infty x^\lambda \varphi(x)\, dx . \qquad (1.3.1)$$

Moreover these integrals are defined for any continuous function φ that decreases rapidly at $+\infty$. The latter means that all the functions $x^k\varphi(x)$ are bounded on the entire half-line, $k = 0, 1, 2, \ldots$. For every such function φ the integral (1.3.1) is a holomorphic function of the parameter λ. It tends to infinity as $\lambda \to -1$ if $\varphi(0) \neq 0$.

Theorem 2 (cf. (Hadamard 1932; M. Riesz 1949; Schwartz 1950/51)). *The family of generalized functions x_+^λ can be continued on the parameter λ to a function that is meromorphic on the entire complex plane with values in $\mathcal{K}'$ and has simple poles at the points $\lambda = -1, -2, \ldots$.*

This means that for any test function φ the integral (1.3.1) has a meromorphic extension with these poles. What has just been said also holds for any function φ that is infinitely differentiable in a neighborhood of zero and rapidly decreasing as $x \to +\infty$.

The residues of this meromorphic function can be computed from the formula

$$\mathrm{res}_{\lambda=-n} x_+^\lambda = \frac{(-1)^{n-1}}{(n-1)!}\delta_0^{(n-1)} .$$

Applying this equality to the test function $\varphi(x) = e^{-x}$, we obtain $x_+^\lambda(e^{-x}) = \Gamma(\lambda+1)$.

If the function φ vanishes in a neighborhood of the point $x = 0$, the right-hand side of (1.3.1) is defined for all λ and is an entire function of this parameter. It follows that this integral coincides with the value of $x_+^\lambda(\varphi)$, i.e., the generalized function x_+^λ is a regularization of the integral (1.3.1), which diverges for $\mathrm{Re}\,\lambda \leq -1$. We shall call the generalized function x_+^λ or the distribution $x_+^\lambda |dx|$ an *Hadamard kernel.* These kernels can be differentiated by the formula

$$\frac{dx_+^\lambda}{dx} = \lambda x_+^{\lambda-1} . \tag{1.3.2}$$

In the case $\mathrm{Re}\,\lambda > 0$ this formula follows from (1.3.1) and consequently holds for all values $\lambda \neq 0, -1, -2, \ldots$, in view of the uniqueness of analytic continuation. Using this formula one can obtain a constructive definition of any Hadamard kernel: if $\mathrm{Re}\,\lambda > -n-1$, then

$$x_+^\lambda(\varphi) = \lim_{\varepsilon \to 0} \left[\int_\varepsilon^\infty x^\lambda \varphi(x)\,dx + \right.$$
$$\left. + \frac{\varepsilon^{\lambda+1}}{\lambda+1}\varphi(0) + \frac{\varepsilon^{\lambda+2}}{\lambda+2}\varphi'(0) + \cdots + \frac{\varepsilon^{\lambda+n}}{\lambda+n}\varphi^{(n-1)}(0) \right] .$$

Another explicit formula follows from (1.3.1) if we approximate the function φ on the closed interval $[0, 1]$ by its Maclaurin series

$$x_+^\lambda(\varphi) = \int_0^1 x^\lambda \left[\varphi(x) - \varphi(0) - x\varphi'(0) - \cdots - \frac{x^n}{n!}\varphi^{(n)}(0)\right] dx +$$
$$+ \sum_{k=0}^n \frac{\varphi^{(k)}(0)}{k!(\lambda+k+1)} + \int_1^\infty x^\lambda \varphi(x)\,dx .$$

The generalized function x_-^λ is defined for $\operatorname{Re}\lambda > 1$ by the convergent integral

$$x_-^\lambda(\varphi) := \int_{-\infty}^0 |x|^\lambda \varphi(x)\,dx \tag{1.3.3}$$

and can be meromorphically continued on the parameter λ so as to have simple poles at the points $\lambda = -1, -2, \ldots$. The residues at these points differ from the resides of x_+^λ only by the coefficient $(-1)^{n-1}$, and the differentiation formula also looks like (1.3.2).

3.3. Regularization of Power Functions with Negative Integer Exponents. For an integer $\lambda = -n$ the divergent integrals (1.3.1) and (1.3.3) have natural regularizations as generalized functions $x_\pm^{-n}$ that are obtained from the family of Hadamard kernels by the formulas

$$x_+^{-n} = \lim_{\lambda\to -n} x_+^\lambda - \frac{(-1)^{n-1}}{(n-1)!}\frac{\delta_0^{(n-1)}}{\lambda+n}, \quad x_-^{-n} = \lim_{\lambda\to -n} x_-^\lambda - \frac{1}{(n-1)!}\frac{\delta_0^{(n-1)}}{\lambda+n}.$$

In other words $x_\pm^{-n}$ is the value of the regular part of the Laurent expansion of the corresponding family of kernels at the point $\lambda = -n$. The differentiation formulas for these generalized functions look different:

$$\frac{d}{dx}x_\pm^\lambda = -nx_\pm^{-n-1} \pm (\mp 1)^n \frac{\delta_0^{(n)}}{n!},$$

but the relation

$$x \cdot x_\pm^\lambda = \pm x_\pm^{\lambda+1} \tag{1.3.4}$$

holds for all $\lambda \in \mathbb{C}$. The generalized function $\{x^{-n}\} = x_+^{-n} + (-1)^n x_-^{-n}$ is the finite part of the divergent integral with power kernel x^{-n}. By definition

$$\{x^{-n}\} = \lim_{\lambda\to -n} \left(x_+^\lambda + (-1)^n x_-^\lambda\right).$$

We note that at the point $\lambda = -n$ each of the two terms on the right-hand side has a simple pole, but their residues differ in sign. Therefore the sum is holomorphic for $\lambda = -n$, and consequently the limit exists and satisfies (1.3.4). These generalized functions can also be defined by the formulas

$$\{x^{-1}\} = \frac{d}{dx}\ln|x|, \quad \frac{d}{dx}\{x^{-n}\} = -n\{x^{-n-1}\}.$$

It can be seen from these formulas that $\{x^{-n}\}$ is the *Cauchy principal value* of the corresponding integral:

$$\{x^{-1}\}(\varphi) = \mathrm{PV}\int \frac{\varphi(x)}{x}\,dx = \lim_{\varepsilon\to 0} \int\limits_{|x|\geq\varepsilon} \frac{\varphi(x)}{x}\,dx = \int_0^\infty \frac{\varphi(x)-\varphi(-x)}{x}\,dx,$$

$$\{x^{-2}\}(\varphi) = \int_0^\infty \frac{\varphi(x)+\varphi(-x)-2\varphi(0)}{x^2}\,dx.$$

We shall consider yet another way of introducing the generalized function $\{x^{-n}\}$ below.

3.4. Cauchy-Sokhotskij Kernels. For an arbitrary complex λ we define the holomorphic function $z^\lambda = e^{\lambda \ln z}$ in the plane of the complex variable $z = x + iy$ by taking the branch $\ln z$ of the logarithm function that assumes real values on the positive half-line and undergoes a discontinuity on the negative half-line. Thus $\ln z = \ln |z| + i \arg z$, $-\pi < \arg z < \pi$. Consider the limit values of the function z^λ on the real axis from above and below

$$(x \pm i0)^\lambda(\varphi) \equiv \lim_{\varepsilon \to 0} \int (x \pm i\varepsilon)^\lambda \varphi(x)\, dx\,. \tag{1.3.5}$$

Theorem 3. *The limits* (1.3.5) *exist for any* $\varphi \in \mathcal{D}(\mathbb{R})$; *consequently* $(x \pm i0)^\lambda$ *is a generalized function on the line for any* λ. *This family depends holomorphically on* λ *on the entire complex plane and for* $\lambda \neq -1, -2, \ldots$ *the following formulas hold*

$$(x + i0)^\lambda = x_+^\lambda + e^{i\lambda\pi} x_-^\lambda, \quad (x - i0)\lambda = x_+^\lambda + e^{-i\lambda\pi} x_-^\lambda\,.$$

The Hadamard kernels $x_\pm^\lambda$ can be expressed in terms of the Cauchy-Sokhotskij kernel if the determinant of the matrix of coefficients in these formulas is nonzero. This determinant equals $2i \sin \lambda\pi$, which explains the appearance of poles in the families of Hadamard kernels at the points $\lambda = -1, -2, \ldots$. For $\lambda = 0, 1, 2, \ldots$ there are no poles because $(x + i0)^\lambda = (x - i0)^\lambda$.

The Cauchy-Sokhotskij kernels are connected with the principal values of the power functions by the formulas

$$(x + i0)^{-n} = \{x^{-n}\} \mp i\pi \frac{(-1)^{n-1}}{(n-1)!} \delta_0^{(n-1)}, \quad n = 1, 2, \ldots\,.$$

When $n = 1$, this relation is essentially the same as the Sokhotskij formulas that connect the Cauchy integral taken over the real axis with the principal values of this integral. For more information see Sect. 2.2, Chapt. 3 and (Gel'fand, Shilov 1958a).

The normalized Hadamard kernels are

$$a_\lambda^\pm = \frac{x_\pm^{\lambda-1}}{\Gamma(\lambda)}\,.$$

When $\lambda \neq 0, -1, -2, \ldots$, the numerator and denominator are holomorphic functions of λ, and the denominator does not vanish; consequently the normalized kernel is also holomorphic. At the points $\lambda = 0, -1, -2, \ldots$ the numerator and denominator have simple poles and the fraction has a removable singularity. Thus the normalized Hadamard kernel is an entire function of the parameter λ, and its values at the negative integers equal the ratios of the residues of the numerator and denominator, i.e., $z_{-n}^\pm = (\pm 1)^n \delta_0^{(n)}$.

Chapter 2
The General Theory

§1. Generalized Functions and Distributions on a Manifold

1.1. Densities and Generalized Functions on a Manifold. Let X be a smooth manifold, i.e., a manifold of class C^∞. A *density bundle* on X is a one-dimensional complex bundle Π_x constructed as follows. Suppose an atlas is distinguished on X consisting of charts $\Phi_\alpha : X_\alpha \to \mathbb{R}^n$, $n = n(\alpha)$, $\alpha \in A$, where $\{X_\alpha\}$ is an open covering of X and let

$$\Phi_{\beta\alpha} : \Phi_\alpha(X_\alpha \cap X_\beta) \to \Phi_\beta(X_\beta \cap X_\alpha)$$

be the transition mappings of these charts. The bundle Π_x is obtained by gluing together the trivial bundles $X_\alpha \times \mathbb{C}$, $\alpha \in A$, using the mappings

$$\Phi_\alpha(X_\alpha) \times \mathbb{C} \ni (x, \lambda) \mapsto (\Phi_{\beta\alpha}(x), a_{\beta\alpha}(x)\lambda) \in \Phi_\beta(X_\beta) \times \mathbb{C}\,,$$

where $a_{\beta\alpha} = \left| \det \dfrac{\partial \Phi_{\beta\alpha}}{\partial x} \right|^{-1}$. A *density* on X is a section of this bundle. In the chart Φ_α with coordinates $(x_1, \ldots, x_{n_\alpha}) = x$ it is convenient to write a density in the form $\rho = \lambda(x)|dx|$, where λ is a function and $|dx|$ denotes a unit section of the bundle $X_\alpha \times \mathbb{C}$. If $\mu(y)|dy|$ denotes the same density in a chart Φ_β that intersects Φ_α, the connection between the functions λ and μ can be expressed by the formula

$$\mu(\Phi_{\beta\alpha}(x)) \left| \det \frac{dy}{dx} \right| = \lambda(x)$$

on the intersection of these charts. The density ρ is said to be *locally integrable, continuous*, or *smooth*, if the functions λ that express ρ in all the charts have these properties. The *support* of a density, denoted $\operatorname{supp} \rho$, is the minimal closed subset of X outside which $\lambda(x) = 0$ almost everywhere. The basic property of densities is that for every locally integrable density ρ with compact support the integral

$$\int_X \rho$$

is well-defined. If ρ is represented as a sum of densities whose supports are contained in charts, then the integral can be found as the sum of the integrals of these terms taken over the domains $\Phi_\alpha(X_\alpha)$ of the coordinate spaces.

On any manifold X there exists a smooth positive density, i.e., a density ρ such that $\lambda > 0$ in all charts. Any other density ρ' has the form $\rho' = c\rho$, where c is a function on X.

The *fundamental space* $\mathcal{K}(X)$ is the set of smooth densities on X with compact supports. It is a vector space over the field of complex numbers. In this space a concept of convergence is defined by the following rule: $\rho_k \to \rho$ in $\mathcal{K}(X)$ if the following conditions hold:

1) for any $\alpha \in A$ and any derivative operator D^i in the space $\mathbb{R}^n$ we have $D^i\lambda_k \Rightarrow D^i\lambda$, where λ_k and λ are the coefficients of the densities ρ_k and ρ in the chart Φ_α and $D^i = \dfrac{\partial^{|i|}}{\partial x_1^{i_1} \cdots}$;
2) there exists a compact set $K \subset X$ such that $\operatorname{supp} \rho_k \subset K$ for all k.

A *generalized function on the manifold* X in the sense of S.L. Sobolev is a linear functional $u : \mathcal{K}(X) \to \mathbb{C}$ that is continuous with respect to the convergence just described. The set of generalized functions on X forms a complex vector space denoted $\mathcal{K}'(X)$ endowed with the topology of weak convergence, and also the strong locally convex topology (cf. Sec. 1.7). Every ordinary function, i.e., every locally integrable function f on X defines a generalized function by the rule

$$[f](\rho) = \int_X f\rho\,.$$

This integral is well-defined, since $f\rho$ is a locally integrable density on X with compact support. As in the case described in Chapt. 1, we have a linear mapping $L \to \mathcal{K}'(X)$ of the space L of ordinary functions; it is an imbedding if we identify functions f and g such that $f = g$ almost everywhere. In other words the space $\mathcal{K}'(X)$ is an extension of L, which justifies the term "generalized function." The theory discussed in §§1 and 2 of Chapt. 1, carries over to the general case with certain modifications.

1.2. The Sheaf of Generalized Functions. Let Y be an open submanifold of X. It follows from the continuity of the imbedding $\mathcal{K}(Y) \to \mathcal{K}(X)$ that there exists a mapping $r_Y{}^X : \mathcal{K}'(X) \to \mathcal{K}'(Y)$ that maps a functional u to its restriction to the subspace $\mathcal{K}(Y)$. The functional $r_Y{}^X(u)$ is called the *restriction* of the generalized function u to the submanifold Y. Let $U \supset V \supset W$ be any open subsets of X. It follows from what has been said that the restriction mappings depicted in the following diagram are defined:

$$\begin{array}{ccc} \mathcal{K}'(\mathcal{U}) & \longrightarrow & \mathcal{K}'(\mathcal{V}) \\ & \searrow \quad \swarrow & \\ & \mathcal{K}'(\mathcal{W}) & \end{array}.$$

This diagram is commutative, and hence the correspondence $U \mapsto \mathcal{K}'(U)$ is a presheaf of vector spaces defined on the topological space X.

Theorem 1. *The presheaf just described is a sheaf.*

This assertion means the following: let $\{Y_\alpha\}$ be any open covering of a subset $Y \subset X$, and for each α a certain generalized function u_α is defined on Y_α such that for every pair of sets Y_α, Y_β with nonempty intersection the restrictions of u_α and u_β to this intersection coincide. Then there exists a generalized function u on Y whose restriction to any subset Y_α is u_α, and such a generalized function u is uniquely determined. This sheaf is called the sheaf of generalized functions on X.

Let u be an arbitrary generalized function on the manifold X. Consider the set of open subsets $U \subset X$ such that the restriction of u to U is zero. Let W be the union of all such subsets. By Theorem 1 the restriction of u to W is also zero. The complement $X \setminus W$ is called the *support* of the generalized function u and is denoted $\operatorname{supp} u$. For generalized functions of compact support the analogues of the assertions of §2.3 of Chapt. 1 hold.

Theorem 2. *Let the manifold X be paracompact. The sheaf $\mathcal{K}'$ of generalized functions on X has zero cohomologies in positive dimensions. More generally, $H_{\Phi^p}(X, K') = 0$ for all $p \geq 1$ and any family of supports Φ. The analogue of Theorem 1 of §2 of Chapt. 1 also holds.*

1.3. Operations on Generalized Functions. The product of a generalized function by a smooth function is defined as in §1 of Chapt. 1. To introduce the definition of differentiation and change of variables it is necessary first to define these operations on the fundamental space. Let $F : X \to Y$ be a diffeomorphism of smooth manifolds and ρ a density on Y. Its inverse image under the mapping F is the density $F^*(\rho)$ on X that satisfies the identity

$$\int_K F^*(\rho) = \int_{F(K)} \rho$$

for any compact $K \subset X$. In order for the right-hand side to be defined it is necessary to assume that ρ is locally integrable. This identity determines the inverse image uniquely; to be specific, if $\mu(y)\,|dy|$ is the expression for ρ in a local chart on Y and $y = f(x)$ is the expression of F in a chart on X, then

$$F^*(\rho) = \mu\big(f(x)\big)\left|\det \frac{\partial f}{\partial x}\right| \cdot |dx|$$

is the expression for the inverse image of the density in the same chart. It is clear that the inverse image operation defines a continuous operator $F^* : \mathcal{K}(Y) \to \mathcal{K}(X)$. It follows from the consistency principle (§1 of Chapt. 1) that the action of the diffeomorphism F on a generalized function $u \in \mathcal{K}'(Y)$ must have the following form:

$$F^*(u)(\rho) = u\big((F^{-1})^*(\rho)\big) \,. \tag{2.1.1}$$

Let t be a smooth vector field on X. It generates a group of automorphisms of X, and consequently its action on a generalized function must be compatible with the action of this group. Hence the following rule of action of t on the generalized function u holds:

$$t(u)(\rho) = -u(L_t\rho) \,, \tag{2.1.2}$$

where L_t is the Lie derivative operator along the field t. The coordinate expression of this operator is as follows:

$$L_t\rho = \sum_{i=1}^{n} \frac{\partial}{\partial x_i}(\lambda t_i)|dx| ,$$

where $\rho = \lambda(dx)$, and $t = \sum t_i \frac{\partial}{\partial x_i}$.

1.4. Schwartz Distributions. Let X be a smooth manifold and $\mathcal{D}(X)$ the space of infinitely differentiable functions on this manifold with compact support. Convergence is defined in it using conditions analogous to requirements 1) and 2) of Sect. 1.1. Any continuous linear functional on $\mathcal{D}(X)$ is called a *distribution on* X by Schwartz (1957). The theory of distributions runs parallel to the theory of generalized functions. Moreover there is an isomorphism of spaces and even sheaves of generalized functions and distributions. To construct this isomorphism we choose any smooth density ρ_0 that never vanishes on X and define the operator $\mathcal{D}(X) \to \mathcal{K}(X)$, mapping the function φ to the density $\varphi \cdot \rho_0$. This mapping is an isomorphism of spaces with convergence. The adjoint mapping $\mathcal{K}'(X) \to \mathcal{D}'(X)$ is also an isomorphism; it maps the distribution r to the generalized function $u = r/\rho_0$. In cases when there is a distinguished density ρ_0, for example Lebesgue measure $|dx|$ on $\mathbb{R}^n$, the isomorphism of spaces of generalized functions and distributions is fixed, and these concepts are sometimes conflated in the literature. In general when there is no distinguished nonzero density, such a conflation is not legitimate. This follows, in particular, from a comparison of the formulas for differentiation and change of variables for distributions.

If $F : X \to Y$ is a diffeomorphism of smooth manifolds and r is a distribution on Y, then the inverse image of F is defined as follows:

$$F^*(r)(\varphi) = r\big((F^{-1})^*(\varphi)\big) \tag{2.1.3}$$

and the mapping $(F^{-1})^*$ acts on the function φ by substitution. The action of the vector field t on the distribution r defined on X is defined by the formula

$$t(r)(\varphi) = -r\big(t(\varphi)\big) . \tag{2.1.4}$$

1.5. Generalized Functions and Distributions on $\mathbb{R}^n$. In this case there is a distinguished density $\rho_0 = |dx|$, and therefore to every generalized function u one can assign a distribution $r = u\,|dx|$ and conversely. A generalized function u or a distribution r is *homogeneous* of order λ if for any $c > 0$ the scaling transformation $F_c(x) = cx$ acts on it by the formulas

$$F_c^*(u) = c^\lambda u, \quad F_c^*(r) = c^\lambda r ,$$

where the left-hand sides are defined using (2.1.1) and (2.1.3). If a generalized function u is homogeneous of order λ, then the corresponding distribution $r = u|dx|$ is homogeneous of order $\lambda + n$.

Example. The *delta-function* δ_0 in $\mathbb{R}^n$ is the generalized function operating on $\rho = \varphi\,|dx|$ according to the formula $\delta_0(\rho) = \varphi(0)$. It is homogeneous of order $-n$. At the same time the delta-distribution $\delta_0\,|dx|$, i.e., the functional on $\mathcal{D}(\mathbb{R}^n)$ defined by the rule $\varphi \mapsto \varphi(0)$ is homogeneous of order 0.

The action of a vector field $t = \sum t_i \dfrac{\partial}{\partial x_i}$ on a generalized function is consistent with its action on ordinary smooth functions, i.e., with differentiation along this field. At the same time the action of t on a distribution r according to formula (2.1.4) may be written in the form

$$t(r) = t(u\,|dx|) = t(u)|dx| + u \cdot L_t|dx| = t^*(u)|dx|\,,$$

where $t^* = \sum t_i \dfrac{\partial}{\partial x_i} + \sum \dfrac{\partial t_i}{\partial x_i}$ is the formally adjoint differential operator.

1.6. The Direct Product. Let X and Y be smooth manifolds and p_X and p_Y the projections of the manifold $X \times Y$ onto the direct factors. For any functions $\varphi \in \mathcal{D}(X)$ and $\psi \in \mathcal{D}(Y)$ the product $\varphi \otimes \psi \equiv p_X^*(\varphi) \cdot p_Y^*(\psi)$ is an element of the space $\mathcal{D}(X \times Y)$. This construction gives a bilinear mapping $\mathcal{D}(X) \otimes \mathcal{D}(Y) \to \mathcal{D}(X \times Y)$, whose image is dense. The *direct product* of the distributions $r \in \mathcal{D}'(X)$ and $s \in \mathcal{D}'(Y)$ (Schwartz 1957) is the distribution $r \otimes s \in \mathcal{D}'(X \times Y)$ for which the following identity holds:

$$r \otimes s(\varphi \otimes \psi) = r(\varphi) \cdot s(\psi)\,.$$

Theorem (Schwartz 1957). *For any distributions r and s the* direct product *$r \otimes s$ is uniquely defined. The corresponding mapping $\mathcal{D}'(X) \times \mathcal{D}'(Y) \to \mathcal{D}'(X \times Y)$ is bilinear and continuous with respect to the strong topologies (cf. Sect. 1.7).*

It follows from this result that if $r_k \to r$ in $\mathcal{D}'(X)$ and $s_k \to s$ in $\mathcal{D}'(Y)$, then $r_k \otimes s_k \to r \otimes s$ in $\mathcal{D}'(X \times Y)$. The construction of the direct product is as follows: for every function $\theta \in \mathcal{D}(X \times Y)$ and point $y \in Y$ the restriction of θ to a fiber $X_y \cong X$ of the projection p_Y is an element of the space $\mathcal{D}(X)$. The functional r can be applied to this restriction and the result is the function $\psi(y) = r(\theta|X_y) \in \mathcal{D}(Y)$. The distribution $r \otimes s$ is defined by the equality $r \otimes s(\theta) = s(\psi)$.

The direct product of generalized functions is defined and constructed analogously.

1.7. The Topologies in the Fundamental and Dual Spaces. For simplicity we restrict ourselves to the case $X \subset \mathbb{R}^n$. Let $K \subset X$ be a compact set and $\{M_p,\, p = 0, 1, 2, \dots\}$ a sequence of positive numbers. We denote by $B_K\{M_p\}$ the subset of $\mathcal{D}(X)$ formed by the functions φ satisfying the following conditions: $\operatorname{supp}\varphi \subset K$ and

$$\left|\frac{\partial^p \varphi(x)}{\partial x_1^{i_1} \cdots \partial x_n^{i_n}}\right| \le M_p, \quad i_1 + \cdots + i_n = p, \quad p = 0, 1, 2, \dots, \quad x \in K\,.$$

The space $\mathcal{D}(X)$ is endowed with the strongest locally convex topology in which all the sets $B_K\{M_p\}$ are bounded, i.e., absorbed by any neighborhood of zero. We remark that every sequence of functions of $\mathcal{D}(X)$ that converges in the sense of Sect. 1.1 belongs to one of the sets $B_K\{M_p\}$ and also converges with respect to this topology. Every linear functional that is continuous with respect to convergence in $\mathcal{D}(X)$ is also continuous with respect to this topology. This implies the following conclusion about the structure of generalized functions (and distributions).

Theorem. *For any open subset $X \subset \mathbb{R}^n$ every generalized function u on X can be represented in the form*

$$u = \sum_{i=0}^{\infty} \Delta^i [f_i] \,, \tag{2.1.5}$$

where $\Delta = \sum \frac{\partial^2}{\partial x_j^2}$ is the Laplacian, all f_l are locally integrable functions in X, and on each compact set $K \subset X$ only a finite number of these functions are nonzero.

The functions f can be chosen to be continuous or have a prescribed number N of continuous derivatives.

The topology in the conjugate spaces $\mathcal{D}'(X) \cong \mathcal{K}'(X)$ is introduced so that the polars[1] of the sets $B_K\{M_p\}$ form a fundamental system of neighborhoods of zero in $\mathcal{D}'(X)$. This topology can be described more explicitly. Let $K \subset X$ be a compact set and $\{m_i\}$ a sequence of positive numbers. We consider the subset $W_K\{m_i\} \subset \mathcal{D}'(X)$ of generalized functions that are representable in the form (2.1.5), where f_i satisfy the inequalities

$$\int_K |f_i(x)|\, dx < m_i, \quad i = 0, 1, 2, \dots \;.$$

The set of subsets $W_K\{m_i\}$ is a fundamental system of neighborhoods of zero.

These topologies can also be constructed using the operation In of inductive (direct) limits of sequences and the operation Pr of projective (or inverse) limits of sequences of locally convex spaces. We choose a sequence of compact sets $K_1 \Subset K_2 \Subset \cdots K_q \Subset \cdots$, that exhaust X and consider the double spectrum $\{\mathcal{D}_{p,q}\}$ of Hilbert spaces that is direct with respect to the index q and inverse with respect to $p = 0, 1, 2 \dots$, where $\mathcal{D}_{p,q}$ is the space of functions of $W_2^p(\mathbb{R}^n)$ with supports contained in K_q. Thus $\mathcal{D}_{p,q} \to \mathcal{D}_{p,q+1}$ is an imbedding and $\mathcal{D}_{p+1,q} \to \mathcal{D}_{p,q}$ is a compact mapping. There are topological isomorphisms

$$\mathcal{D}(X) \cong \operatorname*{In}_q \operatorname*{Pr}_p \{\mathcal{D}_{p,q}\}, \quad \mathcal{D}'(X) \cong \operatorname*{Pr}_q \operatorname*{In}_p \{\mathcal{D}'_{p,q}\} \,.$$

[1] Let G be a subset of a vector space E, and E' the dual space to E. The polar of G in E' is the set of functionals $f \in E'$ such that $|f(e)| \le 1$ for all $e \in G$.

The properties of these topological spaces are described in (Schaefer 1966; Rajkov 1967; Trèves 1967). In particular they are complete, barrelled, bornological, and nuclear, and the Open Mapping Theorem and the Closed Graph Theorem hold for them.

It should be noted that the concept of a nuclear space itself was discovered by Grothendieck (1955) after analyzing the well-known theorem of L. Schwartz on the kernel (Schwartz 1952; Grothendieck 1955; Ehrenpreis 1956a), which asserts that every bilinear coordinatewise continuous functional on the space $\mathcal{D}(X) \times \mathcal{D}(Y)$ is defined by some distribution $r \in \mathcal{D}'(X \times Y)$. The concepts of nuclear space and nuclear mapping have found important applications in many questions of functional analysis. In particular these are the key concepts in the theorem of I. M. Gel'fand and A. G. Kostyuchenko (Gel'fand, Vilenkin 1961) on the existence of a complete system of generalized eigenvectors for a self-adjoint operator. A survey of papers on this topic can be found in (Gel'fand, Vilenkin 1961).

§2. De Rham Currents

2.1. Odd Differential Forms (de Rham 1955). Let X be a smooth not necessarily orientable manifold. An *even form* on X is defined to be any ordinary differential form on this manifold. Such forms are sections of the bundle Λ^* of exterior algebras of fibers of the cotangent bundle $T^*(X)$. We consider a one-dimensional bundle S on X with transition mappings

$$b_{\beta\alpha} = \operatorname{sgn} \det \left\| \frac{\partial \Phi_{\beta\alpha}}{\partial x} \right\|$$

(cf. §1). It is called the bundle of scalars of odd type and $S\Lambda^* \equiv S \otimes \Lambda^*$ is the bundle of differential forms of odd type or, for short, the bundle of odd forms. An *odd form* on X is a section of this bundle. In each chart on X such a section can be written as an ordinary form

$$\alpha = \sum a_{i_1,\dots,i_p}\, dx_{i_1} \wedge \dots \wedge dx_{i_p}\,,$$

where the coefficients $a_{i_1,\dots,i_p}$ depend skew-symmetrically on the indices, but the coefficients $b_{j_1,\dots,j_p}$ of the expression for it in an adjacent chart are related by the formula

$$b_{j_1,\dots,j_p} = \operatorname{sgn} \det \left\| \frac{\partial y}{\partial x} \right\| \cdot \sum a_{i_1,\dots,i_p} \frac{\partial x_{i_1}}{\partial y_{j_1}} \cdots \frac{\partial x_{i_p}}{\partial y_{j_p}}\,,$$

which differs from the usual formula by the coefficient sgn $= \pm 1$. In particular a density on X is a section of the bundle of odd forms of highest order.

Let Ω^* and $S\Omega^*$ be the space of smooth even forms and odd forms respectively on X. The exterior product of an odd form and an even form is defined; it is an odd form, and the product of two odd forms is an even form. Thus the vector space $\Omega^* \oplus S\Omega^*$ is an exterior algebra with a double grading $\mathbb{Z} \otimes \mathbb{Z}_2$. The first grading corresponds to the order of the form and the second to its parity. In this algebra the exterior derivative operator d is defined, which increases the order of a form by one and does not change its parity. The local expression for this operator on even and odd forms is the same, so that $d^2 = 0$ on any forms. This operator is a derivation, which means that the following relation holds:

$$d(\varphi \wedge \psi) = d\varphi \wedge \psi + (-1)^p \varphi \wedge d\psi, \quad p = \deg \varphi .$$

An odd form can be regarded as a multilinear skew-symmetric mapping from the space $T(X)$ of smooth vector fields on X into the space $S\Omega^0$ of functions of odd type. Let t be a vector field and φ an odd form of order $p \geq 1$. Substituting the field t as the first argument into φ, we obtain an odd form of order $p-1$. This defines a linear mapping $i(t) : S\Omega^p \to S\Omega^{p-1}$ called the *inner product* with the field t. Forms of order zero vanish under the inner product. This multiplication is also a differentiation. The *Lie derivative* of the form φ along the field t is defined by the formula

$$L_t(\varphi) = d(i(t)\varphi) + i(t)d\varphi .$$

2.2. Currents. Let Ω_0^* and $S\Omega_0^*$ be respectively the space of smooth even forms and smooth odd forms on the manifold X with compact support. In the space $\Omega_0^* \oplus S\Omega_0^*$ we introduce a convergence that is described by conditions analogous to 1) and 2) of §1. The continuous linear functionals on this space are called *currents* (de Rham 1955). A current is called even (resp. odd) if it vanishes on all even (resp. odd) forms. If a current is equal to zero on all forms of order not equal to p, it is said to be homogeneous of dimension p or degree $n-p$, where n is the dimension of X. Thus the space of currents on X has $\mathbb{Z} \oplus \mathbb{Z}_2$ grading with respect to dimension and parity.

Example I. Every generalized function on X is an even current of degree zero.

Example II. Every distribution on X is an odd current of dimension zero.

Example III. Every differential form α on X with locally integrable coefficients defines a current by the formula

$$u(\varphi) = \int_X \alpha \wedge \varphi .$$

The parity and degree of this current coincide with the parity and order of the form α.

Example IV. An oriented chain $c \subset X$ (an odd chain in the sense of de Rham 1955) defines an odd current on X of the same dimension as c according to the rule

$$u(\varphi) = \int_c \varphi .$$

Every even chain c, in particular, every co-oriented smooth chain, defines an even current by the same rule.

The exterior differential d on the space of currents is defined by the formula $du(\varphi) = w \cdot u(d\varphi)$, where w is an involution that maps the homogeneous current v of degree p into $(-1)^p v$. Currents can be multiplied by smooth forms; when this is done, the parities and the degrees of the factors are added.

Let $f : X \to Y$ be a mapping of smooth manifolds and u an odd current on X such that the mapping $f : \operatorname{supp} u \to Y$ is proper. Then the direct image is defined: $v = \int_f u$, which is an odd current on Y:

$$v(\psi) = u(f^*(\psi)), \quad \psi \in \Omega_0^* .$$

The dimension of the current v is equal to the dimension of u. Similarly we construct the direct image of an even current u under an oriented mapping f. In particular the direct image of a distribution r is defined for any smooth mapping f that is proper on $\operatorname{supp} r$, and this image is again a distribution.

§3. Inverse and Direct Images

3.1. The Direct Image of a Density. Let $f : X \to Y$ be a mapping of smooth manifolds and v a generalized function on Y. The expression $v(f(x))$ is in general not defined on X, even when f is a homeomorphism. This expression can be given a well-defined meaning only under certain conditions connecting f and v. We recall that the mapping f is a *submersion* if at each point $x \in X$ the rank of df equals the dimension of the manifold Y at the point $f(x)$ (the equality $f(X) = Y$ is not assumed). Every fiber X_y of a submersion f is a manifold; its dimension at the point x is $\dim_x X - \dim_{f(x)} Y$.

Let $f : X \to Y$ be a submersion; consider the linear mapping

$$\int_f : \mathcal{K}(X) \to \mathcal{K}(Y) . \tag{2.3.1}$$

operating on the density by integration along the fibers of f. The coordinate form of this mapping is as follows: let $x_0 \in X$, and let $y_1, \ldots, y_n$ be coordinates in a neighborhood of the point $f(x_0)$ and $x_1, \ldots, x_{m+n}$ coordinates in a neighborhood $U \ni x_0$ chosen so that $x_{m+j} = y_j(f(x))$, $j = 1, \ldots, n$. We assume that $\rho \in \mathcal{K}(X)$, $\operatorname{supp} \rho \subset U$, and $\rho = \lambda(x)|dx|$. Then

$$\int_f \rho = \mu(y)|dy| ,$$

where

$$\mu(y) \equiv \int\limits_{X_y} \frac{\rho}{df_1 \wedge \cdots \wedge df_m} = \int\limits_{X_y} \lambda(x)|dx_1 \wedge \cdots \wedge dx_m| \,.$$

The integrand on the right-hand side is a density on the fibers of f, and hence μ is a well-defined function in a neighborhood of $f(x_0)$. The expression $\mu|dy|$ is a density on Y, since when the variables $y_1, \ldots, y_n$ are changed the function μ is transformed using the absolute value of the Jacobian of the change of variables. This function is infinitely differentiable, and the support of the density $\mu|dy|$ is contained in $f(\operatorname{supp} \rho)$, hence is compact. Therefore $\int\limits_f \rho \in \mathcal{K}(Y)$. In the case of an arbitrary density $\rho \in \mathcal{K}(X)$ the same conclusion can be obtained using a suitable partition of unity on Y. It is clear that the mapping (2.3.1) is continuous.

The following invariant definition of this mapping exists: for any compact set $K \subset Y$

$$\int\limits_K \Big(\int\limits_f \rho \Big) = \int\limits_{f^{-1}(K)} \rho \,.$$

We remark that this definition makes sense for any smooth mapping f, but to guarantee the smoothness of the integrated density it is necessary to assume that f is a submersion.

3.2. The Inverse Image of a Generalized Function. The mapping adjoint to (2.3.1)

$$f^* : \mathcal{K}'(Y) \to \mathcal{K}'(X)$$

is called the *inverse image* operation or *pullback* of generalized functions. Thus for any generalized function v on Y

$$f^*(v)(\rho) = v\Big(\int\limits_f \rho \Big) \,.$$

This operation is consistent with the operation of substitution in ordinary functions: if the generalized function v corresponds to a locally integrable function $g(y)$, then $f^*(v)$ corresponds to the ordinary function $g(f(x))$. The operation of inverse image is functorial in the argument f.

A vector field t on X is called *vertical* if $df(t) = 0$. For every vertical field t and every generalized function v on Y we have $t(f^*(v)) = 0$, which means that the inverse image of a generalized function is "locally constant" along fibers.

3.3. The Delta-Function on a Hypersurface. Suppose that on a smooth manifold X there is defined a smooth function $f : X \to \mathbb{R}$ that has no critical points on the manifold of zeros $Z = f^{-1}(0)$. This manifold is a nonsingular hypersurface in X. The *delta-function* on this hypersurface is defined to be the generalized function $\delta_f \equiv f^*(\delta_0)$, the inverse image of the generalized function $\delta_0 \in \mathcal{K}'(\mathbb{R})$. To write an explicit expression for it we represent δ_0 as the weak limit of a family of ordinary functions g_ε, $\varepsilon \to 0$, where $g_\varepsilon = \dfrac{1}{2\varepsilon}$ on the interval $[-\varepsilon, \varepsilon]$ and $g_\varepsilon = 0$ outside this interval. As a result we obtain

$$\delta_f(\rho) = \lim_{\varepsilon \to 0} \frac{1}{2\varepsilon} \int\limits_{-\varepsilon \le f(x) \le \varepsilon} \rho, \quad \rho \in \mathcal{K}(X) . \tag{2.3.2}$$

We now introduce a Riemannian metric on X: let $dV \in S\Omega^*$ be the volume element on X and dS the volume element on Z. We write $\rho = \varphi\, dV$, where $\varphi \in \mathcal{D}(X)$. By passage to the limit, we obtain from (2.3.2)

$$\delta_f(\rho) = \int\limits_Z \frac{\varphi}{|\mathrm{grad}\, f|}\, dS . \tag{2.3.3}$$

More generally, if we write $\rho = df \wedge \omega$, where ω is an odd form on X (the *Léray form*), then

$$\delta_f(\rho) = \int\limits_Z \omega , \tag{2.3.4}$$

and the coorientation of Z is defined by the form df. The delta-function can also be obtained as the derivative of a function $Y(f)$ (equal to 1 when $f \ge 0$ and 0 when $f < 0$):

$$dY(f) = df \cdot \delta_f .$$

We assume that f can be written as a product of smooth functions $f = g \cdot h$ (they also have no critical zeros). According to (2.3.3) or (2.3.4),

$$\delta_f = \frac{\delta_g}{|h|} + \frac{\delta_h}{|g|}$$

(cf. (Gel'fand, Shilov 1958a), where the absolute value signs are omitted).

Example. Let us write an expression for the delta-function on the hyperbola $xy = s$. It is a fiber of the mapping $f(x, y) = xy$, and consequently if $\rho = \varphi|dx \wedge dy|$, then

$$\delta_f(\rho) = \int\limits_{xy=s} \frac{\varphi|dx \wedge dy|}{ydx + xdy} = \int \frac{\varphi(x, \frac{s}{x})}{|x|}\, dx = \int \frac{\varphi(\frac{s}{y}, y)}{|y|}\, dy$$

(cf. (Gel'fand, Shilov 1958a), where the denominators must be replaced by their absolute values).

Let Z be a submanifold of X that is the preimage of the origin under some submersion $f : X \to \mathbb{R}^n$. The *delta-function* $\delta_f = f^*(\delta_0)$ on this submanifold can be defined in analogy with formulas (2.3.2) and (2.3.4).

3.4. Relative Densities and Connection. Let $f : X \to Y$ be any submersion. Consider the quotient space

$$\mathcal{K}_f = \mathcal{K}(X)/f^*(\Omega^{\max}(Y)) \wedge dS\Omega_0^*(X) , \tag{2.3.5}$$

where $\Omega^{\max}(Y)$ denotes the space of smooth even forms on Y having maximal order on each connected component of Y. The denominator is assumed to be zero on each connected component of X whose dimension is equal to the dimension of its image in Y. The elements of the space $\mathcal{K}_f$ are called *relative densities* on X/Y. We now extend the operation defined in Sect. 3.1 to relative densities. To do this it suffices to show that this operation vanishes on the denominator of (2.3.5). Let $\rho = f^*(\omega) \wedge d\sigma$, where $\omega \in \Omega^n(Y)$ and $\sigma \in S\Omega_0(X)$. We choose any chart $V \subset Y$ with coordinates $y_1, \dots, y_n$ and in this chart we write $\omega = b(y)dy_1 \wedge \dots \wedge dy_n$. In $f^{-1}(V)$ we have $\rho = df_1 \wedge \dots \wedge df_n \wedge b(f(x))d\sigma$, where $y_j = f_j(x)$, $j = 1, \dots, n$, is the coordinate expression for the mapping f. Consequently

$$\int\limits_{X_y} \frac{\rho}{df_1 \wedge \dots \wedge df_n} = b(y) \int\limits_{X_y} d\sigma = 0 ,$$

whence $\int\limits_f \rho = 0$. Hence it follows that the operation of integration of relative densities along the fibers of f is well-defined.

We now establish the rule for differentiation of these integrals. Let $\mathcal{E}(Y)$ be the ring of infinitely differentiable functions on Y. The space $\mathcal{K}_f$ is a module over this ring. A *connection* ∇ on this module is by definition a linear mapping of the space $T(Y)$ of smooth fields on Y into the space of differential operators on $\mathcal{K}_f$, satisfying the relation

$$\nabla_t(a\rho) = t(a)\rho + a\nabla_t(\rho), \quad a \in \mathcal{E}(Y), \quad \rho \in \mathcal{K}_f \tag{2.3.6}$$

for any field $t \in T(Y)$. We construct a special connection in $\mathcal{K}_f$, which we shall call the *canonical* connection. Along a given field t we construct a smooth field s on X satisfying the relation $df(s) = t$ (it is called a *lifting* of t). A lifting is easy to construct locally, using the fact that f is a submersion. The local liftings can be combined into a global lifting using a suitable partition of unity. Now consider the *Lie derivative* operator $L_s = d \cdot i(s)$ in $\mathcal{K}(X)$. We verify that it acts invariantly in the denominator of (2.3.5). We have $i(s)f^*(\omega) = f^*(i(t)\omega)$ for every $\omega \in \Omega^n(Y)$, and so

$$i(s)(f^*(\omega) \wedge d\sigma) = f^*(i(t)\omega) \wedge d\sigma + (-1)^n f^*(\omega) \wedge i(s)\, d\sigma .$$

Therefore

$$L_s(f^*(\omega) \wedge d\sigma) =$$
$$= df^*(i(t)\omega) \wedge d\sigma + f^*(\omega) \wedge d(i(s)d\sigma) \in f^*(\Omega^n(Y)) \wedge dS\Omega_0^*(X) ,$$

which proves our assertion. Thus L defines an endomorphism of the space $\mathcal{K}_f$, which we denote by ∇_t. Let us verify that ∇_t is independent of the choice of the lifting s. If s' is another lifting of the field t, then $df(s'-s)=0$ and hence the field $r=s'-s$ is tangent to the fibers of f. Any density $\rho \in \mathcal{K}(X)$ can be written locally in the form $\rho = df_1 \wedge \cdots \wedge df_n \wedge \sigma$. Since $i(r)df_j = 0$, we have

$$d(i(r)\rho) = df_1 \wedge \cdots \wedge df_n \wedge d(i(r)\sigma) \in f^*(\Omega^n(Y)) \wedge dS\Omega_0^*(X)\,,$$

and consequently L_s and $L_{s'}$ defined the same operator on $\mathcal{K}_f$. Thus the canonical connection operator ∇_t is well-defined and obviously depends linearly on the field t. The formula (2.3.6) follows from the relation $s(a)\rho - da \wedge i(s)\rho = 0$.

For any fields t and v on Y we have the equality

$$[\nabla_t, \nabla_v] = \nabla_{[t,v]}\,,$$

whose meaning is that the connection ∇ is flat. The basic relation for the canonical connection is the rule for differentiating an integral along the fibers of a mapping f:

$$L_t\Big(\int_f \rho\Big) = \int_f \nabla_t \rho, \quad \rho \in \mathcal{K}_f, \quad t \in T(Y)\,. \tag{2.3.7}$$

3.5. Differentiation of the Inverse Image of a Generalized Function. Let $f: X \to Y$ be a submersion and v a generalized function on Y. For every smooth vector field t on Y the derivative $t(v)$ is defined by the formula (2.1.2). The following identity relates the inverse images of v and $t(v)$:

$$f^*(t(v))(\rho) = -f^*(v)(\nabla_t(\rho))\,. \tag{2.3.8}$$

It follows from (2.3.7).

Example. Let $f: X \to \mathbb{R}^n$ be a submersion, let Z be the fiber of it over the point $y=0$, and let $\delta_f = f^*(\delta_0)$ be delta-function on this fiber. The generalized function $\dfrac{\partial \delta_f}{\partial f_j} \equiv f^*\Big(\dfrac{\partial \delta_0}{\partial y_j}\Big)$ can be regarded as the derivative with respect to the argument f_j. Let us find an explicit expression for it by setting $t_j = \dfrac{\partial}{\partial y_j}$. By (2.3.8)

$$\frac{\partial \delta_f}{\partial f_j} = -\delta_f(\nabla_{t_j}(\rho))\,.$$

If we write $\rho = df_1 \wedge \cdots \wedge df_n \wedge \sigma$, then in $\mathcal{K}_f$

$$\nabla_{t_j}(\rho) = d(i(t_j)\rho) = (-1)^{j-1} d(df_1 \wedge \cdots \widehat{df}_j \cdots \wedge df_n \wedge \sigma)\,,$$

and consequently

$$\frac{\partial \delta_f}{\partial f_j}(\rho) = (-1)^j \int\limits_Z \frac{d(df_1 \wedge \cdots \widehat{df}_j \cdots \wedge df_n \wedge \sigma)}{df_1 \wedge \cdots \wedge df_n} \,,$$

This formula is due to I.M. Gel'fand and Z.Ya. Shapiro (Gel'fand, Shilov 1958a).

Using (2.3.8) one can obtain a method of differentiating the inverse image along any vector field s on X. Using the fact that f is a submersion, we write

$$s = s_0 + \sum a_i s_i, \quad a_i \in \mathcal{E}(X)\,,$$

where the sum is locally finite, the fields s_i are liftings of the fields $t_i \in T(Y)$, and the field s_0 is vertical. Since the field s_0 annihilates the inverse image of any generalized function, we have

$$s(f^*(v)) = \sum a_i f^*(t_i(v)), \quad v \in \mathcal{K}'(Y)\,.$$

3.6. The Direct Image of a Generalized Function. As we have seen in §2, the operation of direct image has a natural definition in the category of distributions, but not in the category of generalized functions (while the opposite is the case for the operation of inverse image). The direct image of a generalized function under an arbitrary smooth mapping $f : X \to Y$ can be defined in the following more complicated way. Let $u \in \mathcal{K}'(X)$ and let σ_X be any smooth density on X such that the support of the distribution $\sigma_X \cdot u$ is proper over Y. We also choose a smooth nonzero density σ_Y on Y and consider the expression

$$\sigma_Y^{-1} \cdot \int\limits_f (\sigma_X \cdot u)\,.$$

It has an interpretation as a generalized function on Y, since $\int\limits_f (\sigma_X u)$ is a distribution. This generalized function does not change if σ_Y is multiplied by any nonzero smooth function ψ and σ_X is simultaneously multiplied by $f^*(\psi)$. To determine the direct image in a more invariant manner, we consider the pair (σ_X, σ_Y) as the mapping $\Pi(Y) \to \Pi(X)$ operating according to the formula $\rho \mapsto f^*(\sigma_Y^{-1}\rho)\sigma_X$, where $\Pi(Y)$ and $\Pi(X)$ are the spaces of smooth densities on these manifolds. This mapping is linear over the ring $\mathcal{E}(Y)$. Conversely, every element of the space

$$\mathrm{Hom}_{\mathcal{E}(Y)}(\Pi(Y), \Pi(X)) \tag{2.3.9}$$

can be represented in this form. Such an element will be called a *relative density*. In a neighborhood of any point where f is a submersion the elements (2.3.9) can be regarded as smooth densities on the fibers of f. We define the support of an element of (2.3.9) as the support of its value on any nonzero density σ_Y. This support is independent of the choice of σ_Y. We denote by

$\mathcal{K}(X/Y)$ the subspace of (2.3.9) formed by the elements whose supports are proper over Y.

We define the *direct image* of generalized functions defined on X as the bilinear mapping $f_* : \mathcal{K}'(X) \times \mathcal{K}(X/Y) \to \mathcal{K}'(Y)$ operating according to the rule

$$f_*(u, h) = \sigma_Y^{-1} \cdot \int_f h(\sigma_Y) \cdot u, \quad \sigma_Y \in \Pi(Y).$$

We now establish the relation between the operations of direct and inverse images under a mapping f that is a submersion. For every element $h \in \mathcal{K}(X/Y)$ one can define integrals over fibers

$$\int_f h \equiv \sigma_Y^{-1} \int_f h(\sigma_Y).$$

These integrals determine an infinitely differentiable function on an open set $f(X) \subset Y$. The required relation can be expressed by the equality

$$f_*(f^*(v), h) = \left(\int_f h \right) \cdot v | f(X).$$

Example. Let $f : X \to Y$ be a proper imbedding and r a distribution on X. According to Sect. 2.2 $\int_f r$ is a distribution on Y with support contained in $f(X)$, called the "delta-distribution with density r." If u is a generalized function on $f(X)$, then the direct image $f_*(u, h)$ is defined and depends on the argument $h \in \mathcal{K}(X/Y)$. In this case $\mathcal{K}(X/Y)$ is isomorphic to the space of smooth sections of the linear bundle $| \wedge (N_x)|$, where $\wedge(N_X)$ denotes the maximal exterior degree of the normal bundle N_X of the submanifold $f(X)$, and the absolute value sign means that the transition coefficients in this bundle have been replaced by their absolute values. In particular, formula (2.3.3) can be interpreted as the direct image of the generalized function $u \in \mathcal{K}'(Z)$

$$u(\sigma) = \int_Z h\sigma.$$

The coefficient $h = \frac{1}{|\text{grad}\, f|} \frac{dS}{dV}$ in this formula is a section of the bundle $|N_Z|$.

§4. Partial Smoothness of Generalized Functions

4.1. Smoothness Relative to a Mapping. Let $f : X \to Y$ be a mapping of smooth manifolds. The generalized function u defined on X will be called smooth over Y if for every element $\tau \in \mathcal{K}(X/Y)$ the direct image $f_*(u, \tau)$ is an infinitely differentiable function on Y. We shall say that the generalized function u is *fiberwise smooth* if for every vertical field t on X/Y (i.e., every field satisfying the condition $df(t) = 0$) with compact support $\operatorname{supp} t$ there exists a sequence of positive numbers c_k, $k = 0, 1, 2, \ldots$, such that the family of generalized functions $c_k t^k(u)$ is weakly bounded, i.e., for any density $\rho \in \mathcal{K}(X)$ the numerical sequence

$$c_k t^k(u)(\rho), \quad k = 0, 1, 2, \ldots$$

is bounded.

Example 1. Let $u = \delta_Z$ be the delta function on a closed submanifold (cf. §3). It is smooth over Y if and only if the mapping $f : Z \to Y$ is a submersion.

Example 2. Let $g : X \to Z$ be a submersion and $v \in \mathcal{K}'(Z)$. The generalized function $g^*(v)$ is smooth over Y if the restriction of f to any fiber of g is a submersion. It can be seen from Example 1 that this condition cannot be weakened in general. On the other hand, every generalized function of the form $g^*(v)$ is fiberwise smooth if there exists a smooth mapping $h : Y \to Z$ such that $g = h \circ f$. This condition is also necessary if all the fibers of g are connected. From these examples it can be seen that the condition of fiberwise smoothness is comparatively more restrictive than the condition of smoothness over the base of the mapping.

In what follows we limit ourselves to the case when the mapping f is a submersion. In this case the definitions just described are essentially due to L. Schwartz (1957), who considered the case where $X = X_0 \times Y$ and $X_0 \subset \mathbb{R}^m$ and $Y \subset \mathbb{R}^n$ are open sets. A distribution $r \in \mathcal{D}'(X)$ is *semiregular* in the sense of Schwartz over Y if for any function $\varphi \in \mathcal{D}(X_0)$ the distribution on Y defined by the formula

$$r_\varphi(\psi) \equiv r(\varphi \otimes \psi), \quad \psi \in \mathcal{D}(Y) ,$$

is an infinitely differentiable function. If $|dx|$ is a nonzero density on X_0, then $\tau = \varphi|dx|$ can be regarded as an element of the space $\mathcal{K}(X/Y)$, and consequently the requirement of semiregularity is formally weaker than the condition that r be smooth over Y. The equivalence of the two conditions is proved using the following theorem.

Theorem 1. *If the distribution $r \in \mathcal{D}'(X)$ is semiregular with respect to Y, then for any compact sets $K \subset X_0$ and $L \subset Y$ and any integer l there exists an integer k having the property that for every function $\varphi \in C^k(X_0)$ such that $\operatorname{supp} \varphi \subset K$ the distribution r_φ is an ordinary function on L and has continuous derivatives up to order l in the interior of L.*

L. Schwartz takes integral semiregularity of a distribution to mean that the distribution is continuous in a certain topology that is weaker than the standard topology of $\mathcal{D}(X)$ (cf. Sect. 1.7). He shows that this condition is equivalent to the structural representation shown in the theorem below. It follows from this that integral semiregularity of a distribution $r \in \mathcal{D}'(X)$ is equivalent to its being fiberwise smooth with respect to the projection $X \to Y$.

Theorem 2. *Let $f : X \to Y$ be a submersion. For a generalized function $u \in \mathcal{K}'(X)$ to be fiberwise smooth, it is necessary and sufficient that for any compact set $K \subset X$ it may be written as a finite sum*

$$u(\rho) = \sum_{1}^{N} \int_X g_n a_n(\rho), \quad \operatorname{supp} \rho \subset K \, , \, .$$

where a_n are certain differential operators with coefficients in $\mathcal{E}(X)$ and g_n are continuous functions on X such that for every smooth vertical field t all the functions $t^k(g_n)$, $k = 1, 2, \ldots$, are also continuous on K.

4.2. Smoothness Relative to a Foliation. We now transform the definitions just described into more invariant form. Let X be a smooth manifold. A *foliation* on X is a structure $\mathcal{F}$ that is defined by an open covering $\{X_\alpha\}$ of the manifold X and a set of smooth submersions $\psi_\alpha : X_\alpha \to \mathbb{R}^n$ that are consistent with one another. The consistency condition is that for any indices α and β such that $X_\alpha \cap X_\beta \neq \emptyset$ there exists a smooth mapping $\psi_{\beta\alpha} : \psi_\alpha(X_\alpha \cap X_\beta) \to \psi_\beta(X_\beta \cap X_\alpha)$ such that $\psi_{\beta\alpha} \cdot \psi_\alpha = \psi_\beta$. In view of the symmetry between α and β this mapping is necessarily a diffeomorphism. The pre-images of points under the mapping ψ_α are called local fibers and the mapping itself is called a chart of the foliation. Since the charts are consistent, the local fibers can be extended from one set of the covering to another. In particular if a submersion $f : X \to Y$ is given, its fibers form a foliation. However, not every foliation can be defined in this way. The set of charts (X_α, ψ_α) is called an *atlas* of the foliation.

Suppose a foliation $\mathcal{F}$ is defined on the manifold X. We shall say that a generalized function u on X is *smooth across* this foliation if for any chart (X_α, ψ_α) of the foliation the restriction of u to X_α is a smooth generalized function over $\mathbb{R}^n$ with respect to the mapping ψ_α. It is clear that this condition is independent of the choice of the atlas of the foliation.

We shall call the generalized function u *smooth along the foliation* $\mathcal{F}$ if for any α the generalized function $u|X_\alpha$ is fiberwise smooth relative to ψ_α. In other words, u is smooth along $\mathcal{F}$ if for any smooth vector field t with compact support tangent to $\mathcal{F}$ there exists a sequence of positive numbers c_k such that the family of generalized functions $c_k t^k(u)$, $k = 0, 1, 2, \ldots$, is weakly bounded.

Example. The delta-function on a submanifold $Z \subset X$ is smooth along any foliation in which Z coincides locally with a fiber and is smooth across any foliation transversal to Z.

Let $\mathcal{K}'_{\mathcal{F}}(X)$ be the space of generalized functions on X that are smooth along $\mathcal{F}$ and $\mathcal{K}'^{\mathcal{F}}(X)$ the space of generalized functions that are smooth across this foliation. These spaces are invariant with respect to the operations of multiplication by a smooth function and differentiation along a smooth field. For the space $\mathcal{K}'^{\mathcal{F}}(X)$ these assertions are trivial, and for $\mathcal{K}'_{\mathcal{F}}(X)$ they follow from Theorem 2.

4.3. Restriction of Generalized Functions to Submanifolds. Suppose a closed submanifold $Z \subset X$ is a fiber of some foliation $\mathcal{F}$, and u is a generalized function on X that is smooth across $\mathcal{F}$. In this case one can define the restriction of u to Z, which we shall denote $u|Z \in \mathcal{K}'(Z)$. To give a precise meaning to this statement we define an adequate convergence in the space $\mathcal{K}'^{\mathcal{F}}(X)$. By definition $u_k \to u$ in $\mathcal{K}'^{\mathcal{F}}(X)$ if this sequence converges to u in $\mathcal{K}'(X)$ and in addition for every chart (X_α, ψ_α) of the foliation $\mathcal{F}$, any element $\tau \in \mathcal{K}(X_\alpha/\mathbb{R}^n)$, and any continuous functional $h \in \mathcal{E}'(\mathbb{R}^n)$ the numerical sequence

$$h(\psi_{\alpha *}(u_k, \tau))$$

is bounded (a posteriori it converges to $h(\psi_{\alpha *}(u, \tau))$).

Theorem 3. *For any manifold X, any foliation $\mathcal{F}$ on X and any fiber Z of this foliation there exists a linear operator*

$$R_Z^X : \mathcal{K}'^{\mathcal{F}}(X) \to \mathcal{K}'(Z),$$

having the following properties.

1) *It maps every continuous function of $\mathcal{K}'^{\mathcal{F}}(X)$ to its restriction to the manifold Z.*
2) *It is continuous with respect to the convergences in $\mathcal{K}'^{\mathcal{F}}(X)$ and in $\mathcal{K}'(Z)$.*
3) *The operators R_Z^X commute with the restriction mappings of generalized functions to open subsets $X' \subset X$ and to $X' \cap Z$ respectively.*
4) *Let $f : X \to Y$ be a submersion, and let the generalized function u be smooth over Y. Then for every $\tau \in \mathcal{K}(X/Y)$*

$$f_*(u, \tau)(y) = R_{X_y}^X(u)(\tau|X_y), \quad y \in Y.$$

The operators R_Z^X are uniquely determined by these conditions.

We now explain the proof of the theorem. Since the problem is local, we can assume that Z is a fiber of some submersion $f : X \to \mathbb{R}^n$ over the point $y = 0$. For every generalized function u that is smooth over $\mathbb{R}^n$ we define a generalized function u_0 on Z by the formula

$$u_0(\tau|Z) \equiv \delta_0(f_*(u, \tau)), \quad \tau \in \mathcal{K}(X/\mathbb{R}^n),$$

where δ_0 is the delta-distribution at the point $y = 0$. It is clear that every element of $\mathcal{K}(Z)$ has the form $\tau|Z$ for a suitable element $\tau \in \mathcal{K}(X/\mathbb{R}^n)$. To establish that this definition is unambiguous one need only show that the right-hand side is zero if $\tau | Z = 0$. Let $y_1, \ldots, y_n$ be coordinates on $\mathbb{R}^n$. We can write $\tau = \sum_1^n f^*(y_j)\tau_j$ with certain $\tau_j \in \mathcal{K}(X/\mathbb{R}^n)$, and consequently

$$\delta_0(f_*(u,\tau)) = \delta_0\Big(\sum_1^n y_j f_*(u,\tau_j)\Big) = 0\,.$$

Thus we can set $R_Z^X u = u_0$. The verification of properties 1)–4) presents no difficulty.

4.4. General Properties of Partially Smooth Generalized Functions. The following two theorems show that the properties of smoothness along and across a foliation are complementary in a natural sense.

Theorem 4. *Let $\mathcal{F}$ be a foliation on the manifold X. If a generalized function u defined on X is smooth both along and across $\mathcal{F}$, then it is smooth, i.e., coincides with an ordinary infinitely differentiable function.*

The second result is the possibility of multiplying a generalized function along $\mathcal{F}$ by a generalized function that is smooth across $\mathcal{F}$. To do this we introduce convergence in the space $\mathcal{K}'_{\mathcal{F}}(X)$. By definition $u_j \to u$ in this space if this sequence converges in $\mathcal{K}'(X)$ and for any smooth field t with compact support tangent to $\mathcal{F}$ there exists a sequence of positive numbers c_k such that the family of generalized functions

$$c_k t^k(u_j), \quad k = 0, 1, 2, \ldots, \quad j = 1, 2, \ldots$$

is weakly bounded.

Theorem 5. *For any manifold X and foliation $\mathcal{F}$ on it the operation of multiplication of generalized functions*

$$M : \mathcal{K}'_{\mathcal{F}}(X) \times \mathcal{K}'^{\mathcal{F}}(X) \to \mathcal{K}'(X)$$

is well-defined. This means the following.

1) *The mapping M is bilinear.*
2) *On every pair of continuous functions this mapping coincides with the usual multiplication.*
3) *If $u_k \to u$ in $\mathcal{K}'_{\mathcal{F}}(X)$ and $v_k \to v$ in $\mathcal{K}'^{\mathcal{F}}(X)$, then $M(u_k, v_k) \to M(u, v)$ in $\mathcal{K}'(X)$.*

The space $\mathcal{K}'_{\mathcal{F}}(X)$ is maximal in the sense that there exists no larger subspace $L \subset \mathcal{K}'(X)$ such that the mapping M can be extended to $L \times \mathcal{K}'^{\mathcal{F}}(X)$ preserving these properties. The space $\mathcal{K}'^{\mathcal{F}}(X)$ is also maximal in the analogous sense.

§5. The Wave Front

5.1. Definitions. The *singular support* of a generalized function $u \in \mathcal{K}'(X)$ is the smallest closed set $\operatorname{sing\,supp} u \subset X$ outside which u coincides with a smooth function. The complementary set can be characterized as follows: $x \in X \setminus \operatorname{sing\,supp} u$ if and only if there exists a chart $U \ni x$ and a density $\rho \in \mathcal{K}(u)$ different from zero at the point x such that the function $v = F(\rho u)$ is rapidly decreasing at infinity, i.e.,

$$v(\xi) = O(|\xi|^{-q}), \quad |\xi| \to \infty \tag{2.5.1}$$

for any $q = 0, 1, 2, \ldots$. Here F denotes the Fourier transform of the distribution ρu computed in this chart. The property of rapid decrease is independent of the choice of the chart. And if $x \in \operatorname{sing\,supp} u$, the directions in the dual space $\mathbb{R}^n_\xi$ along which u does not have such decrease can be described by the construction of a wave front. We shall give a precise definition of this concept first for the case when X is an open subset of $\mathbb{R}^n$. In this case the cotangent bundle $T^*(X)$ is the direct product of X and an n-dimensional space that can be identified with $\mathbb{R}^n_\xi$. We denote the set of nonzero cotangent vectors by $T^*_0(X)$. The *wave front* (Hörmander 1971) of a generalized function u can be defined as the conic closed subset $\mathrm{WF}(u) \subset T^*_0(X)$ having the following property: $(x, \xi) \notin \mathrm{WF}(u)$ if and only if there exists a density ρ different from zero at the point x such that the function v is rapidly decreasing at infinity in some conic neighborhood of the ray $\{t\xi,\, t > 0\}$. A set $K \subset T^*(X)$ is called *conic* if each of its fibers is invariant with respect to the similarity transformations $\xi \mapsto t\xi$, $t > 0$.

Let u be any generalized function on a smooth manifold X. According to what has been said for every chart X_α on X one can define the wave front of the restriction $u|X_\alpha$ as a conic subset in $T^*_0(X_\alpha)$. These sets coincide over the intersections of the charts and consequently define a single closed conic set $\mathrm{WF}(u) \subset T^*_0(X)$ called the *wave front* of u. This construction is consistent with the operation of restriction of a generalized function to an open submanifold and with the action of diffeomorphisms.

It is convenient to introduce the concept of an extended wave front

$$\widehat{\mathrm{WF}}(u) \equiv \mathrm{WF}(u) \cup \operatorname{supp} u \subset T^*(X),$$

where $\operatorname{supp} u$ is interpreted as a subset of $T^*(X)$ by identification of X with the set of zero cotangent vectors.

Naturally these definitions can also be given for distributions on a manifold in such a way that if $r = \sigma u$, where u is a generalized function and σ is a smooth nonzero density, then $\mathrm{WF}(r) = \mathrm{WF}(u)$.

We now describe the connection of the wave front with the properties of partial smoothness (§4). Let $\mathcal{F}$ be a foliation on X and $N^*_{\mathcal{F}} \subset T^*_0(X)$ its conormal bundle. If a generalized function u is smooth along $\mathcal{F}$, then $\mathrm{WF}(u) \subset N^*_{\mathcal{F}}$ (the converse is not true); and if $\mathrm{WF}(u) \cap N^*_{\mathcal{F}}$ is empty, then u is smooth across $\mathcal{F}$ (the converse is again not true).

5.2. The Wave Front of a Generalized Function of One Variable

Proposition 1. *Let u be a generalized function of compact support on the line. Its wave front is contained in the half-plane $\{(x,\xi), \xi > 0\}$ (resp. $\{(x,\xi), \xi < 0\}$) if and only if the convolution*

$$v = u * \frac{|dx|}{(x \mp i0)}$$

is infinitely differentiable.

Indeed the Fourier transform maps the convolution into the product of the images, and the Fourier image of the distribution $(x \mp i0)^{-1}|dx|$ is zero when $\xi > 0$ (resp. $\xi < 0$) and is constant on the complementary ray. Therefore the convolution with this distribution "cuts off" the part of the wave front contained in the half-plane $\xi > 0$ (resp. $\xi < 0$). Proposition 1 remains true if these distributions are replaced by any Cauchy-Sokhotskij kernels $(x \mp i0)^{-q}|dx|$, $q = 2, 3, \dots$.

Let X be any open subset on the line $\mathbb{R}$. Suppose there is an open subset U in the complex plane $\mathbb{C}$ such that $U \cap \mathbb{R} = X$ and a function h that is holomorphic in $U^+ = U \cap \mathbb{C}^+$ or in $U^- = U \cap \mathbb{C}^-$, where $\mathbb{C}^\pm$ are the upper and lower half-planes. If for any density $\rho = \varphi(x)|dx|$ the limit

$$u(\rho) \equiv \lim_{\varepsilon \to +0} \int h(x \pm i\varepsilon)\varphi(x)\, dx$$

exists, then u is a generalized function in X called the *boundary value of the holomorphic function* h. The function h is determined uniquely by u, and for any compact set $K \subset U$ it satisfies the inequality

$$|h(z)| \leq C|y|^{-q}$$

on $K \cap \mathbb{C}^\pm$ with some constants C and q. Every generalized function on X can be represented as the difference of the boundary values of a holomorphic function h_+ defined on U^+ and a holomorphic function h_- in U^-, the set U being such that $U \cap \mathbb{R} = X$ is arbitrary (Tilmann 1961; Schapira 1970). The functions $h_\pm$ are defined up to a term g that is holomorphic in U. This pair of functions is called an *analytic representation* of the generalized function u.

Theorem 2. *Let $h_\pm$ be an analytic representation of the generalized function $u \in \mathcal{K}'(X)$, and $X_\pm$ the open subset of X consisting of the points x such that $\mathrm{WF}_x(u)$ does not contain the ray $\xi > 0$ (resp. $\xi < 0$). All the derivatives of the function $h_\pm$ can be extended by continuity to $U^\pm \cup X_\pm$. The converse is also true.*

Thus the wave front of a generalized function of one variable can be completely described using the boundary behavior of the functions $h_\pm$ that define its analytic representation. The proof is based on Proposition 1.

5.3. Other Descriptions of the Wave Front

Theorem 3 (cf. (Guillemin, Sternberg 1977)). *Let X be an open set in $\mathbb{R}^n$ and u a generalized function in X. Then $(x_0, \xi_0) \neq 0$ is not contained in* $\mathrm{WF}(u)$ *if and only if there exists a neighborhood $U \ni x_0$ and a density $\rho \in \mathcal{K}(U)$ that is nonzero at x_0 such that for any manifold S with a distinguished point s_0 and any function $f : U \times S \to \mathbb{R}$ whose partial differential $d_x f$ at the point (x_0, s_0) equals ξ_0, the direct image of the convolution $v = \rho u * (f + i0)^{-1}|dx|$ under the mapping $F : U \times S \to \mathbb{R} \times S$ is an infinitely differentiable function in a neighborhood of $\mathbb{R} \times \{s_0\}$. Here $F(x,s) = (f(x,s), s)$.*

Hörmander (1983) describes the wave front u as the intersection of the sets $\gamma(A)$, where A are pseudodifferential operators such that $Au \in \mathcal{E}(X)$ and $\gamma(A)$ is the set of rays in $T^*(X)$ on which the symbol A is asymptotically equal to zero.

5.4. Properties of the Wave Front (Hörmander 1971; Guillemin, Sternberg 1977).

1) The projection of $\mathrm{WF}(u)$ on X is $\operatorname{sing\,supp} u$.
2) For any linear differential operator a on X with smooth coefficients

$$\mathrm{WF}(au) \subset \mathrm{WF}(u) .$$

3) For any generalized functions u and v on X

$$\mathrm{WF}(u) \triangle \mathrm{WF}(v) \subset \mathrm{WF}(u+v) \subset \mathrm{WF}(u) \cup \mathrm{WF}(v) ,$$

where the symbol $\triangle$ denotes the symmetric difference of sets.

4) For any generalized functions u on X and v on Y

$$\widetilde{\mathrm{WF}}(u \otimes v) = \widetilde{\mathrm{WF}}(u) \times \widetilde{\mathrm{WF}}(v) .$$

5) If Z is a submanifold of X such that $N_Z^* \cap \mathrm{WF}(u) = \varnothing$, then the restriction $u\big|Z \in \mathcal{K}'(Z)$ (cf. Sect. 4.3) is defined, and its wave front is contained in the restriction of $\mathrm{WF}(u)$ to $T(Z)$. The following assertion is a consequence of 5) and 4).
6) Let K and L be conic subsets of $T_0^*(X)$ such that $\xi + \eta \neq 0$ when $(x, \xi) \in K$ and $(x, \eta) \in L$. Then for any generalized functions u with $\mathrm{WF}(u) \subset K$ and v with $\mathrm{WF}(v) \subset L$ the product $u \cdot v$ can be defined as in Sect. 4.4, and

$$\widetilde{\mathrm{WF}}(u \cdot v) \subset \widetilde{\mathrm{WF}}(u) + \widetilde{\mathrm{WF}}(v) .$$

The following property generalizes 5).

7) Let $f : X \to Y$ be a smooth mapping and $v \in \mathcal{K}'(Y)$, while for any point $(x, \xi) \in \mathrm{WF}(v)$ the inequality $df_x^*(\xi) \neq 0$ holds. Then the inverse image $f^*(v) \in \mathcal{K}'(X)$ is well-defined, and

$$\mathrm{WF}(f^*(v)) \subset df^*(\mathrm{WF}(v)) ,$$

and the formulas of §3 are preserved for this operation.

8) Again suppose that $f : X \to Y$ is a mapping of smooth manifolds and $u \in \mathcal{K}'(X)$. The wave front of the direct image is described by the following formula

$$\widetilde{\mathrm{WF}}(f_*(u,\tau)) \subset \{(y,\eta) \in T^*(Y),\, y = f(x),\, df_x^*(\eta) \in \widetilde{\mathrm{WF}}(u)\} ,$$

where $\tau \in \mathcal{K}(X/Y)$.

5.5. Trigonometric Integrals. Consider the integral

$$I(x) = \int\limits_{\mathbb{R}^N} e^{i\varphi(x,\theta)} a(x,\theta)\, d\theta , \tag{2.5.2}$$

where $\varphi(x,\theta)$ is a real-valued function that is smooth in $\Omega \times \mathbb{R}^N \setminus \{0\}$ and positive homogeneous of first order with respect to θ, i.e., $\varphi(x,t\theta) = t\varphi(x,\theta)$ if $t > 0$; let a also be a smooth function in $\Omega \times \mathbb{R}^N \setminus \{0\}$ that is positive homogeneous with respect to θ or has the following asymptotic expansion for large θ:

$$a \sim a_m + a_{m-1} + \cdots + a_k + \cdots ,$$

where a_k is a positive homogeneous function of order k and $\Omega \subset \mathbb{R}^n$ is an open set. Such an integral is called *trigonometric* or *oscillatory* with phase function φ and amplitude a. The term "Fourier integral" is also used; the Maslov canonical operator (Maslov, Fedoryuk 1976) has a similar form. The detailed theory of integrals of the form (2.5.2) can be found in (Hörmander 1971; Trèves 1980; Hörmander 1985). A trigonometric integral is divergent in general. It is interpreted as a generalized function depending on the parameters x:

$$I(\rho) = \int\limits_{\Omega \times \mathbb{R}^N} e^{i\varphi(x,\theta)} a(x,\theta)\psi(x)\, d\theta\, dx, \quad \rho = \psi\, dx .$$

The phase function φ is said to be *nondegenerate* if its total differential does not vanish anywhere. If this function is nondegenerate, then the singularities of the generalized function I are determined by the critical points of this function with respect to the variables θ, i.e., by the points (x,θ) where $\varphi'_\theta = 0$. Such points form a manifold $C_\varphi \subset \Omega \times \mathbb{R}^N \setminus \{0\}$ of dimension n, and the mapping

$$\varphi'_x : C_\varphi \to T^*(\Omega)$$

is an immersion. Let Λ_φ be the image of this mapping.

Theorem 4. $\mathrm{WF}(I) \subset \Lambda_\varphi$.

The construction (2.5.2) and its generalizations can be used to study various problems of the theory of linear differential equations (Duistermaat, Hörmander 1972; Maslov, Fedoryuk 1976; Shubin 1978; Trèves 1980).

5.6. The Analytic Wave Front. This concept characterizes the degree of nonanalyticity of a generalized function in a manner similar to that in which a wave front measures the deviation from infinite differentiability. According to Hörmander (1974) the complement to the *analytic wave front* $\mathrm{WF}_A(u) \subset T_0^*(X)$ of a generalized function $u \in \mathcal{K}'(X)$ is the set of points (x_0, ξ_0) such that there exists a neighborhood $V \ni x_0$ and an open cone $\Lambda \ni \xi_0$ in the fiber of $T_0^*(X)$ over x_0 and also a bounded sequence of distributions $\{g_\nu\} \subset \mathcal{E}'(X)$ that coincide with $u\,|dx|$ on V such that

$$|F(g_\nu)(\xi)| \leq C\Big(\frac{C\nu}{|\xi|}\Big)^\nu, \quad \xi \in \Lambda, \quad \nu = 1, 2, \dots ,$$

with some common constant C. The analytic wave front is related to the usual wave front by the obvious inclusion $\mathrm{WF}(u) \subset \mathrm{WF}_A(u)$. On the other hand, for every generalized function u and every open conic subset $\Lambda \subset T_0^*(X)$ such that $\mathrm{WF}(u) \subset \Lambda$ the representation $u = v + g$ is possible, where $g \in \mathcal{E}(X)$ and $\mathrm{WF}_A(v) \subset \Lambda$.

Bony (1976) has shown that $\mathrm{WF}_A(u)$ coincides with the singular support $\mathrm{SS}(u)$ of the generalized function u regarded as a hyperfunction. The role of these concepts is illustrated by the following uniqueness theorem (Sato, Kawai, Kashiwara 1973; Hörmander 1983a): *Let h be a hyperfunction defined in a neighborhood of the point $x_0 \in \mathbb{R}^n$, and $\xi_0 \neq 0$ a cotangent vector at that point. If $\mathrm{SS}(h)$ does not contain the element (x_0, ξ_0) or the element $(x_0, -\xi_0)$ and $\operatorname{supp} h$ is contained in the half-space $\{\xi_0 \cdot (x - x_0) \geq 0\}$, then $h = 0$ in a neighborhood of x_0.* Polyakov and Khenkin (1986) used the concept of an analytic wave front to describe the Radon transform on projective space (the Radon-Penrose transform).

Chapter 3
The Fourier Transform

§1. Elementary Results

1.1. The Invariant Form of the Transform. We write the classical Fourier integral as[2]

$$F(\rho) = g(\xi) \equiv \int_X e^{-2\pi i \xi \cdot x} f(x)\,dx, \quad X = \mathbb{R}^n\,. \tag{3.1.1}$$

We shall say that in this formula it is the density $\rho = f(x)\,|dx|$ rather than the function $f(x)$ that is transformed. If the density is integrable (i.e., the

[2] The choice of the coefficient 2π in the exponent is convenient, since it frees us of numerical coefficients in many basic formulas of the theory.

function is integrable with respect to Lebesgue measure $|dx|$), the integral (3.1.1) defines a continuous function g that tends to zero as $|\xi| \to \infty$. Thus the transform (3.1.1) maps integrable densities defined on the space X into functions on the dual space $\Xi \cong \mathbb{R}^n$. The duality between these spaces is established using the bilinear form $\xi \cdot x = \sum \xi_i x_i$.

Parallel to (3.1.1) we define the *Fourier transform*, which maps integrable functions on X into densities on Ξ according to the rule

$$F(f) = g(\xi)|d\xi| \equiv \int\limits_p e^{-2\pi i \xi \cdot x} f(x)\,|dx\,d\xi|\,, \tag{3.1.2}$$

where $\int\limits_p$ is the operator that integrates densities along the fibers of the projection $p : X \times \Xi \to \Xi$ in the sense of Sect. 3.1 of Chapt. 2 and $|dx\,d\xi|$ is the density on $X \times \Xi$ that is canonically connected with the duality and independent of the choice of bases in X and Ξ. Naturally if both sides of (3.1.2) are divided by $|d\xi|$, we arrive at (3.1.1).

1.2. The Euclidean Structure on Functions and Densities. We introduce the bilinear form

$$\langle g, \rho \rangle_X \equiv \int\limits_X g\rho\,,$$

in which the first argument is a function on X and the second is a density, the product being integrable. If $\rho = h|dx|$, integrability of this product means that the function gh is integrable with respect to Lebesgue measure. We note that both Fourier operators—on functions and on densities—are symmetric with respect to the bilinear forms just described that are constructed on X and Ξ

$$\begin{aligned} \langle F(\rho), \sigma \rangle_\Xi &= \langle F(\sigma), \rho \rangle_X\,, \\ \langle f, F(g) \rangle_X &= \langle g, F(f) \rangle_\Xi\,. \end{aligned} \tag{3.1.3}$$

We introduce the operation $*$ that maps a function f defined on X into the density $*f \equiv \bar{f}|dx|$, and the density $\rho = g|dx|$ into the function $\bar{g}$. Using this operation we define positive definite Hermitian forms on the space of functions

$$(f, g) \equiv \langle f, *g \rangle = \int\limits_X f\bar{g}\,dx$$

and on the space of densities

$$(\rho, \sigma) \equiv \langle *\rho, \sigma \rangle = \int\limits_X \frac{\bar{\rho}}{|dx|}\sigma\,.$$

We denote by $L_2(X)$ the direct sum of the spaces of functions and densities with finite norm $\|a\| = \sqrt{(a,a)}$. The operation $*$ is a unitary involution in this space.

Theorem 1 (Plancherel). *The Fourier transform* (3.1.1) *and* (3.1.2) *can be extended to a bounded operator* $F : L_2(X) \to L_2(\Xi)$. *This operator is unitary, i.e.,*

$$(F(f), F(g))_\Xi = (f, g)_X \;, \quad (F(\rho), F(\sigma))_\Xi = (\rho, \sigma)_X \;.$$

It follows from this theorem that the adjoint operator F^* is also unitary, and the following relations hold:

$$F^* \circ F = E, \quad F \circ F^* = E \;.$$

The explicit expression for F^* differs from F only in the sign in the exponent:

$$F^*(\rho) = \langle e^{2\pi i x \cdot \xi}, \rho \rangle, \quad F^*(g) = \int_q e^{2\pi i x \cdot \xi} g(\xi) |d\xi \, dx| \;.$$

In these formulas the function ρ and the density g are assumed to be integrable.

Theorem 2 (Bochner 1932). *The Fourier transform in* L_2 *can be defined by the formula*

$$F(\rho) = \lim_{R \to \infty} \int_{|x| \le R} e^{-2\pi i \xi \cdot x} \rho(x) \;,$$

where the limit is taken in the L_2*-norm. An analogous relation holds for functions* $f \in L_2$. *The same formula with the opposite sign in the exponential gives an explicit expression for the inverse operator* $F^{-1} = F^*$ *on* L_2.

1.3. Properties of the Fourier Transform

$$F\left(\frac{\partial^j a}{\partial x^j}\right) = (2\pi i \xi)^j F(a) \;,$$

$$F((-2\pi i x)^j a) = \frac{\partial^j}{\partial \xi^j} F(a) \;, \qquad a \in L_2(X) \;,$$

$$F(\tau_k a) = \exp(-2\pi i h \cdot \xi) F(a) \;,$$

$$F(\exp(2\pi i h \cdot x) a) = \tau_h F(a) \;,$$

where τ_h denotes a shift by $h : \tau_h a(x) = a(x - h)$. The same formulas hold for the operator F^* with x and ξ replaced by $-x$ and $-\xi$.

Linear Changes of Variable. Let $A : X \to X$ be any linear automorphism, let $B : \Xi \to \Xi$ be an isomorphism such that the transform $A \times B$ of the space $X \times \Xi$ preserves the form $x \cdot \xi$, and hence also the density $|dx \, d\xi|$, and let $B = A'^{-1}$. We have

$$B^*(F(\rho)) = F(A^*(\rho)), \quad B^*(F(f)) = F(A^*(f)) \;, \tag{3.1.4}$$

where the asterisk denotes the action of the linear transform on a function or a density.

The convolution of two integrable densities ρ and σ is defined by the formula

$$\rho * \sigma = \int_S \rho \times \sigma , \tag{3.1.5}$$

where $S : X \times X \to X$ is the mapping that takes (x, y) to $x + y$. The convolution is also an integrable density. The fundamental property of the Fourier transform is contained in the equalities

$$F(\rho * \sigma) = F(\rho) \cdot F(\sigma), \quad F(f \cdot g) = F(f) * F(g) . \tag{3.1.6}$$

In the second equality the densities $F(f)$ and $F(g)$ must be assumed integrable. In these equalities the operator F can be replaced by F^*.

1.4. The Fourier Transform of Semidensities. A *semidensity* in $X = \mathbb{R}^n$ is an expression

$$s(x) = f(x)|dx|^{\frac{1}{2}} ,$$

which under a change of variable $x = x(y)$ transforms according to the rule

$$s(x(y)) = f(x(y)) \left| \det \frac{\partial x}{\partial y} \right|^{\frac{1}{2}} |dy|^{\frac{1}{2}} .$$

The square of a semidensity, and also the square of its absolute value are densities and consequently, can be integrated if the coefficient f is square-integrable. Let $L^2(X)$ be the space of semidensities on X with square-integrable absolute value. It is a Hilbert space with the inner product

$$(s, t) = \int_X s \cdot \bar{t} .$$

The Fourier transform $F : L^2(X) \to L^2(\Xi)$ acting according to the formula

$$F(s) = \int_X e^{-2\pi i \xi \cdot x} s \cdot |dx\, d\xi|^{\frac{1}{2}} = \int_X e^{-2\pi i \xi \cdot x} f(x)|dx| \cdot |d\xi|^{\frac{1}{2}} ,$$

the integral being interpreted as in Theorem 2. This is a unitary operator whose adjoint can be written by a similar integral with the sign in the exponent reversed. In this respect the theory of semidensities is simpler than the theory of transforms of functions and densities. However the operations of multiplication and convolution lead outside the class of semidensities, since the product of semidensities is a density and their convolution is a function. However the formulas (3.1.6) are preserved if $F(s * t)$ is understood as the Fourier transform of a function and $F(s \cdot t)$ as the Fourier transform of a density.

Semidensities arise naturally as the symbols of trigonometric integrals (Hörmander 1971; Maslov, Fedoryuk 1976).

1.5. Let r be a distribution in X with compact support. It can be extended (uniquely) to a continuous functional on the space $\mathcal{E}(X)$ (cf. Sect. 2.3 of Chapt. 1). Consequently the Fourier transform of r can be defined by the formula

$$\tilde{r}(\xi) = F(r) \equiv r(e^{-2\pi i \xi \cdot x}) .$$

Theorem (Paley, Wiener 1934; Schwartz 1957). *The function $\tilde{r}$ can be extended to an entire function $\tilde{r}(\zeta)$ in $\Xi \otimes \mathbb{C} \cong \mathbb{C}^n$ that satisfies the following inequality with some constants q and C:*

$$|\tilde{r}(\zeta)| \leq C(|\zeta| + 1)^q \exp(2\pi\gamma(\operatorname{Im}\zeta)) , \tag{3.1.7}$$

where γ is the support function of the compact set $K = \operatorname{supp} r$, i.e.,

$$\gamma(\eta) = \max_{x \in K} \eta \cdot x .$$

Conversely, if f is an entire function satisfying (3.1.7), where γ is the support function of some compact convex set $K \subset X$, then $f = F(r)$, where r is a uniquely defined distribution with support in K. This distribution can be found from the formula $r(\varphi) = \langle f, F^*(\varphi) \rangle$. If the generalized function u in the Paley-Wiener-Schwartz theorem is of compact support, the Fourier transform is actually being applied to the distribution $r = u|dx|$. All the properties enumerated in Sect. 1.3 are preserved; the convolution of distributions will be discussed in the next section. Generalizations of the Paley-Wiener Theorem for the case of symmetric spaces can be found in (Gel'fand, Graev, Vilenkin 1962; Helgason 1984).

§2. The Convolution of Distributions

2.1. The convolution of the distributions $u, v \in \mathcal{D}'(\mathbb{R}^n)$ is defined by the formula

$$u * v := s_*(u \otimes v) , \tag{3.2.1}$$

where $s : \mathbb{R}^n \times \mathbb{R}^n \to \mathbb{R}^n$ is the group operation in the additive group $\mathbb{R}^n$: $s(x, y) = x + y$. Here it is usually assumed that the support of the distribution $u \otimes v$ is proper over $\mathbb{R}^n$. The result of the convolution is a distribution on $\mathbb{R}^n$, and

$$\operatorname{supp}(u * v) \subset \operatorname{supp} u + \operatorname{supp} v .$$

This operation is commutative. According to formula (3.2.1) one can also define the convolution of a generalized function u and a distribution v by fixing an orientation of $\mathbb{R}^n$. In this case the direct product $u \otimes v$ is a current of dimension n on $\mathbb{R}^n$. Therefore, according to Sect. 2.2 of Chapt. 2, its direct image $s_*(u \otimes v)$ is a current of the same dimension on the oriented space $\mathbb{R}^n$, i.e., a generalized function. For generalized functions u and v the convolution can be defined as the *convolution of the distributions* $u|dx|$ and $v|dx|$.

The convolution of distributions or generalized functions is defined if one of the factors is of compact support; the other factor is arbitrary. If we do not assume that one of the factors is of compact support, some restrictions must be imposed on the support of each. Let C be a convex closed cone in $\mathbb{R}^n$ with vertex at the origin and containing no lines. Consider the space A_C of distributions in $\mathbb{R}^n$ satisfying the condition $\operatorname{supp} u \subset C + K$, where K is a compact set depending on u. For any two elements of A_C the convolution (3.2.1) makes sense and belongs to A_C. The operation of convolution in A_C is associative and commutative. In the algebra A_C there are no divisors of zero, and for distributions with compact supports a more precise result is known:

Theorem (cf. (Lions 1953; Mikusinski 1960)). *For any* $u, v \in \mathcal{E}'(\mathbb{R}^n)$

$$H(\operatorname{supp}(u * v)) = H(\operatorname{supp} u) + H(\operatorname{supp} v) ,$$

where $H(G)$ *denotes the convex hull of the set* $G \subset \mathbb{R}^n$.

This equality is preserved in certain cases if the supports are replaced by singular supports (sing supp). Such an equality holds in particular if one of the distributions has finite support, for example, if $u*$ is a differential operator (Gårding 1947; Malgrange 1960).

The problem of defining the convolution without a priori assumptions on the supports is discussed in (Antosik, Mikusinski, Sikorski 1973), where examples are given of the lack of associativity with such a convolution.

2.2. Riemann-Liouville Integrals. Let $\mathbb{C}^{\pm}$ be the right or left half-line and $A_{\mathbb{C}^{\pm}}$ the algebra with respect to convolution generated by distributions on the line equal to zero in a neighborhood of $\mp\infty$. The Hadamard kernel $a_\lambda^+ = \frac{x_+^{\lambda-1}}{\Gamma(\lambda)}|dx|$ (Sect. 4.5 of Chapt. 1) belongs to $A_{\mathbb{C}^+}$. The convolution operator with this kernel

$$a_\lambda^+ * f = \frac{1}{\Gamma(\lambda)} \int_{-\infty}^{x} (x-t)^{\lambda-1} f(t)\, dt$$

is called the fractional *Riemann-Liouville integral* of order λ. It converges if $\operatorname{Re}\lambda > 0$ and $f \in A_{\mathbb{C}^+}$ is a continuous function; it can be extended as an entire function of λ in the sense of a convolution of distributions. It follows from the classical formula

$$\frac{x_+^{\lambda-1}}{\Gamma(\lambda)}|dx| * \frac{x_+^{\mu-1}}{\Gamma(\mu)}|dx| = \frac{x_+^{\lambda+\mu-1}}{\Gamma(\lambda+\mu)}|dx| ,$$

in view of the uniqueness of analytic continuation, that

$$a_\lambda^+ * a_\mu^+ = a_{\lambda+\mu}^+ \tag{3.2.2}$$

for all $\lambda, \mu \in \mathbb{C}$. Thus the family of Hadamard kernels forms a group with respect to convolution isomorphic to the group of complex numbers. For an integer $n \geq 0$ the convolution operator with $a_{-n}^{+} = \delta_0^{(n)}|dx|$ is the nth derivative operator. Therefore by (3.2.2) a_n^+ is the nth derivative operator in $A_{\mathbb{C}^+}$, and its action on an ordinary function f is given by the Dirichlet formula

$$F_n(x) = \frac{1}{(n-1)!} \int_{-\infty}^{x} (x-t)^{n-1} f(t)\, dt\,.$$

In the general case a_λ^+ is called the integration operator of order λ (or the differentiation operator of order $-\lambda$). If $-\lambda$ is not a nonnegative integer, then $\operatorname{supp} a_{-\lambda}^{+} = \mathbb{R}^+$, i.e., the operator of fractional differentiation is not local.

The kernels a_λ^- satisfy a relation analogous to (3.2.2).

2.3. The M. Riesz Groups. In $\mathbb{R}^n$ we consider the quadratic form $q(x) = x_1^2 - x_2^2 - \cdots - x_n^2$; let C be a convex cone in the half-space $x_1 \geq 0$, where $q \geq 0$ (the future cone). M. Riesz (1949) studied in detail the family of distributions defined for $\operatorname{Re}\lambda > 0$ by the formula

$$q_+^\lambda(\varphi) = \int_C q^\lambda(x)\varphi(x)\, dx, \quad \varphi \in \mathcal{D}(\mathbb{R}^n)\,.$$

This family has a meromorphic extension to the complex plane with poles at the points $\lambda = -1, -2, \ldots$ and $\lambda = -\frac{n}{2}, -\frac{n}{2} - 1, \ldots$ (cf. also §2 of Chapt. 4), and after normalization

$$Z_\lambda = \frac{q_+^{\lambda - \frac{n}{2}}}{\pi^{\frac{n-2}{2}} \cdot 2^{2\lambda-1} \cdot \Gamma(\lambda)\Gamma\left(\lambda + 1 - \frac{n}{2}\right)} \tag{3.2.3}$$

becomes an entire function of the parameter λ with values in $S'(\mathbb{R}^n)$, and $\operatorname{supp} Z_\lambda \subset C$. Thus the distribution Z_λ belongs to the algebra A_C and therefore their convolution is defined for all values of the parameters. The following important formula holds

$$Z_\lambda * Z_\mu = Z_{\lambda+\mu}\,. \tag{3.2.4}$$

The points $\lambda = 0, -1, -2, \ldots$ are poles of the numerator and denominator in (3.2.3), and consequently the value of Z_λ at these points can be found as a ratio of residues:

$$Z_0 = \delta_0|dx|, \quad Z_{-k} = \Box^k Z_0\,, \tag{3.2.5}$$

where $\Box = \frac{\partial^2}{\partial x_1^2} - \frac{\partial^2}{\partial x_2^2} - \cdots - \frac{\partial^2}{\partial x_n^2}$ is the differential operator dual to the form q. In particular the operator of convolution with Z_0 is the identity, which together with (3.2.4), means that under the operation of convolution the family of Riesz kernels Z_λ is a group isomorphic to $\mathbb{C}$, and (3.2.5) implies the relation

$$\Box^k Z_k = Z_{-k} * Z_k = Z_0 = \delta_0|dx|\,,$$

which means that Z_k is a fundamental solution (cf. Chapt. 4) for the hyperbolic operator $\Box^k$ with support contained in the cone C. We remark that in the case of an even n the point $\lambda = k = \frac{n}{2} - 1$ is a pole of the numerator and the denominator of the fraction (3.2.3), as a consequence of which the support Z_k belongs to the boundary of the cone C. This fact is an expression of the strong Huyghens principle: in an even-dimensional space the wave described by the d'Alembertian operator $\Box^k$ has a back, in particular the wave from a point source is concentrated on the surface of a sphere.

The computation of Z_1 as a ratio of residues gives the explicit formula

$$Z_1 = \frac{Y(x_1)}{2\pi^{\nu+1}}\delta^{(\nu)}(q)|dx| \,, \quad \nu \equiv \frac{n}{2} - 2 \geq 0$$

(the definition of the delta-function $\delta(q)$ on the cone $q = 0$ is given in §1 of Chapt. 4). When n is odd, the support of Z_1 coincides with the entire cone C and

$$Z_1 = \frac{1}{2\pi^{\nu+1}\Gamma(-\nu)} q_+^{-\nu-1} \,.$$

If we replace the quadratic form q by a positive-definite form $s(x) = \sum_1^n x_l^2$, a family of kernels arises that display some analogies. A *Riesz potential* (M. Riesz 1949; Helgason 1980) is a distribution in $\mathbb{R}^n$ defined by the formula

$$I^\lambda = \frac{\Gamma(\frac{n}{2} - \lambda)}{4^\lambda \pi^{\frac{n}{2}} \Gamma(\lambda)} s^{\lambda - \frac{n}{2}} |dx|$$

for values of λ in the half-plane $\operatorname{Re}\lambda > 0$. This family can be extended meromorphically to the whole complex plane with poles at the points $\lambda = \frac{n}{2}, \frac{n}{2} + 1, \ldots$. For every λ the distribution I^λ is slowly increasing (cf. §3). The group property of Riesz potentials can be written as follows:

$$I^\lambda * I^\mu * \varphi = I^{\lambda+\mu} * \varphi, \quad \operatorname{Re}\lambda + \operatorname{Re}\mu < \frac{n}{2} \,.$$

It can be verified directly for $\operatorname{Re}\lambda > 0$ and $\operatorname{Re}\mu > 0$, when the potentials are ordinary functions and, consequently, is true in the domain $\operatorname{Re}\lambda + \operatorname{Re}\mu < \frac{n}{2}$ by the uniqueness of analytic continuation. The following formulas hold:

$$\Delta I^\lambda = -I^{\lambda-1}, \quad I^0 = \delta_0 |dx| \,,$$

where Δ is the Laplacian. Therefore when $k \neq \frac{n}{2}, \frac{n}{2} + 1, \ldots$ the distribution $(-1)^k I^k$ is a fundamental solution for the operator Δ^k.

Analogues of the M. Riesz groups are described in (Gårding 1947; Erdélyi 1975).

§3. The Fourier Transform of Distributions of Unrestricted Growth

3.1. The Schwartz Space. The vector space $S = S(\mathbb{R}^n)$ consists of infinitely differentiable functions φ in $\mathbb{R}^n$ satisfying the following conditions: for any multi-indices $i = (i_1, \ldots, i_n)$ and $j = (j_1, \ldots, j_n)$ the function $x^j \dfrac{\partial^i \varphi(x)}{\partial x^i}$ is bounded, i.e., the quantity

$$p_{i,j}(\varphi) = \sup_{\mathbb{R}^n} \left| x^j \frac{\partial^i \varphi(x)}{\partial x^i} \right|$$

is finite. The function $p_{i,j}$ is a seminorm on S. A sequence φ_k converges to φ in the space S if $p_{i,j}(\varphi_k - \varphi) \to 0$ for all i and j. Since the set of such seminorms is countable, this convergence is metrizable; the space S is complete with respect to this metric. The convergence and topology in S can be defined by the equivalent system of seminorms

$$p_N(\varphi) \equiv \|(|x|^2 - \Delta)^N \varphi\|_{L_2(\mathbb{R}^n)} .$$

Similarly one can define the space of rapidly decreasing smooth densities ρ, i.e., the space of densities of the form $\rho = \varphi\,|dx|$, where $\varphi \in S$. We shall denote this space also by $S(\mathbb{R}^n)$. In the spaces just described the operations of differentiation with respect to any coordinate and multiplication by any polynomial are defined and continuous, as is any linear change of variables. The functions $\varphi(x) = p(x)e^{-x^2}$, where p is a polynomial, form a dense set in S.

We now map $\mathbb{R}^n$ onto the "punctured" sphere $S^n \setminus N$ using stereographic projection (N is a point of S^n). Let $\mathcal{E}_0(S^n)$ be the space of infinitely differentiable functions on S^n that are *flat at the point* N, i.e., equal to zero along with all derivatives at that point. The inverse image of the space $\mathcal{E}_0(S^n)$ under this projection is all of $S(\mathbb{R}^n)$.

The Fourier transform operator F and the inverse Fourier transform operator F^{-1} establish an isomorphism of topological vector spaces

$$S(\mathbb{R}^n_x) \xrightarrow{F} S(\mathbb{R}^n_\xi) \xrightarrow{F^{-1}} S(\mathbb{R}^n_x) , \tag{3.3.1}$$

Here $S(\mathbb{R}^n_\xi)$ is a space of functions and $S(\mathbb{R}^n_x)$ is a space of densities or vice versa.

3.2. Slowly Increasing Generalized Functions. A continuous linear functional on the space S of functions (resp. densities) is called a *slowly increasing distribution* (resp. *generalized function*). The space $\mathcal{K}(\mathbb{R}^n)$ is a part of S and the convergence $\rho_k \to \rho$ in $\mathcal{K}$ implies that this convergence holds in S. Therefore a mapping $S' \to \mathcal{K}'$ is defined; it is an imbedding, since $\mathcal{K}$ is dense in S. The term "slowly increasing generalized function" is justified by the following facts:

A) every ordinary function f that is *slowly increasing*, i.e., such that $|f(x)| = O\big(|x|+1)^q\big)$ for some q, defines, by a formula analogous to (1.1.1), a continuous functional $[f] \in S'$;
B) every element $u \in S'$ can be written as

$$u = (-\Delta + 1)^q [f] \,,$$

where f is a slowly increasing ordinary function and $q \geq 0$ is an integer;
C) for a generalized function $u \in \mathcal{K}'$ to be slowly increasing it is necessary and sufficient that for any $\rho \in \mathcal{K}$ the convolution $u * \rho \in \mathcal{E}(\mathbb{R}^n)$ be slowly increasing.

3.3. The Fourier Transform in S'. In accordance with (3.1.3) the *Fourier transform* of *slowly increasing* generalized functions and distributions is performed according to the formulas

$$F'(u)(\psi) = u(F(\psi)), \quad F^{-1\prime}(u)(\psi) = u(F^{-1}(\psi)) \,.$$

According to (3.3.1) the operators F' and $F^{-1\prime}$ are mutually inverse and establish an isomorphism of the Schwartz spaces $S'(\mathbb{R}^n_x) \cong S'(\mathbb{R}^n_\xi)$. On distributions of compact support the operator F' coincides with the Fourier transform introduced in Sect. 1.5. All the properties of the usual Fourier transform except (3.1.6) are preserved for the operators just described. For interpretation of the first relation (3.1.6) see Sect. 2.1.

3.4. The Poisson Summation Formula. Because of the choice of the coefficient 2π in (3.1.1) this formula can be written very simply:

$$\sum_{k=-\infty}^{\infty} F(f)(k) = \sum_{k=-\infty}^{\infty} f(k), \quad f \in S(\mathbb{R}^1) \,.$$

This equality can be rewritten in the form

$$\sum F(\delta_k) = \sum \delta_k \,.$$

We note the related formula

$$\sum F(f)\left(k + \frac{1}{2}\right) = \sum (-1)^l f(l) \,.$$

More complicated formulas of the same type were obtained by Ramanujan (1914). In these formulas f is an even function and f and $\tilde{f} = F(f)$ decrease sufficiently rapidly at infinity; if $f \in S$, these conditions are satisfied:

$$\tilde{f}\Big(\frac{1}{\sqrt{8}}\Big) - \tilde{f}\Big(\frac{3}{\sqrt{8}}\Big) - \tilde{f}\Big(\frac{5}{\sqrt{8}}\Big) + \tilde{f}\Big(\frac{7}{\sqrt{8}}\Big) + \cdots$$
$$\cdots = f\Big(\frac{1}{\sqrt{8}}\Big) - f\Big(\frac{3}{\sqrt{8}}\Big) - f\Big(\frac{5}{\sqrt{8}}\Big) + f\Big(\frac{7}{\sqrt{8}}\Big) + \cdots \,,$$

where $1, 3, 5, 7, \ldots$ are the odd numbers, and

$$\tilde{f}\Big(\frac{1}{\sqrt{12}}\Big) - \tilde{f}\Big(\frac{5}{\sqrt{12}}\Big) - \tilde{f}\Big(\frac{7}{\sqrt{12}}\Big) + \tilde{f}\Big(\frac{11}{\sqrt{12}}\Big) + \cdots$$
$$\cdots = f\Big(\frac{1}{\sqrt{12}}\Big) - f\Big(\frac{5}{\sqrt{12}}\Big) - f\Big(\frac{7}{\sqrt{12}}\Big) + f\Big(\frac{11}{\sqrt{12}}\Big) + \cdots,$$

where $1, 5, 7, 11, \ldots$ are the numbers relatively prime to 6.

3.5. The Fourier Transform of Power Functions. The transformation formulas for power functions of one variable with integer exponents are shown in the following figure:

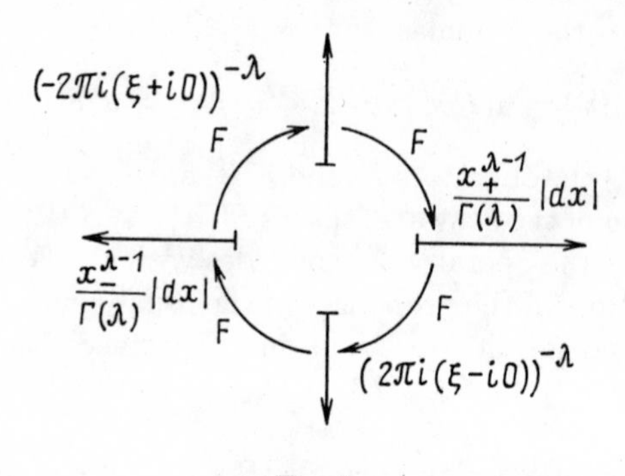

Fig. 1

From this and from §4 of Chapt. 1 there follow formulas for integer exponents:

$$F(\{x^{-n}\}|dx|) = \frac{(2\pi i)^n}{2(n-1)!}\xi^{n-1}\operatorname{sgn}\xi\,.$$

Consider the quadratic form

$$P(x) = x_1^2 + \cdots + x_n^2 - x_{p+1}^2 - \cdots - x_{p+q}^2$$

of rank $p + q = n$ defined on $\mathbb{R}^n$ and the form

$$Q(\xi) = \xi_1^2 + \cdots + \xi_p^2 - \xi_{p+1}^2 - \cdots - \xi_{p+q}^2\,,$$

defined on the dual space. For each complex λ we define a slowly increasing generalized function

$$(P \pm i0)^\lambda(\rho) = \lim_{\varepsilon\to+0}\int_{\mathbb{R}^n}(P \pm i\varepsilon)^\lambda\rho\,.$$

As a function of the parameter λ this generalized function has poles at the points $\lambda = -\frac{n}{2}, -\frac{n}{2} - 1, \ldots$, and the normalized family $(\pi(P \pm i0))^{\lambda - \frac{n}{2}}/\Gamma(\lambda)$ is an entire function of λ. Moreover at the points $\lambda = -k$, $k = 0, 1, \ldots$, it assumes the values (Gel'fand, Shilov 1958a)

$$\lim_{\lambda \to -k} \frac{(\pi(P \pm i0))^{\lambda - \frac{n}{2}}}{\Gamma(\lambda)} = e^{\mp \frac{\pi i q}{2}} \frac{\left(-\frac{\Box}{4\pi}\right)^k \delta_0}{\Gamma(\frac{n}{2} + k)}, \quad k = 0, 1, \ldots,$$

where $\Box = \frac{\partial^2}{\partial x_1^2} + \cdots + \frac{\partial^2}{\partial x_p^2} - \cdots - \frac{\partial^2}{\partial x_{p+q}^2}$ is the dual differential operator (cf. §1 of Chapt. 4). The following equality holds

$$F\left(\frac{(\pi(P \pm i0))^{\lambda}}{\Gamma(\lambda + \frac{n}{2})} |dx|\right) = e^{\mp \frac{\pi i q}{2}} \frac{(\pi(Q \mp i0))^{-\lambda - \frac{n}{2}}}{\Gamma(-\lambda)},$$

In the case of the positive definite form $P(x) = \sum x_i^2$ a more general formula is known: for every homogeneous harmonic polynomial h of degree $k = 1, 2, \ldots$

$$F\left(\frac{h(\pi P)^{\lambda - \frac{k}{2}}}{\Gamma(\lambda + \frac{n+k}{2})} |dx|\right) = t^{-k} \frac{h(\pi Q)^{-\lambda - \frac{n+k}{2}}}{\Gamma(-\lambda + \frac{k}{2})},$$

for $0 < \operatorname{Re} \lambda < \frac{n}{2}$. In the case $\lambda = 0$ the equality is preserved if the integral on the left-hand side is interpreted as a principal value (Stein, Weiss 1971).

A large number of explicit formulas for the Fourier transform can be found in (Gel'fand, Shilov 1958a; Brychkov, Prudnikov 1977).

3.6. The Fourier Transform of Generalized Functions in Convex Domains. According to the *Paley-Wiener Theorem* (Paley, Wiener 1934) the Fourier transform of square-integrable functions of compact support coefficients φ is the set of functions of $L_2(\mathbb{R}^n_\xi)$ that admit an entire analytic extension to $\mathbb{C}^n$ with the estimate $O(\exp(R|\operatorname{Im} \zeta|)$ for some $R > 0$. Let $K \subset \mathbb{R}^n$ be an arbitrary compact set; its support function is defined as the quantity

$$\gamma_K(\eta) \equiv \max_{x \in K} \eta \cdot x,$$

depending on the point of the dual space $\mathbb{R}^n$. Let Z_K denote the space of entire functions in $\mathbb{C}^n$ satisfying the inequality

$$|\psi(\zeta)| \leq C_q(|\zeta| + 1)^{-q} \exp(2\pi\gamma_K(\operatorname{Im} \zeta))$$

with any $q \geq 0$. In this space we introduce a topology using the sequence of norms $p_q(\psi)$, $q = 0, 1, \ldots$, where $p_q(\psi)$ is a lower bound for the constant C_q satisfying the last inequality. Let $\mathcal{K}_K$ be the subspace of $\mathcal{K}(\mathbb{R}^n)$ formed by the densities whose supports belong to K.

Theorem 1 (cf. (Schwartz 1957; Ehrenpreis 1960, 1970)). *If the compact set K is convex, the direct and inverse Fourier transforms establish isomorphisms of topological vector spaces*

$$Z_K \xrightarrow{F^{-1}} \mathcal{K}_K \xrightarrow{F} Z_K ,$$

if the topology of $\mathcal{K}_K$ is defined by the convergence described in Sect. 1.1 of Chapt. 1.

No similar assertion can hold for a nonconvex set K, since its support function γ_K coincides with the support function of its convex hull.

It follows from Theorem 1 that there is an isomorphism of topological spaces $F^{-1} : Z(\Omega) \to \mathcal{K}(\Omega)$ for any convex domain Ω, where $Z(\Omega)$ is the inductive limit of the spaces Z_K taken over some increasing sequence of compact sets $K_i \subset \Omega$ that exhaust the domain. It follows from this that the *inverse Fourier transform* of generalized functions defined in Ω can be defined as a continuous linear functional on $Z(\Omega)$. Similarly the *direct Fourier transform* $F'(u)$ is a functional on $Z(-\Omega)$. The latter can be written as the integral of a test function with a suitable density defined on $\mathbb{C}^n$. More detailed reasoning leads to the following result:

Theorem 2. *For every generalized function u defined in a convex domain Ω, the Fourier transform $F'(u)$ can be written as an integral*

$$F'(u)(\psi) \equiv u(F(\psi)) = \int \psi h \, d\xi \, d\eta, \quad \zeta = \xi + i\eta, \quad \psi \in Z(-\Omega) , \tag{3.3.2}$$

where h is some measurable function in $\mathbb{C}^n$ satisfying the following condition: for every compact set $K \subset \Omega$ there exists a number q such that

$$\int \left| (|\zeta| + 1)^{-q} \exp(2\pi\gamma_K(-\eta)) h(\zeta) \right|^2 d\xi \, d\eta < \infty . \tag{3.3.3}$$

Conversely if the function h satisfies this condition, then the integral (3.3.2) *defines some generalized function on Ω.*

Example 1. The Fourier transform of the exponential

$$u(x) = \exp(2\pi i \theta \cdot x)$$

is, according to the definition just given, the delta-distribution δ_θ at the point $\theta \in \mathbb{C}^n$, i.e., the functional on $Z(\mathbb{R}^n)$ acting according to the formula $\delta_\theta(\psi) = \psi(\theta)$. To write this distribution in the form (3.3.2) one can proceed as follows: using the fact that the function ψ is harmonic, we represent $\psi(\theta)$ as the average of ψ over a sphere of radius r and then average this representation with respect to r with some smooth finite density $\rho(r)$ such that $\int \rho = 1$.

The Fourier transform of an exponential polynomial $u(x) = p(x)\exp(2\pi i\theta \cdot x)$ is a linear combination of the derivatives of the delta-distribution $F'(u) = p\Big(\frac{i}{2\pi}\frac{\partial}{\partial\zeta}\Big)\delta_\theta$.

Example 2. The Fourier transform of the rapidly increasing function $u(x) = e^{\pi x^2}$ can be represented by a measure concentrated on the imaginary subspace

$$F'(u)(\psi) = \int_{\mathbb{R}^n} e^{-\pi\eta^2}\psi(i\eta)\,d\eta\,.$$

This integral can be converted into (3.3.2) by the device described in the previous example.

3.7. Description of the Densities that Define Zero Functionals. As was seen in the examples, the function h in (3.3.2) can be chosen with a great degree of arbitrariness.

Example 3. Let $\theta \in \mathbb{C}^n$, and let μ be a density whose integral is 1 uniformly distributed on a sphere in $\mathbb{C}^n$ with center at θ. The distribution $\delta_\theta - \mu$ defines the zero functional on $Z(\mathbb{R}^n)$.

Example 4. If $y \in \mathbb{R}^n \setminus \Omega$, the density $\exp(-2\pi i y\cdot\xi)|d\xi|$ defines the zero functional on $Z(-\Omega)$, since

$$\int \exp(-2\pi i y\cdot\xi)\psi(\xi)\,d\xi = F(\psi)(y) = 0$$

for every function $\psi \in Z(-\Omega)$ by virtue of Theorem 1.

Theorem 3 (cf. (Palamodov 1962a)). *A necessary and sufficient condition for a function h satisfying (3.3.3) to define the zero functional on $Z(-\Omega)$ via formula (3.3.2) is that for any compact set $K \subset \Omega$ there exist a number $q \geq 0$ and square-integrable functions λ_j on $\mathbb{R}^n \times \mathbb{R}^n$, $j = (j_1, \dots, j_n)$, $|j| \leq q$, such that for every continuous function φ of compact support on $\mathbb{C}^n$*

$$\int \varphi h\,d\xi\,d\eta = \sum_{|j|\leq q}\int_{\mathbb{R}^n\times\mathbb{R}^n}\int_{\Gamma_{y,\tau}} \exp(-2\pi i y\cdot\zeta)\zeta^j\varphi(\xi,\eta)\,d\xi\,\times$$

$$\times\exp(-2\pi(\gamma_K(-\tau) + y\cdot\tau))\lambda_j(y,\tau)\,dy\,d\tau\,,$$

where $\Gamma_{y,\tau}$ is an n-dimensional chain equal to $\mathbb{R}^n_\xi + i\tau$ for $y \notin K$ and equal to the union of $\mathbb{R}^n_\xi + i\tau$ and the manifold $\mathbb{R}^n_\xi$ taken with the reverse orientation in the case $y \in K$.

For an application of this result cf. Sect. 3.3 of Chapt. 4.

The Fourier transform operator $F' : \mathcal{K}'(\mathbb{R}^n) \to Z'(\mathbb{R}^n)$ and the inverse operator $(F^{-1})'$ have all the properties enumerated in Sect. 1.3; in the first equality of (3.1.6) one of the distributions, for example ρ, has compact support. By the Paley-Wiener-Schwartz theorem the function $F'(\rho)$ is an entire function and satisfies (3.1.7), and therefore is a multiplier in $Z'(\mathbb{R}^n)$.

§4. The Fourier Transform of Ultradistributions and Smooth Functions

4.1. Other Test Spaces. Suppose fixed a sequence $M = \{M_p\}$ of positive numbers. In $\mathcal{D}(\mathbb{R}^n)$ we consider the subspace $\mathcal{D}^M$ consisting of the functions φ that satisfy the inequalities

$$\left|\frac{\partial^i \varphi(x)}{\partial x^i}\right| \leq CA^p M_p, \quad p \equiv |i| = 0, 1, \ldots . \tag{3.4.1}$$

In one of the versions of the definition A and C are constants depending on φ, while in the other $A > 0$ is arbitrary and C depends on A. It is usually assumed that the sequence M is *logarithmically convex*, i.e., $M_p^2 \leq M_{p-1}M_{p+1}$, and conditions are imposed that are necessary for the space $\mathcal{D}^M$ to be an algebra with respect to multiplication and for the operations of differentiation with respect to the coordinate fields to be defined on it. The space $\mathcal{D}^M$ is endowed with a natural convergence; continuous linear functionals on $\mathcal{D}^M$ are called *ultradistributions.* Since there is an embedding $\mathcal{D}^M \hookrightarrow \mathcal{D}$ that preserves convergence, a mapping $i : \mathcal{D}' \to \mathcal{D}^M$ arises from the space of distributions into the space of ultradistributions. Different versions of this idea, which leads to an extension of the concept of distribution (or generalized function), were developed by Gel'fand and Shilov (1958b), Sebastião a Silva (1958), Roumieu (1960), and Beurling; detailed expositions can be found in (Lions and Magenes 1968/70) and (Komatsu 1973).

The fundamental question of the theory is whether the mapping i is an imbedding, or at least whether the space $\mathcal{D}^M$ contains a function not identically zero. The answer is contained in the following theorem.

Theorem (Denjoy-Carleman-Mandelbrojt, cf. (Komatsu 1973)). *A necessary and sufficient condition for the existence of a nontrivial function $\varphi \in \mathcal{D}^M$ is the inequality*

$$\sum_{p=1}^{\infty} \frac{M_{p-1}}{M_p} < \infty$$

or the equivalent condition $\sum M_p^{\frac{1}{p}} < \infty$. If this condition holds, then $\mathcal{D}^M$ is dense in $\mathcal{D}(\mathbb{R}^n)$, and, consequently the mapping i is an imbedding. For any covering of $\mathbb{R}^n$ one can construct a partition of unity subordinate to it formed by functions satisfying (3.4.1).

For every open set $\Omega \subset \mathbb{R}^n$ one can define the space $\mathcal{D}^M(\Omega) = \mathcal{D}^M \cap \mathcal{D}(\Omega)$ with the following convergence: a sequence $\{\varphi_k\}$ converges in $\mathcal{D}^M(\Omega)$ if it converges in $\mathcal{D}^m$ and in $\mathcal{D}(\Omega)$. The dual spaces $\mathcal{D}^M(\Omega)'$ form a sheaf on $\mathbb{R}^n$. This sheaf is flabby. The structure of its elements can be described as follows.

Proposition (cf. (Roumieu 1960; Komatsu 1973)). *Let $u \in \mathcal{D}^M(\Omega)'$, and let $K \subset \Omega$ be a compact set. The values of the functional u on the functions $\varphi \in \mathcal{D}^M$ with $\operatorname{supp}\varphi \subset K$ can be written in the form*

$$u(\varphi) = \sum_i [f_i]\left(\frac{\partial^i \varphi}{\partial x^i}\right),$$

where f_i are integrable functions on K subject to the inequality

$$\sum_{p=|i|=0}^{\infty} A^p M_p \|f_i\|_{L_1(K)} < \infty$$

for any A (resp. for some A).

The requirement of compact support on the test functions can also be replaced by the condition that all the derivatives be rapidly decreasing, that is, by the condition

$$\left| x^j \frac{\partial^i \varphi(x)}{\partial x^i} \right| \leq C A^p M_p \cdot B^q N_q, \quad p = |i|, \quad q = |j| = 0, 1, \dots\,, \tag{3.4.2}$$

where $M = \{M_p\}$ and $N = \{N_q\}$ are given positive sequences and the constants A, B, and C can depend on φ. These inequalities distinguish a space, denoted $\mathcal{D}_N^M$, where a suitable convergence is introduced. The conjugate space $(\mathcal{D}_N^M)'$ can be naturally mapped into $\mathcal{D}^{M\prime}$. The image consists of ultradistributions with bounded growth at infinity. This growth is controlled by the choice of the sequence N. A comparatively simple scale $\{S_\beta^\alpha\}$ of this type was first considered by Shilov (cf. Gel'fand, Shilov 1958b). The space S_β^α is defined using the inequalities (3.4.2) in which $M_p = (p!)^\alpha$ and $N_q = (q!)^\beta$. In the case $\alpha < 1$ the functions $\varphi \in S_\beta^\alpha$ can be extended to entire functions in $\mathbb{C}^n$ admitting the estimate

$$|\varphi(z)| \leq C \exp\left(-\beta \left|\frac{x}{B}\right|^{\frac{1}{\beta}} + \alpha' |Ay|^{\frac{1}{\alpha'}} \right), \quad \alpha' = 1 - \alpha\,.$$

If $\alpha + \beta < 1$, i.e., $\beta < \alpha'$, the rate of decrease along the real space that is implied by this estimate is incompatible with any order of increase not greater than $\frac{1}{\alpha}$ by virtue of the Phragmén-Lindelöf theorem, which implies that S_β^α is trivial. If $\alpha + \beta \geq 1$, excluding the cases $\alpha = 0$, $\beta = 1$ and $\alpha = 1$, $\beta = 0$, this space is nontrivial and contains "sufficiently many" functions. A study of the nontriviality of the spaces $\mathcal{D}_N^M$ in general was conducted in (Babenko 1956).

The scale of spaces S_β^α was used in the study of the characteristic Cauchy problem for systems of equations of evolution type (Gel'fand, Shilov 1958c). An important property of this scale is that the Fourier transform preserves it.

Theorem (cf. (Ehrenpreis 1955, 1956bc; Sebastião a Silva 1958; Gel'fand, Shilov 1958b). *For any α and β the Fourier transform (and also the conjugate transform) defines an isomorphism*

$$F : S^{\alpha}_{\beta} \widetilde{\rightarrow} S^{\beta}_{\alpha} .$$

It is a topological isomorphism with respect to the natural topologies in these spaces.

It follows from this result that the dual operator of the Fourier transform maps $\left(S^{\beta}_{\alpha}\right)'$ isomorphically onto $\left(S^{\alpha}_{\beta}\right)'$.

For a more general scale of spaces of analytic functions a similar theorem was obtained by Hörmander (1955) in terms of the Legendre dual majorants. A similar study of the spaces $\mathcal{D}^M_N$ and their duals was conducted by Roumieu (1960); see also (Sebastião a Silva 1958).

4.2. Hyperfunctions. The theory of hyperfunctions can be regarded as an extension of the theory of generalized functions and ultradistributions. However this extension is based on different ideas. In the theory of Sato (1958) the space of *hyperfunctions* defined on an open set $U \subset \mathbb{R}^n$ is defined as the cohomology group $H^n_U(\widetilde{U}, \mathcal{O})$, where $\widetilde{U}$ is any complex neighborhood of U such that $\widetilde{U} \cap \mathbb{R}^n = U$, and $\mathcal{O}$ is the sheaf of holomorphic functions in $\mathbb{C}^n$. The connection of this construction with the theory of generalized functions is that every generalized function in U can be represented as the sum of the boundary values of a holomorphic $n-1$ cochain defined on a suitable covering of $\widetilde{U} \setminus U$. Thus the generalized functions in U correspond precisely to the cohomology classes $h \in H^n_U(\widetilde{U}, \mathcal{O})$, which can be represented by cochains having such boundary values. Such cochains are characterized by power growth near U. In the general case a hyperfunction can be interpreted only as an element of this cohomology group. The theory of hyperfunctions has some peculiarities that distinguish it from the theory of ultradistributions. In particular, the sheaf of hyperfunctions is flabby, i.e., every hyperfunction defined on an arbitrary open subset $U \subset \mathbb{R}^n$ can be extended to a hyperfunction on $\mathbb{R}^n$. On the other hand, the space of hyperfunctions defined on a noncompact set is absolutely inseparable in its natural topology.

An exposition of the elements of the theory of hyperfunctions can be found in (Schapira 1970); in (Sato, Kawai, Kashiwara 1973) hyperfunctions are applied to problems of the theory of differential equations. The construction of Fourier hyperfunctions (Kawai 1970; Kaneko 1972; Palamodov 1977) can serve as a connecting link between the theory of generalized functions and the theory of hyperfunctions. The space of Fourier hyperfunctions in $\mathbb{R}^n$ is isomorphic to the Shilov space $(S^1_1)'$. This space maps onto the group $H^n_{\mathbb{R}^n}(\mathbb{C}^n, \mathcal{O})$ of hyperfunctions defined on $\mathbb{R}^n$.

4.3. The Fourier Transform of Slowly Increasing Functions. A construction of the Fourier transform for slowly increasing functions of one variable was proposed by Bochner (1932) before the axiomatics of generalized functions appeared. If a function f defined on the real axis is integrable with the weight $(|x|+1)^{-k}$, its "k-transform" in the sense of Bochner is defined by the formula

$$E(\alpha, k) = \frac{1}{2\pi} \int f(x) \frac{e^{-i\alpha x} - L_k(x)}{(-ix)^k}\, dx\,,$$

where L_k is the segment of the MacLaurin series of the function $e^{-i\alpha x}$ up to terms of order $k-1$. Bochner wrote the inverse correspondence in the form

$$f(x) \sim \int e^{ix\alpha} \frac{d^k E(\alpha, k)}{d\alpha^{k-1}}\,.$$

This essentially means that the function f can be recovered as the inverse Fourier transform of the distribution equal to the product of the density $d\alpha$ and the kth derivative of the bounded continuous function $E(\alpha, k)$. Bochner (1932) also anticipated the statement of problems that later became central in the theory of generalized functions: convolution equations with a generalized kernel, the representation of the solutions, and the division problem.

The class of functions considered by Bochner can be imbedded in the Schwartz space S', and the Fourier transform operation in this space described in §3 provides a replacement for Bochner's construction and a generalization of it to the case of several variables. In addition L. Schwartz (1950/51) describes the Fourier image of the subspace $\mathcal{O}_M \subset S'$ consisting of smooth slowly increasing functions. The condition of slow increase means that the function and each of its derivatives $\dfrac{\partial^i f}{\partial x^i}$ admits the estimate $O((|x|+1)^q)$ at infinity, where q may depend on i. The space $\mathcal{O}_M$ is the set of multipliers in S', i.e., functions f such that the operator $\varphi \mapsto f\varphi$ is defined and continuous in S. A distribution r is *rapidly decreasing* if for every function $\varphi \in \mathcal{D}$ the convolution $r * \varphi$ tends to zero faster than any negative power of $|x|$. The set of rapidly decreasing distributions forms a space denoted $\mathcal{O}'_C$. It coincides with the space of *convolvers* in S, i.e., operators that commute with translations. The Fourier transform $F : S \to S$ maps the multipliers defined on the first of these spaces into convolvers acting on the second. Hence Schwartz concludes that the adjoint operator $F' : S' \to S'$ maps the space $\mathcal{O}_M$ into $\mathcal{O}'_C$ and establishes an isomorphism of these spaces that preserves in a natural manner the given convergences. Qualitatively the operator f' turns increase of functions of $\mathcal{O}_M$ into singularity of the distributions belonging to $\mathcal{O}'_C$, and smoothness into decrease.

Example. In the case $n = 1$ the function $f(x) = \exp(i\pi x^2)$ belongs simultaneously to $\mathcal{O}_M$ and $\mathcal{O}'_C$. The first assertion can be verified directly by differentiating this function and the second by integrating it. However, the second follows from the first by virtue of the relation

$$F'(\exp(i\pi x^2)) = \frac{1+i}{\sqrt{2}} \exp(-i\pi\xi^2)\,.$$

4.4. The Transform of Rapidly Increasing Smooth Functions. Consider a scale of spaces $\mathcal{E}_{\alpha,A}^{\beta,B}$ depending on four indices assuming positive values. The space $\mathcal{E}_{\alpha,A}^{\beta,B}$ consists of functions φ defined on $\mathbb{R}^n$ satisfying the system of inequalities

$$\left|\frac{\partial^i \varphi(x)}{\partial x^i}\right| \leq CB^p (p!)^\beta \exp\left(\frac{1}{\alpha}\left|\frac{x}{A}\right|^{\frac{1}{\alpha}}\right), \tag{3.4.3}$$
$$p = |i|, \quad \alpha > 0\,, \quad \beta \geq 0.$$

In the case $\alpha = 0$ this space consists of the infinitely differentiable functions defined in the ball $|x| \leq A$ and satisfying there the estimate

$$\left|\frac{\partial^i \varphi}{\partial x^i}\right| \leq CB^p (p!)^\beta\,.$$

For fixed α and β the spaces $\mathcal{E}_{\alpha,A}^{\beta,B}$ form a family depending on two indices, increasing on B and decreasing on A. From this family one can construct six limiting spaces using the operations of inductive and projective limits over each of the indices. Let $\mathcal{E}_\alpha^\beta$ denote the set of these limiting spaces. The spaces $\mathcal{E}_{\alpha,A}^{\beta,B}$ are naturally always nontrivial, but there is another question for them: for which values of the indices is such a space degenerate, i.e., imbedded in another space of the same scale $\mathcal{E}_{\alpha',A'}^{\beta',B'}$, in which it is not formally imbedded, i.e., at least one of the inequalities $\alpha' \leq \alpha$, $A' \leq A$, $\beta' \geq \beta$, $B' \geq B$ does not hold? Such an imbedding can arise because for $\beta < 1$ the estimate on the growth of the derivatives occurring in (3.4.3) restricts the growth of the holomorphic function φ on $\mathbb{R}^n$ more strongly than the exponential factor. The spaces $\mathcal{E}_{\alpha,A}^{\beta,B}$ are degenerate in the case $\alpha + \beta < 1$ and in the case $\alpha + \beta = 1$, $AB <$ const. Hence it follows that the spaces $\mathcal{E}_\alpha^\beta$ are degenerate in the cases $\alpha + \beta < 1$ or $\alpha + \beta = 1$, $\alpha\beta = 0$ and depend on the type of limiting passage. Thus a nearly complete parallelism exists between the cases when these spaces are degenerate and the cases when the spaces S_α^β are trivial.

It follows in particular from this that every nondegenerate space $\mathcal{E}_\alpha^\beta$ except $\mathcal{E}_0^1$ and $\mathcal{E}_1^0$ can be imbedded in the conjugate to the nontrivial space $S_{\alpha'}^\beta$, and consequently the Fourier transform operator described in Sect. 4.1 is applicable to its elements. On $\mathcal{E}_0^1$ and $\mathcal{E}_1^0$ the Fourier operator is also defined. The Fourier images of these spaces are described by the following theorem:

Theorem (cf. (Schwartz 1950/51; Ehrenpreis 1954, 1956b; Sebastião a Silva 1958; Ehskin 1961; Palamodov 1962b)). *For every nondegenerate space $\mathcal{E}_\alpha^\beta$ the Fourier transform defines an isomorphism that can be symbolically written as*

$$F : \mathcal{E}_\alpha^\beta \rightrightarrows (\mathcal{E}_\beta^\alpha)'\,.$$

More precisely, the Fourier image of such a space can be obtained by interchanging the upper and lower indices, replacing the inductive limit by the projective limit and the projective limit by the inductive limit and passing to the dual space.

§5. The Radon Transform

5.1. Expansion of the Delta-function in Plane Waves (John 1955). A *plane wave* in $\mathbb{R}^n$ is a function of the form $f(x) = g(\omega \cdot x)$, where g is a function of one variable and $\omega \cdot x = \sum \omega_i x_i$, ω being a fixed vector. When n is odd, the generalized function $\delta_0 \in \mathcal{D}'(\mathbb{R}^n)$ has the following expansion in plane waves

$$\delta_0 = \frac{1}{2(2\pi i)^{n-1}} \int\limits_{S^{n-1}} \delta^{(n-1)}(\omega \cdot x)\, d\omega\,, \tag{3.5.1}$$

where S^{n-1} is the unit sphere and $d\omega$ is its volume element. In the even-dimensional case

$$\delta_0 = \frac{(n-1)!}{(2\pi i)^n} \int\limits_{S^{n-1}} \{(\omega \cdot x)^{-n}\}\, d\omega\,. \tag{3.5.2}$$

(See Sect. 4.3, Chapt. 1 for the definition of $\{t^{-n}\}$.) The two cases can be combined in the single formula

$$\delta_0 = \frac{(n-1)!}{(\mp 2\pi i)^n} \int \frac{d\omega}{(\omega \cdot x \pm i0)^n}\,, \tag{3.5.3}$$

if the contributions of the antipodal points of the sphere are added (Gel'fand, Shilov 1958b). Formulas (3.5.1) and (3.5.2) follow from the expansion

$$2\pi^{\frac{n-1}{2}} \frac{|x|^\lambda}{\Gamma(\frac{\lambda+n}{2})} = \int \frac{|\omega \cdot x|^\lambda}{\Gamma(\frac{\lambda+1}{2})}\, d\omega\,.$$

The left- and right-hand sides can be extended as entire functions of the parameter λ (cf. §4 of Chapt. 1), and the left-hand side becomes the delta-function when $\lambda = -n$.

5.2. The Radon Transform. Let f be a continuous function in $\mathbb{R}^n$; its *Radon transform* is defined as the set of integrals

$$\check{f}(\omega, p) = \int\limits_{\omega \cdot x = p} f(x)\, dS\,,$$

taken over all hyperplanes in $\mathbb{R}^n$ (under the assumption that these integrals exist). The image of this transform is a function on the manifold $U = S^{n-1} \times \mathbb{R}$ satisfying the relation $\check{f}(-\omega, -p) = \check{f}(\omega, p)$. To recover a function from its Radon transform it suffices to write it in the form of a convolution $f = f * \delta_0\, |dx|$ and express the delta-function δ_0 by (3.5.1) and (3.5.2). Under suitable conditions of smoothness and decrease of f at infinity the *Radon formula* ($n = 2$) and the *John formula* (John 1955) are valid:

$$f(x) = \frac{1}{2(2\pi i)^{n-2}} \int\limits_{S^{n-1}} \left(\frac{\partial}{\partial p}\right)^{n-1} \check{f}(\omega, \omega \cdot x)\, d\omega$$

for odd n and

$$f(x) = \frac{(n-1)!}{(2\pi i)^n} \int \{(\omega \cdot x - p)^{-n}\}(\check{f}(\omega, p)\, dp)\, d\omega$$

for even n. When $n = 2$, this formula is the basis of a very effective method of computerized tomography (Natterer 1986).

Theorem (cf. (Helgason 1980)). *The Radon transform of the space $S(\mathbb{R}^n)$ coincides with the space S_K of infinitely differentiable functions $g(\omega, p)$ on U that are rapidly decreasing at infinity together with all derivatives satisfying the evenness condition $g(-\omega, -p) = g(\omega, p)$ such that for any $k = 0, 1, \ldots$ the integral*

$$\int_{\mathbb{R}^n} g(\omega, p) p^k \, dp$$

is a homogeneous polynomial of degree k.

An invariant definition of the Radon transform, which relates to densities ρ in $\mathbb{R}^n$ and is independent of the choice of metric, is more natural. In the manifold $\mathbb{R}^n \times U$ the hypersurface H defined by the equation $p = \omega \cdot x$ is considered. A density ρ is lifted to an odd form ρ^* on $\mathbb{R}^n \times U$ and then the quotient $\rho' = \dfrac{\rho^*}{d(\omega \cdot x)}$ is restricted to H. This restriction is an odd form, which can be integrated over the fibers of the submersion $h : H \to U$. The direct image

$$R\rho = \int_h \rho' ,$$

which is an odd 0-form on U, is the Radon transform of ρ.

The Radon transform of a generalized function can be defined starting from the Plancherel formula, which in the odd-dimensional case can be written as

$$\int fg \, dx = \frac{1}{2(2\pi i)^{n-1}} \iint \check{f}(\omega, p)(\partial/\partial p)^{n-1} \check{g}(\omega, p)\, dp\, d\omega \, .$$

If $u \in S'(\mathbb{R}^n)$, the Radon image $\check{u}$ is defined in (Gel'fand, Graev, Vilenkin 1962) as a functional on S_K, acting according to the formula

$$\check{u}((\partial/\partial p)^{n-1}\check{\varphi}) = 2(2\pi i)^{n-1} u(\varphi) \, .$$

This functional can be extended to the space of smooth rapidly decreasing functions on $S^{n-1} \times \mathbb{R}$, but such an extension is far from unique. For a generalized function with compact support the Radon transform with compact support is uniquely determined and there is an inversion formula (Helgason 1980). The Radon transform on symmetric spaces has also been studied in (Gel'fand, Graev, Vilenkin 1962) and (Helgason 1984).

5.3. Some Applications to Differential Equations. Let $p(\partial/\partial x)$ be a homogeneous linear differential operator of order m with constant coefficients. In application to plane waves it becomes an ordinary homogeneous operator:

$$p(\partial/\partial x)g(\omega \cdot x) = p(\omega)g^{(m)}(\omega \cdot x) .$$

This remark opens a simple route for constructing a fundamental solution for such an operator using the expansion of the delta-function in plane waves. Following this route, one obtains formulas for fundamental solutions of the Cauchy problem in the half-space $x_1 \geq 0$ for strictly hyperbolic operators

$$E(x) = -\frac{1}{(2\pi i)^{n-1}(m-n)!} \int_{\Gamma} (x \cdot \xi)^{m-n} \ln |x \cdot \xi| \frac{d\xi}{dp} , \quad p = p(\xi)$$

in the case of odd n and

$$E(x) = \frac{\pi i}{2(2\pi i)^{n-1}(m-n)!} \int_{\Gamma} (x \cdot \xi)^{m-n} \operatorname{sgn}(x \cdot \xi) \frac{d\xi}{dp}$$

in the case of even n. Here Γ is any $n-2$-dimensional cycle belonging to the real cone $K = \{p(\xi) = 0\}$ intersecting each ray in this cone exactly once, and oriented by the projection on the unit sphere in the hyperplane $\xi_1 = 0$ (cf. John 1955). After the integral over Γ is transformed into a sum of integrals over the components of the manifold $K \cap \{\xi_1 = 1\}$ (ovals) and the form $\frac{d\xi}{dp}$ is replaced by $\pm\frac{dS}{|\operatorname{grad} p|}$, the Herglotz-Petrovskij formulas arise (Petrovskij 1945; Atiyah, Bott, Gårding 1970). Generalizations of these formulas for hyperbolic operators with multiple characteristics were given in (Atiyah, Bott, Gårding 1970).

5.4. Other Transforms. The Radon transform in the complex space $\mathbb{C}^n$ is defined for functions $f \in S(\mathbb{C}^n)$ using integrals over complex hyperplanes (Gel'fand, Graev, Vilenkin 1962). The Radon transform of generalized functions defined in $\mathbb{C}^n$ can be defined using an analogue of the Plancherel formula. Another complex analogue of the Radon transform is the Penrose transform, which includes integration of $\bar{\partial}$-closed differential forms defined in a domain of complex projective space over projective lines. It is used to transfer the wave equations into the space of twistors. A survey of the results connected with the Radon-Penrose transform can be found in (Khenkin 1985).

Palamodov (1975) has constructed an expansion of generalized functions in "holomorphic waves," which is an analogue of the expansion in plane waves (Sect. 5.1), and also a resolvent of the space of generalized functions that begins with this expansion.

Properties and tables of the Fourier, Laplace, Mellin, Hilbert, Hankel, and Stieltjes transforms are given in (Zemanian 1968; Brychkov, Prudnikov 1977).

Chapter 4
Special Problems

§1. The Inverse Image of Generalized Functions in the Presence of Critical Points

1.1. Let $f : X \to Y$ be a mapping of smooth manifolds. The point $x \in X$ is said to be *critical* for f, and $f(x)$ is a *critical value* of f if $\operatorname{rang} df(x) < \dim_{f(x)} Y$. In particular an imbedding $X \hookrightarrow Y$ is critical at all points $x \in X$ where the dimension of X is less than the dimension of Y. In §3 of Chapt. 2, we defined the inverse image of any generalized function defined on Y under a submersion $f : X \to Y$, i.e., under a mapping that has no critical points. We now consider the following general problem: how should the inverse image of a generalized function be defined in the case when f has critical points? This problem combines several problems of the theory:

1) Define the value of a generalized function $u \in \mathcal{K}'(X)$ at a given point x (cf. Antosik, Mikusinski, Sikorski 1973). The value at a point can be interpreted as the inverse image under the imbedding $\{x\} \hookrightarrow X$. A more general problem is to define the restriction of a generalized function on a given submanifold $Z \subset X$ (cf. also §4 of Chapt. 2).
2) Give a meaning to the product of generalized functions u and v defined on a manifold X. This product can be interpreted as the restriction of the direct product $u \otimes v \in \mathcal{K}'(X \times X)$ to the diagonal $\Delta \subset X \times X$.
3) The problem of division by a function $F : X \to R$. This requires solving the equation $F \cdot u = v$ in $\mathcal{K}'(X)$ for every right-hand side $v \in \mathcal{K}'(X)$. In particular if $v = 1$, to solve this equation it suffices to construct the inverse image of the generalized function $\left[\frac{1}{y}\right] \in \mathcal{K}'(R)$ (cf. §2 of Chapt. 4).
4) Description of invariant generalized functions. Suppose a Lie group G operates on X and $F_1, \ldots, F_n$ is the complete set of invariant functions. They define a smooth mapping $F : X \to \mathbb{R}^n$. The inverse image of every generalized function v on $\mathbb{R}^n$ is an invariant generalized function on X. However not all invariant generalized functions are obtained this way. Other invariant generalized functions arise as the inverse images of functionals over some extension of the space $\mathcal{K}(\mathbb{R}^n)$ obtained by applying the operation of integration along the fibers F to $\mathcal{K}(X)$ (cf. below).

In general the inverse image is defined for ordinary, for example continuous, functions; its inverse image is again a continuous function. A generalized function has an inverse image under an arbitrary smooth mapping if sufficiently strong smoothness conditions are imposed on it.

Proposition 1 (cf. (Guillemin, Sternberg 1977)). *Let $f : X \to Y$ be a mapping of smooth manifolds and $W \subset T^*(Y)$ a closed subset such that $\lambda \circ df \neq 0$*

for any $\lambda \in W$. *The inverse image operation* f^* *can be uniquely defined on the subspace* $\mathcal{K}'_W \subset \mathcal{K}'(Y)$ *formed by the generalized functions* v *for which* $\mathrm{WF}\,(v) \subset W$ *and*

$$\mathrm{WF}\,(f^*(v)) \subset df^*(\mathrm{WF}\,(v)) = \{\mu = \lambda \circ df,\, \lambda \in \mathrm{WF}\,(v)\}\,.$$

This construction is unique in the following sense: the operation $f^* : \mathcal{K}'_W \to \mathcal{K}'(X)$ is continuous with respect to the natural topology in $\mathcal{K}'_W$ and coincides with the usual inverse image operation on smooth functions (which are dense in $\mathcal{K}'_W$).

A more complete study of the inverse image problem requires a detailed analysis of the operation (cf. §3 of Chapt. 2):

$$\int_f : \mathcal{K}(X) \to (?) \supset \mathcal{K}(Y)\,.$$

Here two problems arise:

I) Describe the maximal subspace $\mathcal{K}(X)$ whose image is contained in $\mathcal{K}(Y)$.
II) Describe the image of the whole space $\mathcal{K}(X)$.

In the general case very little is known about the solution of these problems. We note the following fact.

Proposition 2. *Let* $f : X \to Y$ *be an analytic mapping of real-analytic manifolds and* Σ *the set of its critical points. For every density* $\rho \in \mathcal{K}(X)$ *that is flat on* Σ *the density* $\int_f \rho$ *belongs to* $\mathcal{K}(Y)$ *and is flat on the set* $f(\Sigma)$ *of critical values. The operator of integration defined on the space of flat densities* $\rho \in \mathcal{K}(X)$ *is continuous.*

A smooth density ρ is called flat on the set $G \subset X$ if in any chart that intersects G its coefficient is zero on G together with all its derivatives. The proof of this proposition is based on the formula (2.3.7) and a known inequality of Lojasiewicz (1959).

By virtue of this proposition for every generalized function u on Y the inverse image $f^*(u)$ can be defined as a continuous functional on the space of smooth densities. This functional can be extended to all of $\mathcal{K}(X)$ using the Hahn-Banach theorem, which, however, leaves too much arbitrariness in the choice of the inverse image. In particular situations this arbitrariness can be significantly restricted.

1.2. Let $f : X \to \mathbb{R}$ be a nonconstant analytic function on a connected analytic manifold X with a unique critical value, say $x = 0$. Its level sets $X_s = f^{-1}(s)$ are nonsingular hypersurfaces (some of them may be empty).

Theorem 3 (cf. (Jeanquartier 1970; Malgrange 1974/75)). *For every density* $\rho \in \mathcal{K}(X)$ *the integral* $\int_f \rho$ *is a smooth density everywhere except the point* $s = 0$, *where it has the asymptotic expansion*

$$\frac{\int_f \rho}{ds} \equiv \int_f \frac{\rho}{df} \sim \sum_{\beta \in B} \sum_{q<n} b^{\pm}_{\beta,q} |s|^{\beta-1} \ln^q \frac{1}{|s|} + \sum_0^{\infty} b_k s^k \; ; \tag{4.1.1}$$

in which the signs $\pm$ *correspond to the semineighborhoods* $s \gtrless 0$. *Here* B *is the set of rational numbers of the form* $\frac{k}{r}$, $k = 1, 2, \ldots$, *and* r *is a natural number independent of* ρ.

The essential step in the proof consists of the use of a theorem of Hironaka on the resolution of the singularities of analytic sets. To solve the problems I and II detailed information about this expansion is needed. We consider the first term on the right-hand side of (4.1.1); by transferring certain terms to the second term, one can assume that in this expansion $b^{-}_{\beta,0} = 0$ for all integers β. In such a case all the coefficients $b^{\pm}_{\beta,q}$ and b_k are uniquely determined and are continuous linear functionals of the density ρ, i.e., generalized functions on X. The supports of all generalized functions $b^{\pm}_{\beta,q}$ are contained in the set Σ of critical points f. Hence a solution of Problem 1 follows: a necessary and sufficient condition for $\int_f \rho$ to belong to $\mathcal{K}(Y)$ is that $b^{\pm}_{\beta,q}(\rho) = 0$ for all β and q. Let $\mathcal{K}_f(X)$ be the subspace of $\mathcal{K}(X)$ where these equations hold.

Definition. The inverse image of $v \in \mathcal{K}'(Y)$ is any generalized function $u \in \mathcal{K}'(X)$, connected with v by the relation

$$u(\rho) = v\left(\int_f \rho\right), \quad \rho \in \mathcal{K}_f(X) .$$

According to what has just been said the inverse image exists for any generalized function and is determined up to finite linear combinations of functionals $b^{\pm}_{\beta,q}$.

In particular the delta-function δ_f on the critical level set of f can be defined according to the formula $\delta_f(\rho) = b_0(\rho)$. The generalized function $\left[\frac{1}{f}\right]$ can be defined as the inverse image of $\left[\frac{1}{s}\right]$. A closely related method was used in (Bernstein, Gel'fand 1969; Atiyah 1970).

Analysis of the expansion (4.1.1) leads to the following construction.

Definition (cf. (Palamodov 1985)). We write (4.1.1) in the form

$$\frac{\int\limits_f \rho}{ds} \sim \sum_{\beta\in B}\sum_q \operatorname{Re}\left(\tilde{b}_{\beta,q}(s+i0)^{\beta-1}\ln^q(s+i0)\right) + \sum_0^\infty c_k s^k\,,$$

where $\tilde{b}_{\beta,0} = 0$ if β is an integer. For every $\beta \in B$ we denote by q_β the largest number q such that the functional $\tilde{b}_{\beta,q-1}$ is not zero if β is a noninteger and such that $\tilde{b}_{\beta,q} \neq 0$ if β is an integer. It follows from the relation $\tilde{b}_{\beta,q}(f\rho) = \pm\tilde{b}_{\beta-1,q}(\rho)$ that the function q_β is nondecreasing on any arithmetic progression with difference 1. The *real critical value spectrum* of the function f is the set $S_B(f)$ of points where the function q_β is increasing, i.e., the numbers β such that $q_\beta > q_{\beta-1}$. The quantity q_β is called the *order* of the point β of the spectrum.

This definition is closely connected with the local Picard-Lefschetz monodromy corresponding to the holomorphic extension of the function. Assume that f has a unique real critical point x_0 and U is a small ball with center at x_0, where the extension f is holomorphic and Z_s is the intersection of the complex level hypersurface $f(z) = s$ with this ball. The cohomology groups $H^{n-1}(Z_s, \mathbb{C})$ form a vector bundle over the punctured sphere $S = \{0 < |s| < \varepsilon\}$ of sufficiently small radius ε, where $n = \dim X$. In this bundle there is a canonical connection; the section σ is horizontal with respect to the connection if $\sigma(\gamma_s) = \text{const}$ for any continuous family of cycles $\gamma_s \subset Z_s$, $s \in V \subset S$. Suppose the point $s \in S$ is fixed; for any element $c \in H^{n-1}(Z_s, \mathbb{C})$ one can construct a horizontal section passing through the point (s, c) over any simply connected subset $V \subset S$. Extending this section around the center of S to the point s we obtain a new element $h^*(c) \in H^{n-1}(Z_s, \mathbb{C})$. As a result there arises a linear transformation

$$h^* : H^{n-1}(Z_s, \mathbb{C}) \to H^{n-1}(Z_s, \mathbb{C})\,,$$

called the local *Picard-Lefschetz monodromy.* All of its eigenvalues are roots of unity.

Theorem 4 (cf. (Malgrange 1971; Palamodov 1985)). *Suppose the function f has a unique real critical point and that the number β belongs to its spectrum and has order q. Then the Jordan form of the operator h^* contains a cell of order at least q corresponding to the eigenvalue $\lambda = \exp(2\pi i\beta)$.*

1.3. The Case of a Quadratic Form. Let $f(x) = \dfrac{1}{2}\sum a^{ij}x_i x_j$ be a nondegenerate quadratic form in $\mathbb{R}^n$ of signature (p, q). This function has a unique critical point $x = 0$. In this case the expansion (4.1.1) can be computed completely.

Theorem 5. *If q is even, then*

$$\int\limits_{X_s} \rho/df \sim (-1)^{q/2}\sum_{k=0}^\infty F_k(\rho)s_+^{n/2+k-1} + \sum_0^\infty c_k(\rho)s^k\,, \tag{4.1.2}$$

where $s_+ = \max(s, 0)$ *and*

$$F_k(\rho) = \sqrt{\frac{\pi^n}{|\Delta|}} \frac{\tilde{f}^k(D)\varphi(0)}{k!\Gamma(n/2+k)}, \quad \varphi\,|dx| = \rho\,,$$

where Δ *is the determinant of the form* f, *and*

$$\tilde{f}(D) = \frac{1}{2}\sum a_{ij}\partial^2/\partial x_i \partial x_j\,, \quad \sum_j a^{ij} a_{jk} = \delta^i_k$$

is the dual differential operator. The case of an even p *reduces to* (4.1.2) *by replacing* f *with* $-f$.

In the case when p *and* q *are odd*

$$\int_{X_s} \rho/df \sim (-1)^{\frac{q+1}{2}} \frac{1}{\pi} \sum_0^\infty F_k(\rho) s^{n/2+k-1} \ln|s| + \sum_0^\infty c_k(\rho) s^k\,. \tag{4.1.3}$$

These expansions can be differentiated termwise with respect to s up to any order.

The coefficients c_k in (4.1.2) can be computed as follows:

$$c_k(\rho) = \frac{1}{k!} c_0(\nabla_v^k(\rho)), \quad c_0(\rho) = \int_{X_0} \rho/df\,,$$

where ∇_v is the value of the canonical connection (§3 of Chapt. 2) on the field $v = d/ds$, i.e., $\nabla_v(\rho) = d(\rho/df)$. The integral expressing c_0 converges absolutely in the case $n > 2$. In the case $n = 2$, $p = q = 1$, we can write $f = xy$, $\rho = \varphi(x, y)\,|dx\,dy|$. We have

$$c_0(\rho) = \int \frac{\varphi(x,0)\,dx}{|x|} + \int \frac{\varphi(0,y)\,dy}{|y|}\,, \tag{4.1.4}$$

and each integral can be understood as the action of a generalized function of the form $|t|^{-1} = t_+^{-1} + t_-^{-1}$ (cf. Palamodov 1985).

In all cases the critical point spectrum consists of a single element $\frac{n}{2}$ of order 1. The assertions of Theorem 5 can be interpreted as the asymptotic expansion of the delta-function $\delta_{f-s} \equiv f^*(\delta_s)$ with support X_s in a series of powers (with logarithms) whose coefficients are generalized functions on the singular hypersurface X_0, and also generalized functions concentrated at a critical point. A method of proving (4.1.2) and (4.1.3) is to find the Laurent expansions at the poles of the family of generalized functions $f_\pm^\lambda$. This family is the Mellin transform of the family δ_{f-s}. The Laurent expansion can be found by means of (4.2.2). See (Gel'fand, Shilov 1958a) for details.

1.4. Generalized Functions that are Invariant with Respect to the Pseudo-Orthogonal Group. Let f again be a nondegenerate quadratic form in $\mathbb{R}^n$ of signature (p, q) and G_f the group of linear transformations of $\mathbb{R}^n$ that preserve this form. This group is called the *pseudo-orthogonal* group. Let SG_f be the subgroup of it consisting of transformations with determinant 1. The group SG_f acts transitively on each hypersurface $X_s = \{f(x) = s\}$. Consider the operator of integration along these hypersurfaces

$$\int_f : \mathcal{K}(\mathbb{R}^n) \to \mathcal{K}^f(\mathbb{R}), \quad \rho \mapsto \int_{X_s} \rho/df \cdot ds \,.$$

According to Theorem 5 the image $\mathcal{K}^f(\mathbb{R})$ of this operator is the space of densities of compact support on $\mathbb{R}$ that are infinitely differentiable away from 0 and have expansions of the form (4.1.2) or (4.1.3) in a neighborhood of 0 (this expansion can be differentiated termwise any number of times). Every generalized function v of one variable can be lifted to $\mathcal{K}^j(\mathbb{R}) + \mathcal{K}(\mathbb{R})$ according to the formula

$$\check{v}(\sigma) = v(\sigma - \Sigma_N(\sigma)e) \,,$$

where $e \in \mathcal{K}(\mathbb{R})$ is a fixed density equal to ds in a neighborhood of 0, $\Sigma_N(\sigma)$ is an interval of length N in the first sum in the right-hand side of (4.1.2)–(4.1.3) and the number N may depend on v. To be specific, let $V \subset \mathbb{R}$ be a neighborhood of 0 and M a number such that the restriction of the functional v to the space $\mathcal{K}(V)$ is continuous in the norm of the space $C^M(V)$. In such a case N can be chosen so large that $\sigma - \Sigma_N(\sigma)e \in C^M(V)$. It is clear that the generalized function

$$u(\rho) = \check{v}\Big(\int_f \rho\Big), \quad \rho \in \mathcal{K}(\mathbb{R}^n) \tag{4.1.5}$$

is invariant with respect to the group G_f.

Theorem 6. *The subspace spanned by generalized functions of the form* (4.1.5) *along with the generalized functions*

$$\tilde{f}^k(D)\delta_0, \quad k = 0, 1, 2, \dots \,,$$

coincides with the space of generalized functions in $\mathbb{R}^n$ *that are invariant with respect to the group* SG_f *(or* G_f*).*

Let G_f^0 be the connected component of the identity element of the group SG_f. In the case when $p > 1$ and $q > 1$ the group G_f^0 acts transitively on each hypersurface X_s and consequently the invariant generalized functions for G_f^0 are the same as for SG_f.

Let us consider the case when $p = 1$ and G_f^0 is the proper orthochronal Lorentz group acting on $\mathbb{R}^n$. If $n > 2$, the hypersurfaces X_s are connected when $s < 0$ and decompose into two components each of which is a separate orbit of the group G_f^0 when $s > 0$. When $n = 2$, each curve X_s, $s \neq 0$, decomposes into two orbits. Let $I \in G_f$ be the involution that maps one component of the cone $\{f > 0\}$ into the other. This involution permutes the components of S. Every generalized function u can be written in the form $u = u_+ + u_-$, where u_+ and u_- are generalized functions that are even and odd with respect to I. If u is invariant with respect to G_f^0, then u_+ is invariant with respect to SG_f and consequently can be described by Theorem 6. The generalized function u_- vanishes on all densities ρ that are even with respect to I. Let us describe the action of the functional u_- on the space $\mathcal{K}_-(\mathbb{R}^n) \subset \mathcal{K}(\mathbb{R}^n)$ formed by the odd densities. We choose one component C of the cone $\{f > 0\}$ and consider the integration mapping

$$\int_{f+} : \mathcal{K}_-(\mathbb{R}^n) \to \mathcal{K}(\mathbb{R}_+)$$

along the components of $X_s \cap C$, $s \geq 0$. The result belongs to the space $\mathcal{K}(\mathbb{R}_+)$ of infinitely differentiable densities of compact support defined on the half-line $\{s \geq 0\}$.

Theorem 7. *The image of the adjoint mapping*

$$\mathcal{K}'(\mathbb{R}_+) \to \mathcal{K}'_-(\mathbb{R}^n) \subset \mathcal{K}'(\mathbb{R}^n)$$

is the set of odd generalized functions on $\mathbb{R}^n$ that are invariant with respect to the group G_f^0. Every such generalized function can be written in the form

$$u_-(\rho) = v_+\left(\int_{f+} \rho_-\right), \quad v_+ \in \mathcal{K}'(\mathbb{R}_+), \quad \rho_- = \frac{1}{2}(\rho - I^*\rho),$$

and conversely.

The description of the invariant distributions reduces immediately to describing the invariant generalized functions, since every distribution on $\mathbb{R}^n$ can be written in the form $u\,|dx|$, where u is a generalized function and the density $|dx|$ is invariant with respect to the whole group G_f. The results of Sect. 1.4 for the case $p = 1$ are essentially contained in (Methée 1954) (cf. also (Trèves 1961)). In this paper there is also a description of the invariant solutions of equations connected with the d'Alembertian $\Box = \tilde{f}(D)$.

The invariant slowly increasing generalized functions can also be described by Theorems 6 and 7 if v and v_+ are interpreted as slowly increasing generalized functions. The Fourier transform maps the space of slowly increasing G_f^0-invariant distributions onto the space of G_g^0-invariant slowly increasing generalized functions on the dual space, where g is the quadratic form dual to f. The computation of these transforms and closely related questions are considered in (Methée 1955; Gel'fand, Shilov 1958a).

§2. The Division Problem

2.1. Division by a Function. Let a be an infinitely differentiable function on the manifold X. To divide a generalized function u by a means to solve the equation

$$a \cdot v = u$$

in the space $\mathcal{K}'(X)$. Following (Schwartz 1950/51), we say that division of generalized functions on X is possible if this equation has a solution for any $u \in \mathcal{K}'(X)$. The quotient v is defined up to a generalized function satisfying the equation $a \cdot w = 0$. In the case $\dim X = 1$ division by a is possible if and only if all the zeros x of the function a have finite multiplicity $n(x)$. In this case the quotient is defined up to a generalized function of the form

$$w = \sum_{a(x)=0} \sum_{i=0}^{n(x)-1} c_i(x) \delta_x^{(i)} .$$

The roots of a form a discrete set, and consequently the sum on the right-hand side defines a generalized function for any choice of the quantities $c_i(x)$. By the localization principle it suffices to carry out the construction of the quotient for functions $a(x) \equiv x$, $x \in U \subset \mathbb{R}$. We choose any function $e \in \mathcal{D}(U)$, $e(0) = 1$, and write an arbitrary density $\varphi \in \mathcal{K}(U)$ in the form $\varphi = e \cdot \varphi(0) + x \cdot \psi$. Here $\psi \in \mathcal{K}(U)$ and the operator $\varphi \mapsto \psi$ is continuous. The solution of the equation $x \cdot v = u$ can be obtained from the formula $v(\varphi) = u(\psi)$.

On an analytic manifold X division is possible by every analytic function a if $a \not\equiv 0$ on each connected component of X. In view of the local nature of the problem, this assertion reduces to the case when X is an open subset of $\mathbb{R}^n$. In this case it is established in (Lojasiewicz 1959) and (Malgrange 1966). By a theorem of Whitney (1948) this assertion is equivalent to the following: The function $\varphi \in \mathcal{D}(X)$ belongs to the ideal $a \cdot \mathcal{D}(X)$ if and only if at each point $x \in X$ the Taylor series of φ at this point is divisible by the Taylor series of the function a.

The special case of the problem when the function a is a polynomial on $\mathbb{R}^n$ is important for applications to differential equations with constant coefficients. As shown in (Hörmander 1958; Lojasiewicz 1959; Malgrange 1966), division by any polynomial $a \not\equiv 0$ is possible in the space $S'(\mathbb{R}^n)$. After the Fourier transform is taken, this assertion becomes the following: for any differential operator $a(D)$ with constant coefficients in $\mathbb{R}^n$ the differential equation

$$a(D)\check{v} \equiv \check{u} \tag{4.2.1}$$

is solvable in $S'(\mathbb{R}^n)$ for any right-hand side $\check{u}$ in this space.

2.2. The Method of M. Riesz. After the Fourier transform is applied, the quotient v in division of the generalized function $\check{u} = 1$ by the polynomial a becomes a fundamental solution for the dual operator, i.e., it becomes a solution of (4.2.1) with $\check{u} = \delta_0 \cdot |d\xi|$. The choice of this solution and its properties is essential for studying this equation. M. Riesz ((1949), cf. also (Schwartz 1950/51)) proposed a method of performing such a division using analytic continuation. Let a be a nondegenerate quadratic form in $\mathbb{R}^n$. M. Riesz considers a family of generalized functions $a_+^\lambda = \theta(a)|a|^\lambda$ holomorphically dependent on the parameter $\lambda \in \mathbb{C}$. These functions are continuous if $\operatorname{Re}\lambda > 0$ and satisfy the equation

$$\tilde{a}(D)a_+^{\lambda+1}(x) = (\lambda+1)\Big(\lambda+\frac{n}{2}\Big)a_+^\lambda(x)\,,$$

where $\tilde{a}$ is the dual differential operator (cf. §1). Using this equation one can extend the family to the half-plane $\operatorname{Re}\lambda > -1$ by setting

$$a_+^\lambda = \frac{1}{(\lambda+1)(\lambda+\frac{n}{2})}\tilde{a}(D)a_+^{\lambda+1}\,. \tag{4.2.2}$$

Using this method again, we obtain an extension of the family a_+^λ to the half-plane $\operatorname{Re}\lambda > -2$, etc. As a result a meromorphic function arises in $\mathbb{C}$ with values in the space $S'(\mathbb{R}^n)$ and poles contained in the two arithmetic progressions

$$\lambda = -1, -2, \ldots, -k, \ldots; \quad \lambda = -\frac{n}{2}, -\frac{n}{2}-1, \ldots, -\frac{n}{2}-k, \ldots\,.$$

At points belonging to only one progression the family a_+^λ has a pole of first order (or is holomorphic); the residue is a generalized function with support equal to $X_0 = a^{-1}(0)$ for the first progression and equal to the point $x = 0$ for the second one. At points belonging to two progressions at once, the order of the pole is at most two.

The families of generalized functions

$$(a \pm i0)^\lambda = \lim_{\varepsilon\to+0}(a+i\varepsilon)^\lambda$$

have simpler behavior. Their poles are of first order and located only in the second progression connected with the critical point $x = 0$. Therefore for $n > 2$ one can define the quotient $\dfrac{1}{a}$ as the value of this family for $\lambda = -1$.

For any analytic function $a \not\equiv 0$ the family of generalized functions a_+^λ defined for $\operatorname{Re}\lambda > 0$ has a meromorphic continuation with respect to the parameter λ on the whole complex plane with poles located in a finite number of arithmetic progressions with rational terms. This assertion (the *Gel'fand conjecture*) was established in the case $n = 2$ by M.V. Fedoryuk (1959), who used the local resolution of the singularities of the curve $X_0 = \{a = 0\}$. This assertion was proved in general form (Bernstein, Gel'fand 1969; Atiyah 1970) on the basis of *Hironaka's theorem* on the existence of a local resolution of the singularities of any analytic hypersurface X_0. The idea of the proof is as follows. Let $U \subset \mathbb{R}^n$ be a domain in which the function a is defined, and let $\pi : X \to U$ be a mapping of a real-analytic manifold that resolves the singularities of X_0. This means that the function $b = \pi^*(a)$ vanishes on a set having only *normal singularities*, i.e., in each local chart on X it can be written in the form $b(y) = \pm y_1^{m_1} \cdots y_n^{m_n} \cdot g(y)$, where $y_1, \ldots, y_n$ is a suitable coordinate system and g is a smooth positive function. Therefore the generalized function b_+^λ admits a meromorphic extension with respect to λ with poles at the points $\lambda = -k/m$, $k = 1, 2, \ldots$, $j = 1, \ldots, n$, constructed over all charts on X. The required family a_+^λ is the direct image of b_+^λ under the mapping π, i.e., the restriction of b_+^λ to $\pi^*(\mathcal{K}(U))$. In this "descent" some of the poles drop out and hence the actual set of poles of a_+^λ is only a part of the set.

For a more detailed description of the poles it is convenient to distinguish the sequence $\lambda = -1, -2, \ldots$, which corresponds to the nonsingular points of X_0. To do this one must pass to the families $(a \pm i0)^\lambda$, whose poles depend only on the critical points lying in X_0. These families are the inverse images of families of generalized functions $(s \pm i0)^\lambda$ of one variable that are entire functions of λ (cf. §4 of Chapt. 1). Therefore the poles $(a \pm i0)^\lambda$ are determined by the exponents in the expansion (4.1.1) alone. The following conclusion can then be drawn: these families have meromorphic extensions to the complex plane with poles at the points $\lambda = -\alpha, -\alpha - 1, \ldots, -\alpha - k, \ldots$, where α are elements of the real spectrum $S(a)$ of the isolated critical point of the function a (cf. 1.2). The order of the pole at the point λ of each of these families is the larger of the orders of the elements $\alpha \in S(a)$ such that $\lambda = -\alpha - k$, where $k \geq 0$ is a certain integer.

I.N. Bernstein (1972) generalized the method of M. Riesz to the case of an arbitrary polynomial p. He showed that there exists a polynomial $b(\lambda)$ of one variable and a differential operator $B(x, \lambda, D)$ in $\mathbb{R}^n$ whose coefficients are polynomials in x and λ such that when $\operatorname{Re} \lambda > \deg B$

$$B(x, \lambda, D) p_+^\lambda = b(\lambda) p_+^{\lambda - 1} .$$

This relation makes it possible to construct a meromorphic extension by the method of M. Riesz, proceeding step by step into the left half-plane. Thus the poles of p_+^λ form an arithmetic progression with difference -1, starting at the zeros of the polynomial $b(\lambda + 1)$. More information on the polynomial b can be found in (Malgrange 1971; Yano 1978; Björk 1979; Varchenko 1980).

2.3. Division by a Matrix. Let $a = \{a_{ij}\}$ be a matrix of size $t \times s$ formed of smooth functions on an open set $U \subset \mathbb{R}^n$ (or on a manifold). Consider the equation

$$a \cdot v = u \,, \tag{4.2.3}$$

where $u \in [\mathcal{K}'(U)]^t$ (u is a vector with t components). The problem of division by the matrix can be posed as follows: for which right-hand sides u does there exist a solution $v \in [\mathcal{K}'(U)]^s$ of this equation? There is an obvious necessary condition for existence of a solution: for any row-matrix $b = (b_1, \ldots, b_t)$, where the b_i are smooth functions, such that $b \cdot a = 0$ (b is an annihilator of a), the relation $b \cdot u = 0$ must also hold.

Theorem 1 (cf. (Malgrange 1966)). *If all the functions a_{ij} are analytic in U, then this condition is sufficient for (4.2.3) to be solvable. Moreover for this equation to be solved it suffices that $b \cdot u = 0$ for any annihilator b formed by functions analytic in U. If the matrix a is formed of polynomials and $u \in [S']^t$, then for (4.2.3) to be solvable it suffices that $b^{(k)} \cdot u = 0$ for a suitable finite set of annihilators $b^{(1)}, \ldots, b^{(q)}$ formed by polynomials.*

2.4. Division of Analytic Functionals and Differential Equations. The solution of the differential equation (4.2.1) in $\mathcal{K}'(\Omega)$ reduces to the problem of division by the polynomial $a(2\pi iz)$ in the space $Z'(\Omega)$ if the domain Ω is convex. The following result was one of the first in the division problem.

Theorem 2 (cf. (Malgrange 1956, 1960; Ehrenpreis 1960, 1970)). *For every convex domain $\Omega \subset \mathbb{R}^n$ in the space $Z'(\Omega)$ division by any polynomial $a \not\equiv 0$ is possible. Consequently Eq. (4.2.1) is solvable in the space $\mathcal{K}'(\Omega)$ for any right-hand side from this space.*

The quotient in division of the functional $u \in Z'(\Omega)$ by the polynomial a can be constructed by representing u using the density $r = h\,d\xi\,d\eta$ in accordance with formula (3.3.2). Using the nonuniqueness of this representation, one can change this density so that its support is disjoint from the 1-neighborhood of the set of zeros of a. After that the density $s = r/a$ is unambiguously defined, and using it the required quotient is written

$$v(\varphi) = \int \varphi s\,.$$

A more general result holds:

Theorem 3. *Again let Ω be a convex domain in $\mathbb{R}^n$ and a a matrix formed by polynomials in $\mathbb{C}^n$ having size $t \times s$. A sufficient (and of course necessary) condition for solvability of the system of equations (4.2.3) in $[Z'(\Omega)]^s$ is that the right-hand side satisfy the equation $b \cdot u = 0$ for any polynomial vector $b = (b_1, \ldots, b_t)$ such that $b \cdot a = 0$.*

This result is established in (Palamodov 1967, 1968a) and in weaker forms in (Ehrenpreis 1970) and (Malgrange 1964). The following assertion is one of the stages in the proof.

Theorem 4. *The subspace $a'[Z(\Omega)]^t$ in $[Z(\Omega)]^s$ is localizable (a' is the transposed matrix). This means that every vector $\psi \in [Z(\Omega)]^s$ belongs to this subspace if at each point $z \in \mathbb{C}^n$ the Taylor series of ψ is divisible over the ring of formal power series by the Taylor series of the matrix a'.*

Corollary to Theorem 3. *For any polynomial matrix a the system of differential equations (4.2.1) is solvable in the space $[\mathcal{D}'(\Omega)]^s$ for any right-hand side $\check{u} \in [\mathcal{D}'(\Omega)]^t$ satisfying the formal condition of solvability of $b(D)\check{u} = 0$, where b is any polynomial vector that annihilates this matrix: $b \cdot a = 0$.*

The problem of solvability of more general equations and systems in convolutions leads to the problem of division by holomorphic matrix-valued functions (Malgrange 1956; Ehrenpreis 1960; Martineau 1967; Ehrenpreis 1970; Hörmander 1985).

§3. Translation-Invariant Spaces of Generalized Functions

3.1. A subspace $E \subset \mathcal{K}'(\mathbb{R}^n)$ is *translation-invariant* if for each of its elements u all the translates $\tau_h^*(u)$, $h \in \mathbb{R}^n$, also belong to E; $\tau_h(x) = x + h$. In other words E is a subrepresentation of the group of translations in $\mathbb{R}^n$ in the representation $\mathcal{K}'(\mathbb{R}^n)$.

Example 1. The space of solutions of a differential equation with constant coefficients

$$p\Big(\frac{\partial}{\partial x}\Big)u = 0$$

belonging to $\mathcal{K}'(\mathbb{R}^n)$ or $\mathcal{E}(\mathbb{R}^n)$.

Example 2. The set of generalized functions that are periodic with respect to a fixed lattice $W \subset \mathbb{R}^n$.

Example 3. A generalized function u is *mean-periodic* if it satisfies an equation

$$r * u = 0\,, \tag{4.3.1}$$

where $r \in \mathcal{E}'(\mathbb{R}^n)$ is a given distribution. The space of mean-periodic generalized functions is translation-invariant.

Example 4. The space of solutions of a finite or infinite system of convolution equations

$$r_1 * u = \cdots = r_t * u = \cdots = 0 \tag{4.3.2}$$

is also invariant. This class of spaces contains all the spaces considered in the preceding examples.

Example 5. A minimal nontrivial translation-invariant space has the form

$$E_\zeta = \{\lambda \cdot \exp(\zeta \cdot x),\, \lambda \in \mathbb{C}\}\,,$$

where ζ is a fixed point of $\mathbb{C}^n$. We consider the more general construction:

$$E = L\{p_j(x)\exp(\zeta \cdot x),\, j \in J\}\,, \tag{4.3.3}$$

where p_j are polynomials in $\mathbb{R}^n$ and the symbol L denotes the linear span. A necessary and sufficient condition for the space E to be translation-invariant is that the space of polynomials $L\{p_j,\, j \in J\}$ be invariant under differentiation on all coordinates $x_1, \ldots, x_n$. Another criterion for invariance can be stated as follows: consider the space F of formal power series in $\zeta_1, \ldots, \zeta_n$. To every polynomial $p(x)$ there corresponds a linear functional on F according to the formula

$$\varphi(\zeta) \mapsto p\Big(\frac{\partial}{\partial\zeta}\Big)\varphi(\zeta)|\zeta = 0 \,.$$

The space E is characterized by its *polar*, i.e,. the intersection I of the kernels of the functionals p_j, $j \in J$. A necessary and sufficient condition for E to be invariant is that the space I be an ideal in the ring F. If I is an ideal, we shall denote the space (4.3.3) by $E_{\zeta,I}$. This space is an indecomposable representation of the group of translations. Conversely every finite-dimensional indecomposable representation of this group is isomorphic to some representation of $E_{\zeta,I}$. The representations $E_{\zeta,I}$ are not isomorphic to one another.

Let E be some closed subspace in $\mathcal{K}'(\mathbb{R}^n)$ or in another translation-invariant functional space. The problem of describing its structure includes the following problems:

I) *Spectral analysis.* If $E \neq 0$, does E contain an exponential?
II) *Spectral synthesis.* Are the exponential polynomials (i.e., functions of the form (4.3.3)) contained in the space E dense in E?
III) *The exponential representation.* Can every element of E be represented as an integral over the set of exponential polynomials belonging to E?

3.2. The Case of One Variable. A classical theorem (Euler 1743) contains a positive answer to Problem III, which involves the space E of solutions of an ordinary differential equation with constant coefficients

$$u^{(m)} + c_1 u^{(m-1)} + \cdots + c_m u = 0 \,.$$

The answer can be written in the form $E = \oplus E_{\zeta,f(\zeta)}$, where the direct sum is taken over the roots ζ of the characteristic polynomial $p(\zeta) = \zeta^m + c_1\zeta^{m-1} + \cdots + c_m$ and $I(\zeta)$ is the ideal in the ring F generated by the Taylor series of p at the point ζ.

Another classical case is the expansion of periodic functions in Fourier series, i.e., in convergent series of periodic exponentials. For periodic generalized functions such an expansion was described in §2 of Chapt. 1.

L. Schwartz (1951, 1947) showed that Problem II has a positive solution for any closed invariant subspace in $\mathcal{E}(\mathbb{R})$ and in $\mathcal{K}'(\mathbb{R})$. A.F. Leont'ev (1949) established that the answer to Problem III is in general negative. He studied the following convolution equation (4.3.1) on the line

$$r * u \equiv u\Big(\frac{\pi}{\lambda} - \pi + x\Big) + u\Big(\pi - \frac{\pi}{\lambda} + x\Big) - u\Big(\pi + \frac{\pi}{\lambda} + x\Big) - \\ - u\Big(-\pi - \frac{\pi}{\lambda} + x\Big) = 0 \,.$$

The characteristic function of the kernel r has the form $\tilde{r}(\zeta) = 4\sin\pi\zeta \cdot \sin\frac{\pi\xi}{\lambda}$, λ being a Liouville number. As a result there is an infinite sequence of rapidly convergent pairs of roots $\tilde{r}$. For each such pair ζ_1, ζ_2 the linear combination of exponentials

$$\frac{\exp(\zeta_1 x) - \exp(\zeta_2 x)}{\zeta_1 - \zeta_2}$$

is a solution of (4.3.1), and consequently even an infinite series of such combinations with suitable coefficients converges in $\mathcal{E}(\mathbb{R})$. However, if this series is expanded in individual exponentials, it becomes divergent because of the small denominators $\zeta_1 - \zeta_2$.

Ehrenpreis (1960) has shown that for a large class of equations (4.3.1) Problem III nevertheless has a positive solution if one uses a grouping of the exponential polynomials. To be specific, if the function $\tilde{r}$ is slowly decreasing, every solution $u \in \mathcal{E}(\mathbb{R})$ can be written in the form

$$u(x) = \sum_{k=1}^{\infty} \sum_{\zeta \in V_k} p_\zeta(x) \exp(-i\zeta x) , \tag{4.3.4}$$

where V_k are fixed finite subsets of $\mathbb{C}$, the outer sum converges in $\mathcal{E}(\mathbb{R})$, and all the terms are exponential polynomials satisfying the equation. The condition of slow decrease, which is also important in other problems connected with Eq. (4.3.1), consists of the following: Do there exist positive constants A, B, and ε such that for every $\xi \in \mathbb{R}$

$$\max\{|\tilde{r}(\eta)|;\ \eta \in \mathbb{R},\ |\xi - \eta| \leq A \ln(2 + |\xi|)\} \geq \varepsilon(|\xi| + 1)^{-B} ?$$

3.3. Invariant Spaces of Functions of Several Variables. Malgrange (1956) has shown that Problem II has a positive answer for the space E of solutions of any convolution equation (4.3.1). Hörmander (1983b) obtained a generalization of this result. Problem III has been solved for the space of solutions of an arbitrary system of differential equations with constant coefficients

$$p\Big(\frac{\partial}{\partial x}\Big) u = 0, \quad u \in [\mathcal{K}'(\Omega)]^s . \tag{4.3.5}$$

Here $p\Big(\dfrac{\partial}{\partial x}\Big)$ is any rectangular matrix formed by polynomials p_{ij} in the vector fields $\dfrac{\partial}{\partial x_1}, \ldots, \dfrac{\partial}{\partial x_n}$. A description of this space in various forms (the exponential representation, the fundamental principle) was given in (Malgrange 1964; Palamodov 1967, 1968a; Ehrenpreis 1970); other expositions and versions can be found in (Berenstein, Dostal 1972; de Roever 1977; Björk 1979). Note that in the case of Example 1 a solution follows from Theorem 3 in §3 of Chapt. 3.

We give the statement of a theorem on exponential representation in the case of the system (4.3.5) with one unknown function u and a column matrix $p = (p_1, \ldots, p_t)$. On the dual space to $\mathbb{C}^n$ we consider the polynomials $p_j(-i\zeta)$, $j = 1, \ldots, t$. They generate an ideal J in the ring A of polynomials in $\mathbb{C}^n$. Let $J = \cap J_k$ be its decomposition into primary components and N_k the set of roots of the ideal J_k. For every k there exist differential operators q_{kj}, $j = 1, \ldots, m_k$ in $\mathbb{C}^n$ with polynomial coefficients having the following property: a polynomial $a \in A$ belongs to J_k if and only if

$$q_{kj}\Big(\zeta, \frac{\partial}{\partial \zeta}\Big) a(\zeta) \big| N_k = 0, \quad j = 1, \ldots, m_k \, .$$

The set $q_k = \{q_{kj}, j = 1, \ldots\}$ is called the *Noetherian operator* for the A-module A/J_k (in connection with the fact that these equations correspond to the Lasker-Noether conditions).

Theorem (cf. (Palamodov 1967, 1968a)). *For every convex domain $\Omega \subset \mathbb{R}^n$ every solution $u \in \mathcal{K}'(\Omega)$ of the system* (4.3.5) *can be written as an integral*

$$u(x) = \sum_{j,k} \int_{N_k} q_{k,j}(\zeta, x) \exp(-i\zeta \cdot x) \mu_{k,j} \, , \tag{4.3.6}$$

where $\mu_{k,j}$ are complex-valued densities concentrated on N_k and having finite integrals

$$\int (|\zeta| + 1)^q \exp(\gamma_K(\operatorname{Im} \zeta)) |\mu_{k,j}| \, ,$$

where $K \subset \Omega$ is any compact set and q is a number depending on K. Conversely for any densities $\mu_{k,j}$ possessing these properties the integrals in (4.3.6) *converge absolutely in the space $\mathcal{K}'(\Omega)$ (and uniformly on any bounded set $B \subset \mathcal{K}(\Omega)$) and are solutions of* (4.3.5).

In the case of a general system (4.3.5) it is necessary to consider the A-module

$$M = \operatorname{Cok} \{p' : A^t \to A^s\} \, ,$$

and to represent the zero submodule in M as the intersection of primary submodules $\{0\} = \cap M_k$, then for each component M_k to choose a Noetherian operator $q_k = \{q_{kj}\}$, where q_{kj} are rows of length s and N_k is the support of the module M/M_k. The rest of the statement of the theorem on exponential representation remains unchanged. Generalized or smooth solutions of a general system of the form (4.3.5) satisfying restrictions on the growth at infinity of the type $u(x) = O(\exp(m(x)))$ admit a similar description, where M is a convex function, as do solutions in the spaces of ultradistributions and other spaces (Palamodov 1967).

We remark that by no means all the exponential polynomials satisfying (4.3.5) occur in the representation (4.3.6). Indeed for every point $\zeta \in N = \cup N_k$ in which the set N has positive dimension, the space of exponential polynomials of the form $p(x) \exp(-i\zeta \cdot x)$ satisfying (4.3.5) is infinite-dimensional, while (4.3.6) contains only those of the form $q_{kj}(\zeta, x) \exp(-i\zeta \cdot x)$, where $\zeta \in N_k$. This representation also solves Problem II for (4.3.5), but there are also other versions of the solution. In particular in the space of solutions of (4.3.5) defined in a convex domain Ω the space $\oplus E_{\zeta, I}$ is dense, where the direct sum is taken over any finite set of points ζ that has a nonempty intersection with each component N_k and the local ideal $I \subset F$ is generated by the ideal J.

In the case of a system (4.3.2) of convolution (not differential) equations the space E of solutions can have a more complicated structure. D.I. Gurevich (1975) constructed a system of two such equations in $\mathbb{R}^2$ for which Problem II has a negative solution and a system of six convolution equations for which even Problem I has a negative answer. Thus the theorem of L. Schwartz described above cannot be generalized to the case $n > 1$. On the other hand Gurevich (1975) has shown that in the case $n = 2$ Problem II has a positive solution for any system of differential-difference equations (4.3.2) if the supports of all the kernels r belong to the integer lattice.

General systems of convolution equations in $\mathbb{R}^n$ are studied in (Ehrenpreis 1970; Gurevich 1974; Berenstein, Taylor 1980; Struppa 1983). In particular a multidimensional generalization has been obtained for Ehrenpreis' theorem that it is possible to represent solutions in the form (4.3.4) with terms grouped. Surveys of the papers in this area can be found in (Berenstein and Taylor 1980, 1983; Berenstein and Struppa 1989).

3.4. Function Spaces Invariant with Respect to a Lattice. Let $\mathbb{Z}^n$ be the subgroup of $\mathbb{R}^n$ formed by the points with integer coordinates. This subgroup acts on $\mathbb{R}^n$ and consequently on $\mathcal{K}'(\mathbb{R}^n)$ via translations by integer vectors. Let $E \subset \mathcal{K}'(\mathbb{R}^n)$ be any subspace that is invariant with respect to such an operation. For this subspace questions similar to Problems I, II, and III of Sec. 3.1 make sense, with the difference that instead of exponential polynomials one must consider functions of the form

$$q(x)g(x)\exp(2\pi i\zeta \cdot x) , \tag{4.3.7}$$

where q is a polynomial and g an n-periodic function in $\mathbb{R}^n$, i.e., a function that is invariant with respect to the action of $\mathbb{Z}^n$. The vector ζ is an element of the dual space $\mathbb{C}^n$ (in the physical context it is called a *quasimomentum*). It can be assumed that ζ is an element of the quotient space $\Lambda := \mathbb{C}^n/\mathbb{Z}^{n*}$, where $\mathbb{Z}^{n*}$ is the lattice dual to $\mathbb{Z}^n$, i.e., again the subgroup of integer vectors. Indeed when ζ is translated by an integer vector k, the exponential in (4.3.7) is multiplied by an n-periodic function that can be included in g.

Let $p(x, \partial/\partial x)$ be a differential operator in $\mathbb{R}^n$ with smooth n-periodic coefficients. The space of generalized solutions in $\mathbb{R}^n$ of an equation

$$p(x, \partial/\partial x)u = 0 , \tag{4.3.8}$$

or of a system of such equations, is invariant with respect to the action of $\mathbb{Z}^n$. In the case when this system is elliptic an analogue of the exponential representation holds, and the analogues of exponential polynomials are the solutions of the form (4.3.7). If (4.3.8) is a system of s equations with s unknown functions $u = (u_1, \dots, u_s)$, a solution u of it is called a *Floquet solution* provided all the components u_i have the form (4.3.7). In the case when all the polynomials q are constant u is called a *Bloch solution.* Let $N \subset \Lambda$ be the set of quasi-momenta for which (4.3.8) has at least one Bloch solution. This set is complex-analytic and moreover supports a characteristic coherent sheaf $\mathcal{M}$ defined on the complex manifold Λ. To determine it we consider the scale of Sobolev spaces H^l, $l = 0, 1, \ldots$ formed by the functions defined on the torus $\mathbb{T}^n = \mathbb{R}^n/\mathbb{Z}^n$ and for each l we consider the family of differential operators

$$p_\zeta := p(x, \partial/\partial x + 2\pi i\zeta), \quad \zeta \in \mathbb{C}^n ,$$

acting on this torus. The operator p'_ζ formally adjoint to p_ζ defines for each l and ζ a continuous mapping

$$p'_\zeta : [H^{l+m}]^s \to [H^l]^s ,$$

where m is the order of the system (4.3.8). This mapping depends holomorphically on ζ and in view of the ellipticity is a Fredholm operator for each ζ. Consider now the corresponding mapping of analytic sheaves on $\mathbb{C}^n$

$$p' : E_1 \to E_2 ,$$

where E_1 and E_2 are sheaves of germs of sections of trivial bundles with base $\mathbb{C}^n$ and fibers $[H^{l+m}]^s$ and $[H^l]^s$ respectively. Since the operators are Fredholm operators, the sheaf $\operatorname{Cok} p'$ is a coherent analytic sheaf on $\mathbb{C}^n$, and in view of the ellipticity of p it is independent of l. Under the translation $\zeta \mapsto \zeta + k$ by an integer vector the operator p'_ζ maps to the equivalent operator

$$p'_{\zeta+k} = \exp(-2\pi ik \cdot x) \cdot p'_\zeta \cdot \exp(2\pi ik \cdot x)$$

and consequently the mapping p' goes to an equivalent mapping. Therefore the sheaf $\operatorname{Cok} p'$ is subordinate to the identity automorphism and consequently can be lifted to a coherent sheaf $\mathcal{M}$ on the manifold Λ. The role of this sheaf is analogous to the role of the module M in the situation of Sect. 3.3, while the support coincides with N.

Using the Lasker-Noether theorem, we represent the zero submodule of $\mathcal{M}_\zeta$ at every point $\zeta \in N$ as a finite intersection of primary $\mathcal{O}_\zeta(\Lambda)$-submodules: $\{0\} = \cap L_{\zeta,i}$. The supports $V_{\zeta,i} = \operatorname{supp} \mathcal{M}_\zeta/L_{\zeta,i}$ are well-defined germs of analytic sets at the point ζ. The set N has a unique representation as a locally finite union of irreducible closed analytic sets $N_k \subset \Lambda$ such that at each point $\zeta \in N$ each of the germs $V_{\zeta,i}$ is a component of the germ of one of the sets N_k. We further use the following general fact: for each k there exists a finite number of differential operators in the $\mathcal{O}(\Lambda)$-sheaves

$$q_{kj} : \mathcal{M} \to \mathcal{O}(N_k), \quad j = 1, \ldots, d_k ,$$

having the following property: an element $a \in \mathcal{M}$ is zero if $q_{kj}a = 0$ for all j and k. This means $q_k =_j \oplus q_{kj}$, $k = 1, 2, \ldots$, are Noetherian operators for the sheaf $\mathcal{M}$. The local existence of such operators was established by Palamodov

(1967, 1968b). Global Noetherian operators were constructed for an arbitrary coherent analytic sheaf on a Stein space in (Palamodov1993).

Furthermore, for every k and every point $\zeta \in N_k$ we consider the composition

$$\mathcal{M} \xrightarrow{q_{kj}} \mathcal{O}(N_k) \xrightarrow{r_\zeta} \mathbb{C}\,, \tag{4.3.9}$$

where r_ζ is the operator that computes the value of a function at this point. If the order of the differential operator q_{kj} is zero, this composition may be lifted to a continuous linear functional over E_2, i.e., an n-periodic vector-valued function g with components in H^{-l}, which corresponds to the Bloch solution $g(x)\exp(2\pi i\zeta\cdot x)$ (and, consequently, $g \in C^\infty$). In the general case the Floquet solution with quasi-momentum ζ corresponds to a functional (4.3.9), and the linear span of these solutions over all $j = 1, \dots, d_k$ is a subspace $F_{k,\zeta}$ of the space of solutions (4.3.8) that are invariant with respect to the action of $\mathbb{Z}^n$. The *exponential representation* in the periodic case is an expression for the space of solutions of (4.3.8) as an integral over the manifold of finite-dimensional representations of $F_{k,\zeta}$.

Theorem (Palamodov 1993). *Every solution of* (4.3.8) *with the estimate of growth at infinity*

$$u(x) = O(\exp(a|x|)) \text{ for some } a > 0 \tag{4.3.10}$$

can be expanded into an integral over the Floquet solutions described above, to be specific:

$$u(x) = \sum_{j,k} \int q_{kj}(\zeta, 2\pi i x) g_{kj}(\zeta, x) \exp(2\pi i \zeta x) \mu_{kj}\,,$$

where μ_{kj} are certain complex-valued measures with compact supports contained in N_k, $k = 1, 2, \dots$ and $q_{kj}(\zeta, \partial/\partial\zeta)$ is the composition (4.3.9). *Conversely for any such measures each of these integrals is a solution of* (4.3.8) *and satisfies* (4.3.10).

A closely related result (but more complicated in form) was obtained by P.A. Kuchment (1982), who applied only local Noetherian operators. His reasoning follows the outline of (Palamodov 1967), and make use of the following fact: any smooth function that satisfies the estimate $\varphi(x) = O(\exp(-a|x|))$ for each $a > 0$ admits a harmonic expansion of the form

$$\varphi(x) = \int_{0 \le \xi_j \le 1,\, j=1,\dots,n} \exp(2\pi i \xi \cdot x) \psi(\xi, x)\, d\xi\,,$$

$$\psi(\xi + k, x) = \exp(-2\pi i k x)\psi(\xi, x)\,,$$

where k is any integer vector and $\psi(\xi, x)$ is a function that is n-periodic on x and has an extension as an entire function of ξ. This is a corollary of the Fourier expansion of φ. In (Kuchment 1981) a generalization of this theory to the case of several noncommutative groups is shown.

§4. Causal Generalized Functions

4.1. The Jost-Lehmann-Dyson Representation. In quantum field theory the generalized functions that arise as the commutators of the field operators vanish in the domain $x^2 < 0$, where $x^2 = x_0^2 - x_1^2 - x_2^2 - x_3^2$ is the Lorentz quadratic form. This property is called *causality*, since it relates to the independence of measurements in domains separated by spacelike intervals. Such in particular is the Green's function of the scalar field

$$D^c(x) = \frac{1}{(2\pi)^4} \int \frac{e^{-ipx}\,dp}{m^2 - p^2 - i\varepsilon}, \quad p^2 = p_0^2 - p_1^2 - p_2^2 - p_3^2 \,.$$

This property is possessed by the Green's functions of the spinor and vector fields and other functions of field theory (Bogolyubov and Shirkov 1976).

Definition. A generalized function $u \in S'(\mathbb{R}_p^4)$ is *causal* if the support of its Fourier transform $\tilde{u} \in S'(\mathbb{R}_x^4)$ belongs to the cone $K = \{x : x^2 \geq 0\}$.

Example 1. The Green's function in the momentum representation $D^c(p) = \dfrac{dp}{m^2 - p^2 - i\varepsilon}$ is a causal distribution.

Example 2. The Pauli-Jordan permutation function

$$D(p) = \frac{1}{(2\pi)^3} \int\limits_{\mathbb{R}^3} \frac{dk}{\sqrt{k^2 + m^2}} e^{ikp} \sin(p_0\sqrt{k^2 + m^2})$$

is also causal, since it equals the Fourier transform of the distribution $\operatorname{sgn} k_0 \delta(k^2 - m^2)$, whose support is the cone $k^2 - m^2 = 0$.

Theorem (Jost-Lehmann-Dyson, cf. (Jost, Lehmann 1957; Dyson 1958ab; Bogolyubov, Shirkov 1976)). *Every causal generalized function u can be written as a fourfold integral*

$$u(p) = \int\limits_{\mathbb{R}^3} d\xi \int_0^\infty d\lambda \cdot \operatorname{sgn} p_0 \cdot \delta(p_0^2 - \lambda - (p' - \xi)^2)(v_0 + p_0 v_1)\,, \quad (4.4.1)$$

$$p' = (p_1, p_2, p_3)\,,$$

where v_0 and v_1 are slowly increasing generalized functions defined in the half-space $\{(\xi, \lambda),\ \lambda \geq 0\}$. It is uniquely determined by u. There is an analogous representation as a fivefold integral (without uniqueness).

These representations are based on the following fact:

Theorem (cf. (Dyson 1958ab; Vladimirov 1964)). *With any $n \geq 4$ for every causal generalized function u there exists a generalized function $v \in S'(\mathbb{R}^{n+1})$ satisfying the wave equation*

$$\left(\frac{\partial^2}{\partial q_0^2}-\frac{\partial^2}{\partial q_1^2}-\cdots-\frac{\partial^2}{\partial q_n^2}\right)v=0$$

and such that

$$v(p_1,p_2,p_3,0,\cdots,0)=u(p)\,,$$
$$\frac{\partial v(p_0,p_1,p_2,p_3,0,\ldots,0)}{\partial q_i}=0\,,\quad i>3\,.$$

The converse is also true.

In other words causal functions are the traces on the timelike subspace of solutions of the wave equation that are even with respect to the other variables. The expression for such a solution using the general Kirchhoff formula leads to different versions of the Jost-Lehmann-Dyson representation (Dyson 1958ab; Bogolyubov, Shirkov 1976). In particular when $n=4$, the solution v is uniquely determined by u. In formula (4.4.1) the quantities v_0 and v_1 are traces of this solution and its normal derivative on a spacelike hyperplane. A connection between the supports of u, v_0, and v_1 follows from the uniqueness theorems for the wave equation.

4.2. The Expansion of Causal Functions in Frequency Components. Let $\Gamma\subset\mathbb{R}^n$ be an open cone with vertex at the origin. The domain $T(\Gamma)=\mathbb{R}^n+i\Gamma$ in $\mathbb{C}^n$ is said to be a *tube domain.* Let $U\subset\mathbb{R}^n$ be an open set and Ω an open part of $\mathbb{C}^n$ such that $\Omega\cap\mathbb{R}^n=U$. The *boundary value* of a function f that is holomorphic in $T(\Gamma)\cap\Omega$ is the functional

$$\operatorname{bv} f(\rho):=\lim_{\Gamma\ni y\to 0}\int f(x+iy)\varphi(x)\,dx$$

defined on densities $\rho=\varphi\,dx\in\mathcal{K}(U)$ for which this limit exists. If this limit exists for all elements $\mathcal{K}(U)$, then this functional is continuous and, consequently is a generalized function in U.

Proposition. *Every causal function u can be written as a sum*

$$u=\operatorname{bv} f_+ + \operatorname{bv} f_-\,,$$

where $f_\pm$ are holomorphic functions in the tube domains $T(K^\pm)$ and $K^\pm=\{x^2>0,\ \pm x_0>0\}$ are the future and past cones. These boundary values belong to the space $S'(\mathbb{R}^4)$ and are also causal generalized functions.

Example. The Pauli-Jordan function of Sect. 4.1 can be expanded into frequency terms $D(p)=D_+(p)+D_-(p)$ corresponding to the dual representation of the decomposition of the hyperboloid $k^2-m^2=0$ into connected components. The positive-frequency component D_+ is the boundary value of the holomorphic function $f_+(z)=F(z^2)$ defined in the future tube $T(K^+)$. Here

$$F(\zeta)=\frac{m^2}{8\pi}\frac{H_1^{(2)}(-m\sqrt{\zeta})}{-m\sqrt{\zeta}},\quad H_1^{(2)}(\lambda)=-\frac{d}{d\lambda}H_0^{(2)}(\lambda)\,,$$

and $H_0^{(2)}$ is the Hankel function. The function $F(\zeta)$ is holomorphic on the entire complex plane with a cut along the positive half-line. A similar formula $f_-(z) = F(z^2)$ defines the holomorphic function in the past tube $T(K^-)$ and $D_- = -\text{bv} f_-$. As a result we can write (cf. (Akhiezer, Berestetskij 1969))

$$D(p) = F(p^2 + i\varepsilon 0) - F(p^2 - i\varepsilon 0), \quad \varepsilon = \operatorname{sgn} p_0 .$$

4.3. Properties of Causal Functions. The analytic wave front of every causal function u at any point p is contained in the cone $\{x^2 \geq 0\}$. This fact is an immediate corollary of the definition (cf. §5 of Chapt. 2). It follows from this that u has a well-defined restriction to every smooth timelike curve l, i.e., every curve, all of whose tangent vectors t satisfy the inequality $t^2 > 0$ (cf. Property 5 of Sect. 5.4., Chapt. 2). The following uniqueness theorem holds:

Theorem (cf. (Vladimirov 1964)). *Let l be a timelike curve with endpoints p and q, and let $p_0 > q_0$. The set $B_K(l) = p + K^- \cap q + K^+$ is called the K-envelope of this curve. If a causal function u is zero in a neighborhood of l, then $u = 0$ in a neighborhood of $B_K(l)$.*

This assertion is connected with the trace property of the solution of the wave equation on a timelike subspace (Courant, Hilbert 1962). A generalization in terms of the analytic wave front is described in Sect. 5.6, Chapt. 2.

4.4. The "Edge of the Wedge" Theorem

Theorem. *Let U be a nonempty open subset of $\mathbb{R}^n$ and $K \subset \mathbb{R}^n$ an open cone such that $K \cap (-K) \neq \varnothing$, and let Ω be a complex neighborhood of U. If the function f is holomorphic in $T(K) \cap \Omega$ and has the boundary value $\text{bv} f \in \mathcal{K}'(U)$, then f has an extension to a holomorphic function in Ω, where Ω, where Ω is a complex neighborhood of U depending only on U and K.*

The set K can be taken as the interior of the light cone. In such a case the condition of existence of a boundary value of the function f means that functions $f^{\pm}$ equal to the restriction of f to the cones $T(K^{\pm}) \cap \Omega$ have such values, and these values coincide. A similar theorem was established by N. N. Bogolyubov in 1956 in connection with the derivation of the dispersion relation for scattering amplitudes. The appearance of holomorphic functions in tube domains in this context is connected with the causal character of scattering amplitudes. In its original form the theorem assumed that $U = \mathbb{R}^n$ and that the boundary values are slowly increasing generalized functions. The proofs of the theorem in the general case can be found in (Vladimirov 1964) and (Browder 1963). The "edge of the wedge" theorem gave rise to several generalizations, as well as studies in mixed domains, see in particular (Martineau 1967; Bros, Iagolnitzer 1975).

§5. Multiplication of Generalized Functions

It is impossible to introduce the structure of an associative and commutative algebra in the space of generalized functions $\mathcal{K}'(\mathbb{R})$ that is consistent with the operation of multiplying generalized functions by smooth functions. This is a consequence of the following reasoning due to L. Schwartz (1950/51). Suppose that such an algebra exists and contains the generalized functions δ_0 and $\left[\frac{1}{x}\right]$. Then

$$(\delta_0 \cdot x) \cdot \left[\frac{1}{x}\right] = (x \cdot \delta_0) \cdot \left[\frac{1}{x}\right] = 0, \quad \delta_0 \cdot \left(x \cdot \left[\frac{1}{x}\right]\right) = \delta_0 \neq 0\,,$$

contradicting the assumption of associativity. However, if we relax some of the requirements, multiplicative structures may exist on certain subspaces of $\mathcal{K}'(\mathbb{R}^n)$. Such possibilities are important for field theory (multiplication of singular functions) and nonlinear analysis (generalized solutions of nonlinear equations). There are several approaches to the problem of multiplying generalized functions.

I. The product of generalized functions u and v can be unambiguously defined if they satisfy interconnected conditions on the nature of their singularities. For example if u is an ordinary locally bounded function and v is a measure, the product $u \cdot v$ is defined. If u has locally bounded derivatives up to some order k and v is the sum of the derivatives of measures of order at most k, then the product $u \cdot v$ is again defined, etc. In the case of generalized functions of several variables there are many more combinations of such conditions. In particular the product is well-defined if the generalized function u is smooth along a certain foliation and v is smooth across the foliation (§4 of Chapt. 2), or the sum $\mathrm{WF}\,(u) + \mathrm{WF}\,(v)$ contains no zero vectors (§5 of Chapt. 2), etc.

II. In $S'(\mathbb{R})$ we consider the subspace S'_+ formed by the generalized functions that are boundary values of holomorphic functions $f(z)$ defined in the strip $0 < \operatorname{Im} z < b$ and admit a power estimate of growth at infinity and as $\operatorname{Im} z \to 0$. The function f is uniquely determined by the generalized function $u = \operatorname{bv} f$. If $\operatorname{bv} g = v \in S"_+$, then the product can be defined from the formula $u \cdot v = \operatorname{bv}(f \cdot g) \in S'_+$. As a result S'_+ becomes a commutative and associative algebra. In particular $(x+i0)^\lambda \cdot (x+i)^\mu = (x+i0)^{\lambda+\mu}$ in this algebra. In this construction the upper plane can of course be replaced by the lower plane, yielding an algebra S'_-; however it is not possible to multiply generalized functions in different algebras in this way.

In the case of an arbitrary domain $U \subset \mathbb{R}^n$ from any open convex cone $\Gamma \subset \mathbb{R}^n$ one can distinguish the subspace $\mathcal{K}'_\Gamma(U) \subset \mathcal{K}'(U)$ consisting of the generalized functions of the form $u = \operatorname{bv} f$, where f is a function holomorphic in the tube domain $T(\Gamma) \cap \Omega$, and $\Omega \supset U$ may depend on f.

If we introduce a multiplication in $\mathcal{K}'_{\Gamma}(U)$ by multiplying the corresponding holomorphic functions, there also arises an associative commutative algebra, and this operation is consistent with the ordinary rule for multiplication by functions that are holomorphic in a neighborhood of U. In the case when Γ is a future or past cone in $\mathbb{R}^4$ we thereby obtain the algebras of positive or negative-frequency generalized functions. A substantive discussion of these possibilities can be found in §19 of (Bogolyubov, Shirkov 1976).

III. Every generalized function u is the weak limit of a sequence of ordinary continuous functions. We choose such sequences for given generalized functions $f_n \to u$, $g_n \to v$, and we set

$$u \cdot v = \lim_n [f_n \cdot g_n] ,$$

if this limit exists. Such an approach is ambiguous, since even in the simplest cases it gives different results depending on the choice of these sequences (Antosik, Mikusinski, Sikorski 1973). For example in the case $n = 1$ the generalized function δ is the weak limit of the sequence of functions $f_n^+(x) = n$ on the interval $-\lambda \leq n^2 x \leq n$, and $f_n^+(x) = 0$ outside this interval. Let $f_n^-(x) = f_n^+(-x)$; we also have $f_n^- \to \delta$, but $f_n^+ f_n^- \to 2\lambda\delta$. Thus we arrive at the formula $\delta \cdot \delta = 2\lambda\delta$ with an arbitrary $\lambda \geq 0$. On the other hand if we set

$$f_n(x) = \frac{n}{\pi(1 + n^2 x^2)} ,$$

then $f_n \to \delta$, but f_n^2 has no finite limit in the space of generalized functions.

Let us consider another example. The product $u = \left[\frac{1}{x}\right] \cdot \delta$ can be defined by interpreting u as a solution of the equation $x \cdot u = \delta$. In such a case $u = -\delta' + c\delta$, where c is arbitrary. One can define u differently, namely as the weak limit of the sequence of functions $\left(f_n * \left[\frac{1}{x}\right]\right) \cdot f_n$. This limit exists and leads to the equality $\left[\frac{1}{x}\right] \cdot \delta = -\frac{1}{2}\delta'$!

Despite the ambiguity of the product $\delta^2 = \delta \cdot \delta$, such an expression is encountered in the physical literature, even in a textbook (Kushnirenko 1983, p. 169). In the formula for the retarded Green's function in $\mathbb{R}^4$

$$D(x) = \frac{1}{2\pi}\theta(x_0)\delta(x^2)$$

the product of a discontinuous function θ by the generalized function $\delta(x^2)$ is well-defined, since the latter is a measure. But if u is a more complicated generalized function, the product $\theta(x_0)u$ may fail to have an unambiguous meaning, cf. (Bogolyubov, Shirkov 1976).

IV. *Other Approaches.* König (1955) constructed an algebra of formal products of generalized functions of one variable. Yu.M. Shirokov (1979) constructs a special algebra of generalized functions of three variables, solving the problem of extension of an unbounded quantum-mechanical operator. Bogoliubov and Parasiuk (1957) define the product of causal functions of field theory using subtraction of suitable singular functions. Colombeau (1984) extends the space of generalized functions to an algebra by replacing each generalized function by a suitable family of smooth functions approximating it.

Chapter 5
Contact Structures and Distributions

Introduction

The most important source of the theory of generalized functions has been problems that arise in the theory of hyperbolic differential equations (including analytic questions of field theory). In particular the generalized functions that we call Hadamard kernels (Chapt. 1) were introduced under the name "nonsingular integrals of a new type" to construct a fundamental solution of a hyperbolic equation (Hadamard 1932). The work of Sobolev (1936) is devoted to the solution of this same problem. In the paper of M. Riesz (1949) the important idea of analytic continuation of a family of generalized functions arose in the same context. A new class of singular functions arises in the analysis of a wave field in a neighborhood of a caustic. Among the pioneering works in this area is the paper of V.M. Babich (1961a), where new special functions are introduced to study a field near an elementary caustic.

In this chapter we discuss an approach to the study of such singularities, based on the concept of a distribution connected with a wave (= conic Lagrangian) submanifold in the contact bundle. The problems of integral geometry, in particular the problem of describing the discontinuities of a (generalized) Radon transform, belong to the same class. The construction of distributions that we are using is close to the method of Hadamard and Gårding (1977) and parallel to the constructions of the canonical Maslov operator and the Lagrangian distributions of Hörmander. In §4 we study geometrical properties of distributions of this class are studied which are to a great degree new. The purpose of §5 is to show an application of nonclassical special generalized functions connected with points of hypersurfaces. These series include the Airy and Pearcey functions and their more complicated analogues (Guillemin, Sternberg 1977), as well as the special functions of Babich and their higher analogues.

§1. Contact Structures

1.1. A *contact manifold* is a smooth manifold M of dimension $2n-1$ ($n \geq 1$) with a distinguished $2n-2$-dimensional nondegenerate distribution $H \subset T(M)$ (i.e., at each point $m \in M$ a hyperplane $H_m \subset T_m(M)$ is distinguished that depends smoothly on m). Such a distribution can be defined at least locally by the equation $\alpha = 0$, where α is a nonzero form of first order, called a contact form. Another form that defines the same distribution differs from α by a smooth nonzero factor. The condition of nondegeneracy of the distribution H means that the form $\alpha \wedge d\alpha \wedge \cdots \wedge d\alpha$ of order $2n-1$ is everywhere nonzero on M.

Let X be any smooth manifold; consider the manifold $K(X)$ of co-oriented contact elements of X. A co-oriented element at the point $x \in X$ is a hyperplane in the tangent space $T_*(X)$ with a distinguished side. Such elements are in one-to-one correspondence with the rays in the cotangent bundle $T^*(X)$. Thus $K(X)$ is isomorphic to the quotient bundle of $T^*(X)$ with respect to the action of the group $\mathbb{R}_+$ of positive numbers, and also to the spherical subbundle $S^*(X) \subset T^*(X)$ formed by the unit spheres $S_x^* \subset T_x^*$ with respect to some metric on $T^*(X)$. The bundle $K(X)$ has a canonical contact structure that can be given by the contact form $\alpha = \sum \xi_i \, dx_i$, where $x_1, \ldots, x_n$ are the local coordinates on X and $\xi_1, \ldots, \xi_n$ are the corresponding coordinates in a fiber of $T^*(X)$. The same contact structure can be defined in an invariant manner by setting $H_{x,K} = dp^{-1}(K)$ at every point $(x, K) \in K(X)$, where $p : K(X) \to X$ is a canonical projection.

A submanifold W of a contact manifold M is *integral* if at each of its points x it is tangent to the distinguished hyperplane H_x. Its dimension never exceeds $n-1$ (otherwise the condition that the distribution H be nondegenerate would be violated). A closed integral manifold W of dimension $n-1$ in M will be called a *wave* manifold. W is a wave manifold if and only if its pullback L under the canonical projection $T^*(X) \to K(X)$ is a conic Lagrangian manifold. This notion is closely related to wave fronts and wave propagation (see below). Arnol'd, Varchenko and Gusejn-Zade (1982) use the term *Legendre manifold* for any integral submanifold of dimension $n-1$ of a contact manifold of dimension $2n-1$. The *front* of the wave manifold W is its image $p(W) \subset X$ under the canonical projection.

Example 1. Every closed smooth submanifold $Y \subset X$ is the front of the wave manifold $S^*(Y)$ of its conormal unit vectors.

Example 2. Let Θ be any smooth manifold, f a smooth function on $X \times \Theta$ with zero set Z and $df(x,\theta) \neq 0$ for every point $(x,\theta) \in Z$. The projection $X \times \Theta \to X$ defines a mapping of smooth manifolds $q : Z \to X$. Let C_f be critical set of this mapping. It is defined in $X \times \Theta$ by the system of equations

$$f(x,\theta) = 0, \quad \frac{\partial f(x,\theta)}{\partial \theta_i} = 0, \quad i = 1, \ldots, N, \tag{5.1.1}$$

where $\theta_1, \ldots, \theta_N$ are local coordinates in Θ. We impose the condition

$$(x,\theta) \in C_f \Rightarrow \text{the forms } df, d\frac{\partial f}{\partial \theta_1}, \ldots, d\frac{\partial f}{\partial \theta_N} \text{ are independent}. \tag{5.1.2}$$

In this case C_f is a smooth manifold of dimension $n-1$. Consider the mapping

$$f^* : C_f \to K(X); \quad (x,\theta) \to (x, K = \operatorname{Ker} d_x f(x,\theta)).$$

By (5.1.2) this mapping is a local diffeomorphism. The image of this mapping, if it has no self-intersections, is a wave submanifold $W_f \subset K(X)$. To verify this it is necessary to show that for every tangent vector u to W_f at a point (x, K) the form $d_x f$ vanishes at $dp(u)$. We can write $u = df^*(v)$, where v is a vector tangent to C_f at a point (x,θ). Using (5.1.1), we find

$$d_x f(dp(u)) = d_x f(dp(df^*(v))) = d_x f(dq(v)) = df(v) = 0.$$

The front equals the projection $q(C_f)$, which in general has singular points. We shall study below the structure of the front in a neighborhood of such points.

1.2.

Proposition. *Let $W \subset K(X)$ be a wave manifold and $w = (x, K)$ a point of it. Assume that the rank of the mapping $p : W \to X$ at the point w is r and $x_1, \ldots, x_r$ are functions in a neighborhood of the point x such that the inverse images of the forms $dx_1, \ldots, dx_r$ are independent on W at the point w. Then one can choose a coordinate system $x_1, \ldots, x_r, x_{r+1}, \ldots, x_n$ in a neighborhood of the point x and a coordinate system $\xi_1, \ldots, \xi_{n-1}$ in a neighborhood of the point K on the fiber $K(X)$ such that the functions $x_1, \ldots, x_r, \xi_{r+1}, \ldots, \xi_{n-1}$ form a coordinate system on W, and a contact form can be written as $\alpha = \sum_1^{n-1} \xi_i \, dx_i + dx_n$ in a neighborhood of w.*

Proof. We can assume that the functions $x_1, \ldots, x_r$ vanish at the point x. We have $r < n$, and we choose a coordinate function x_n so that $\alpha = dx_n$ at this point. The forms $dx_1, \ldots, dx_r, dx_n$ are independent at the point x, since by hypothesis the inverse images of $dx_1, \ldots, dx_r$ on $T_w(W)$ are independent and dx_n vanishes on $T_w(W)$. We choose the functions $x_{r+1}, \ldots, x_{n-1}$ so as to obtain altogether a coordinate system in a neighborhood of the point x. In the corresponding coordinates on $T^*(X)$ the contact form can be written $\alpha = \sum \xi_i \, dx_i$, and in passing to $K(X)$ we can set $\xi_n = 1$. It remains only to verify that the forms $ds_1, \ldots, dx_r, d\xi_{r+1}, \ldots, d\xi_{n-1}$ are independent on W in a neighborhood of w. To do this we consider the submanifold $Y \subset X$ defined by the equations $x_1 = \cdots = x_r = 0$, and the bundle $K(Y)$ of its contact elements. Every such element is the intersection of some contact element of the manifold X with $T(Y)$. Consider the set V of elements of the form $(y, L \cap T_y(Y))$, where $(y, L) \in W$. Each of them is a contact element in Y, as follows from the independence of the forms $dx_1, \dots, dx_r$ on W. For the same reason V is a smooth manifold of dimension $n - r - 1$ (in fact a wave manifold). The rank of its projection on Y at the point w is zero, and consequently the rank of the projection on the fiber $K(Y)$ is $n - r - 1$.

We shall call the functions $x' = (x_1, \dots, x_r)$, $\xi'' = (\xi_{r+1}, \dots, \xi_{n-1})$ constructed in this theorem *Darboux coordinates* on the germ (W, w) of the wave manifold.

The Theorem on the Generating Function (compare (Hörmander 1971)). *Let $W \subset K(X)$ be a wave manifold and x' and ξ'' the Darboux coordinates for its germ at the point w. There exists a neighborhood $U \subset X \times \mathbb{R}^{n-r-1}$ of the point $(p(w), 0)$ and a smooth function $f = f(x, \xi''))$ on U such that the germ constructed from f as in Example 2 coincides with the germ of W.*

Proof. We can define W by the equations

$$x_j = X_j(x', \xi''), \quad j = r+1, \dots, n; \quad \xi_i = \Xi_i(x', \xi''), \quad i = 1, \dots, r,$$

where X_j and Ξ_i are certain smooth functions. We set

$$f(x, \xi'') = \sum_{j=r+1}^{n-1} \xi_j (x_j - X_j) + (x_n - X_n).$$

It is obvious that $df \neq 0$. We claim that

$$C_f = \{x_j = X_j(x', \xi''), \quad j = r+1, \dots, n\}. \tag{5.1.3}$$

It is obvious that the right-hand side is contained in the left-hand side. Moreover, since W is an integral manifold, it follows that

$$\alpha \big| W = \sum_1^r \Xi_i \, dx_i + \sum_{r+1}^{n-1} \xi_j \, dX_j + dX_n = 0,$$

whence $\sum \xi_j \, d_\xi X_j + d_\xi X_n = 0$ and

$$\sum \Xi_i \, dx_i + \sum \xi_j \, d_x X_j + d_x X_n = 0. \tag{5.1.4}$$

It follows from the first relation that the left-hand side of (5.1.3) is contained in the right-hand side, and consequently, they coincide. Using (5.1.4), we find that

$$d_x f = \sum \xi_j (dx_j - d_x X_j) + (dx_n - d_x X_n) = \sum \xi_j \, dx_j + \sum \Xi_i \, dx_i + dx_n,$$

and therefore $f^*(x, \xi'') = (\Xi_1, \dots, \Xi_r; \xi_{r+1}, \dots, \xi_{n-1}, 1)$, which proves the theorem.

The function f is called a *generating* function for the germ (W, w). Thus any germ of a wave manifold has at least one generating function. There is some arbitrariness in the choice of the generating function. In particular, one can consider a function

$$g(x, \theta, \eta) = f(x, \theta) + h(\eta) ,$$

depending on m additional variables η, where h is an arbitrary nondegenerate quadratic form. It is clear that $C_g = C_f \times \{0\}$ and $d_x g\big|_{\eta=0} = d_x f$; consequently g is also a generating function of the germ of the wave manifold generated by the function f. We call the function g an *inflation* of f using the quadratic form h.

The Theorem on Equivalence of Generating Functions (cf. (Hörmander 1983a)). *Let f and $\tilde{f}$ be functions that generate germs (W, w) and $(\widetilde{W}, w)$ of wave submanifolds in $K(X)$. These functions are defined in a neighborhood of the point $(x, \theta_0) \in X \times \mathbb{R}^N$ (resp. in a neighborhood of $(x, \tilde{\theta}_0) \in X \times \mathbb{R}^M$), where $x = p(w)$. Necessary and sufficient conditions for the existence of a diffeomorphism φ of a neighborhood of (x, θ_0) onto a neighborhood of $(x, \tilde{\theta}_0)$ over X (i.e., one that commutes with projections on X) and a smooth function $a > 0$ in a neighborhood of the first point such that $\varphi^*(\tilde{f}) = af$ are the following:*

I) *the germs of W and $\widetilde{W}$ coincide,*
II) *$M = \widetilde{M}$,*
III) *the quadratic forms $d^2_\theta f(x, \theta_0)$ and $d^2_{\tilde{\theta}} \tilde{f}(x, \tilde{\theta}_0)$ have the same signature.*

Corollary. *If the functions f and $\tilde{f}$ satisfy only I), then there exist inflations of them g and $\tilde{g}$, a smooth function $a > 0$, and a diffeomorphism over X that maps $\tilde{f}$ to af.*

§2. Distributions Connected with a Wave Manifold

2.1. Let $(W, *)$ be the germ of a wave manifold over a manifold X and f a generating function of this germ defined in the domain $U \subset X \times \mathbb{R}^N$; we denote the coordinates in this domain by $u = (x, \theta)$. Consider the auxiliary space Σ of distributions defined on $U \times \mathbb{R}$ that are infinitely differentiable when $t \neq 0$, where t is a coordinate in this second factor, and are rapidly decreasing as $|t| \to \infty$ together with all their derivatives. For such a distribution $a = a(u, t)$ the Fourier transform is defined

$$\tilde{a}(u, \tau) = F_{t \to \tau} a(u, t) ,$$

and is a smooth function on $U \times \mathbb{R}^*$. We further define the numbers $0 \leq \rho \leq 1$, $0 \leq \delta \leq 1$, and $m \in \mathbb{R}$ arbitrarily, and we consider the subspace $\Sigma^m_{\rho,\delta} \subset \Sigma$ of distributions a for which the following inequalities hold:

$$|D^\alpha_\tau D^\beta_u \tilde{a}(u, \tau)| \leq C_{K,\alpha,\beta} (|\tau| + 1)^{\frac{N}{2} + m + \delta|\beta| - \rho|\alpha|} \tag{5.2.1}$$

on every compact set $K \subset U$ for any multi-indices α and β with certain constants $C_{K,\alpha,\beta}$. (If U is a point, we simplify the notation $\Sigma^m_{\rho,\delta}$ to Σ^m.) We define a distribution on U that can be symbolically written as

$$a_f(\cdot) = a(\cdot, f(\cdot)) . \tag{5.2.2}$$

Proposition 1. *If $\delta < 1$, the distribution a_f is well-defined as an element of $\mathcal{D}'(U)$.*

To give a meaning to (5.2.2), we choose a sequence of functions $\varphi_k \in \mathcal{D}(\mathbb{R})$ tending to δ_0, and we set

$$a_f(\psi) = \lim_{k\to\infty} a(\psi(u)\varphi_k(t - f(u))) .$$

We shall show that this limit exists for any such sequence and any $\psi \in \mathcal{D}(U)$. If $m < -\dfrac{N}{2} - 1$, the function $\tilde{a}$ is integrable with respect to τ uniformly over $u \in K$ for every compact set $K \subset U$. Consequently the distribution a is continuous and this limit exists. Further, reasoning by induction on the number m, we assume that our assertion has been proved for every distribution $b \in \sum^{m-1+\delta}_{\rho,\delta}$. We find a distribution $a_1 \in \sum^{m-1}_{\rho,\delta}$ such that $(2\pi i\tau + 1)\tilde{a}_1 = \tilde{a}$, and we construct a smooth vector field v in U such that $v(f) = 1$. For any smooth function φ we have $\left(\dfrac{\partial}{\partial t} + v\right)\varphi(t - f(u)) = 0$. Therefore, taking into account the relation $\dfrac{\partial a_1}{\partial t} + a_1 = a$, we find

$$a(\psi \cdot \varphi_k(t - f)) = -a_1\left(\psi\frac{\partial}{\partial t}\varphi_k(t - f)\right) + a_1(\psi \cdot \varphi_k(t - f)) .$$

The second term has a limit as $k \to \infty$, since $a_1 \in \sum^{m-1}_{\rho,\delta}$, and the first can be transformed by integrating by parts into the form

$$\begin{aligned} -a_1\left(\psi \cdot \frac{\partial}{\partial t}\varphi_k(t - f)\right) &= a_1(\psi \cdot v(\varphi_k(t - f))) = \\ &= -a_1(v(\psi) \cdot \varphi_k(t - f)) + v^*(a_1)(\psi \cdot \varphi_k(t - f)) , \end{aligned}$$

where v^* is the differential operator adjoint to v. As $k \to \infty$ both terms on the right-hand side have a limit since $v^*(a_1)$ belongs to $\sum^{m-1+\delta}_{\rho,\delta}$.

Definition. Assume that $a \in \Sigma$ and the projection $p : \operatorname{supp} a \to X$ is a proper mapping). We construct a_f using Proposition 1 and define the direct image

$$w = p_*(a_f) = \int_p a(x, \theta, f(x, \theta)) . \tag{5.2.3}$$

The function f is called the *phase* and the function a or $a/|d\theta|$ the *amplitude* of w. In the case $N = 0$ we shall also call the amplitude the *profile* of w. We denote by $\sum^m_{\rho,\delta}(f)$ the space of germs at the point $x_0 = p(*)$ of distributions that can be represented modulo the space $\mathcal{O}_{x_0}(X)$ of germs of infinitely smooth functions in the form (5.2.3) with phase function f and amplitude of class $\sum^{m}_{\rho,\delta}$.

Proposition 2. *For any $\rho > 0$, $\delta < 1$, and m the wave front (§5 of Chapt. 2) of every distribution $w \in \Sigma^m_{\rho,\delta}(f)$ is contained in the union $W \cup (-W)$. If the amplitude is such that the function $\tilde{a}$ is rapidly decreasing as $\tau \to -\infty$ (resp. as $\tau \to +\infty$), then the wave front of w is contained in W (resp. $-W$).*

The first assertion is a consequence of the second. To prove the second we choose a coordinate system $x_1, \dots, x_n$ in a small neighborhood X of the given point x_0 and a closed conic subset $\Lambda \subset T^*(X)$ that does not intersect the manifold $W \cap S^*(X)$. We need to show that the integral

$$I(\xi) = \iiint e^{2\pi i(\tau f(x,\theta) - (x,\xi))} \tilde{a}(x,\theta,\tau)\, d\tau\, d\theta\, dx$$

tends rapidly to zero as the point $\xi \in \Lambda$ tends to infinity. In doing this we assume that the support of a is contained in $X' \times \Theta$, where X' is this neighborhood of the point x_0. The domain $X' \times \Theta$ can be covered by parts in each of which either the inequality $|f| \geq \varepsilon$ holds or for some i the inequality $\left|\frac{\partial f}{\partial \theta_i}\right| \geq \varepsilon$ holds, or $|\xi - \tau f'_x| \geq \varepsilon(|\xi| + |\tau|)$ with some $\varepsilon > 0$. Introducing a suitable partition of unity, we represent I as a sum of integrals over such domains. In a domain of the first type we integrate by parts on the variable τ, in the second type on the variable θ_i, and in the third type on x. In the first two cases the result is an integral of the same type with a new amplitude contained in $\sum^{m-\rho}_{\rho,\delta}$ or $\sum^{m-1+\delta}_{\rho,\delta}$. After repeated application of this device the inner integrals with respect to τ and θ acquire more and more smoothness with respect to x, and after integration on x more and more rapid decrease with respect to ξ. In the third case, after integration by parts with respect to x we obtain an additional decreasing factor of the form $|\tau|^\delta(|\xi| + |\tau|)^{-1}$ in the estimate of the new amplitude. When $|\xi| \geq 1$, it can be estimated by the quantity $(|\tau| + 1)^{-1+\delta}$, which as a result of the repetition of the procedure will give us the required decrease in amplitude. At the same time it can be estimated by $(|\xi| + 1)^{-1+\delta}$, which after iteration guarantees a rapid decrease in the integral. Thus the proposition is proved.

Theorem 3. *If $1 - \rho = \delta < \frac{1}{2}$, the space $\Sigma^m_{\rho,\delta}(f)$ is an invariant of the germ of the wave manifold W.*

By the equivalence theorem (§1), the proof follows from three assertions:

1) the space $\sum^m_{\rho,\delta}(f)$ maps into itself under the action of the diffeomorphism $x = x$, $\theta' = \theta'(x,\theta)$ on the function f;
2) the same is true when f is replaced by hf, where $h \in \mathcal{O}_{x_0}(X)$ and $h > 0$, and
3) if $\delta < \frac{1}{2}$ and g is an inflation of f, then $\sum^m_{\rho,\delta}(g) = \sum^m_{\rho,\delta}(f)$.

The first assertion is obvious, and the third will follow from Theorem 4. Let us prove the second. Let $a \in \sum^m_{\rho,\delta}$; we shall verify that the amplitude $b(u,t) =$

$a(u,h(u)t)$ also belongs to the class $\sum^m_{\rho,\delta}$. We have $\tilde{b}(u,\tau) = g(u)\tilde{a}(u,g(u)\tau)$, where $g = h^{-1}$. Consequently for every vector field v in U, using (5.2.1), we obtain

$$|v(\tilde{b})(u,\tau)| = \left|v(g)(\tilde{a}) + g\cdot v(\tilde{a}) + g\cdot v(g)\tau\frac{\partial \tilde{a}}{\partial \tau}\right| \le C(|\tau|+1)^{\frac{N}{2}+m+\delta}\,.$$

Therefore the amplitude b satisfies the estimate (5.2.1) with $\alpha = 0$, $|\beta| = 1$. All the other derivatives of b are estimated similarly.

Definition. Let $1-\rho = \delta < \frac{1}{2}$ and let W be the germ of a wave manifold. By Theorem 3 the notation $\sum^m_{\rho,\delta}(W) \equiv \sum^m_{\rho,\delta}(f)$ is unambiguous, where f is any generating function of this germ. If W is a closed wave manifold over X, we denote by $\sum^m_{\rho,\delta}(W)$ the space of distributions on X that can be written as a finite sum

$$w = \sum_\alpha w_\alpha, \quad w_\alpha \in \Sigma^m_\delta(W_\alpha)$$

in a neighborhood of any point x, where W_α is a germ of the manifold W at some point lying over x. Then we call w a *W-distribution* and the generalized function $w/|dx|$ a *W-function.*

2.2. Examples

Example 1. The fiber $K_0(\mathbb{R}^n)$ of the contact bundle over $\mathbb{R}^n$ is a wave manifold and has the generating function $f(x,\theta) = \sum_1^n x_i\theta_i$ defined on $\mathbb{R}^n \times S^{n-1}$. We choose the amplitude

$$a(u,t) = \frac{(n-1)!}{(-2\pi i)^n}(t+i0)^{-n}\,|dt\,dx|\,dS\,,$$

where dS is the standard volume element on the sphere S^{n-1}. The corresponding generalized function is found by the formula

$$w = p_*(a_f) = \frac{(n-1)!}{(-2\pi i)^n}\int \frac{|dx|\,dS}{((x,\theta)+i0)^n} = \delta_0(x)\,|dx|\,.$$

This is the expansion of the delta-function in plane waves (Gel'fand, Shilov 1958a). Here one can pass to the complex-conjugate amplitude and also to integration over projective space, adding the values of the integrand density at antipodal points (cf. Sect. 3.5.2)

Example 2. Consider the Fourier integral (Hörmander 1971)

$$I(x) = \int_{\mathbb{R}^N} e^{2\pi i\varphi(x,\theta)}a(x,\theta)\,d\theta$$

with nondegenerate phase function φ and amplitude a of Hörmander class $S^m_{\rho,\delta}(X)$. The condition $a \in S^m_{\rho,\delta}(X)$ means that the function $a(x,\theta)$ is defined in $X \times \mathbb{R}^N$ ($X \subset \mathbb{R}^n$ is an open set), and for any compact set $K \subset X$ and multi-indices α and β the inequality

$$|D_\theta^\alpha D_x^\beta a(x,\theta)| \le C_{K,\alpha,\beta}(|\theta|+1)^{m+\delta|\beta|-\rho|\alpha|} \tag{5.2.4}$$

holds with some constants $C_{K,\alpha,\beta}$. The phase function gives a Lagrangian manifold $L \subset T^*(X)$ that is the image of the set $C_\varphi = \{(x,\theta) : d_\theta\varphi(x,\theta) = 0\}$ under the mapping

$$C_\varphi \to T^*(X) : (x,\theta) \mapsto d_x\varphi(x,\theta) .$$

The condition of nondegeneracy means that C_* is a nonsingular submanifold in $X \times (\mathbb{R}^N \times \{0\})$. The function φ is by hypothesis positive-homogeneous, i.e., $\varphi(x,\lambda\theta) = \lambda\varphi(x,\theta)$ for $\lambda > 0$. It follows from this that the set C_φ is conic, i.e., is the union of rays of the form $\{(x,\lambda\theta), \lambda > 0\}$. Therefore the manifold L is also conic and corresponds to a wave manifold W with generating function $\varphi(x,\omega)$, $x \in X$, $\omega \in S^{N-1}$.

We shall show that the Fourier integral I belongs to $\Sigma_{\rho,\delta}^{m+(N-1)/2}(\varphi)$. To do this we pass to polar coordinates $\theta = r\cdot\omega$ in it and integrate over the radial variable r:

$$I(x) = \int\limits_{S^{N-1}} d\omega \int_0^\infty e^{2\pi i r\varphi(x,\omega)} a(x, r\cdot\omega) r^{N-1}\, dr =$$

$$= \int d\omega\, \tilde{b}(x,\omega,\varphi(x,\omega)) = p_*(\tilde{b}_\varphi) .$$

Here $\tilde{b}$ denotes the inverse Fourier transform of a function $b(x,\omega,\tau) := a(x,\tau\omega)\tau_+^{N-1}$. Inequality (5.2.1) for the amplitude $\tilde{b}$ with the constant m replaced by $m+(N-1)/2$ follows from inequality (5.2.4) for a. This proves our assertion.

It follows from this assertion that every Lagrangian distribution in the sense of (Hörmander 1985) corresponding to an arbitrary conic Lagrangian manifold L is a W-distribution for the wave manifold $W := L \cap S^*(X)$ ($S^*(X)$ is the spherical cotangent bundle). Conversely every W-distribution (when $\rho = 1$, $\delta = 0$) is the Lagrangian distribution connected with the conical Lagrangian manifold L generated by W. To verify this it suffices to conduct the preceding reasoning in the reverse order and use 25.1.5 in (Hörmander 1985).

Example 3. Let $p(x,D)$ be a strictly hyperbolic linear differential operator with respect to the time variable t on the manifold X and E_y a *parametrix* of it, i.e., a solution of the differential equation

$$p(x,D)E_y(x) \equiv \delta_y(x) \pmod{C^\infty(X)}$$

with support belonging to the half-space $t(x) \ge t(y)$, $y \in X$. Indeed the support of E_y is contained in a closed domain bounded by a conoid consisting of rays emanating from the point y and lying in this half-space. This conoid is the projection on X of the wave manifold W formed by the set of zero bicharacteristics of the principal symbol $p_m(x,\xi)$ emanating from the fiber $K_y(X)$ and directed into the future. We assert that in the half-space $t(x) > t(y)$ the parametrix is a W-distribution. Indeed, according to (Lax 1957; Duistermaat, Hörmander 1972; Maslov, Fedoryuk 1976) the parametrix can be written as a sum of $m = \deg p$ Fourier integrals with $N = n-1$, $n = \dim X$, in which the amplitudes a have asymptotic expansions of the form $a \sim a_\nu + a_{\nu-1} + a_{\nu-2} + \cdots$, $\nu = 1-m$, where all $a_i = a_i(x,\theta)$ are positive-homogeneous functions of θ of order i that are smooth jointly in the variables when $\theta \neq 0$. This expansion holds as $|\theta| \to \infty$ and is applicable by means of termwise differentiation to the derivatives of the amplitude. It follows from this that $a \in S^{1-m}_{1,0}(X)$, and, by what was said in Example 2, $E_y \in \Sigma_0^{-m+\frac{n}{2}}(W)$.

Example 4. Suppose that in the domain $\Omega \subset \mathbb{R}^n$ there is given a smooth function f and that S is the set of its zeros, while $df(x) \neq 0$ when $x \in S$. Consider in Ω the density $\rho = a(f(x))b(x)\,dx$ having a discontinuity on the smooth hypersurface S, and suppose $bdx \in \mathcal{K}(\Omega)$. We shall show that its Radon transform (Sect. 5.2, §5, Chapt. 3) is the generalized function connected with some wave manifold. We consider the Radon transform as a function on the space $U = S^{n-1} \times \mathbb{R}$ of cooriented hyperplanes in $\mathbb{R}^n$; here the hyperplane $H_{\omega,p}$ corresponding to the point $(\omega,p) \in U$, is defined by the equation $(\omega,x) - p = 0$. This same equation distinguishes a hypersurface in $\mathbb{R}^n \times U$ which we denote by H. We regard the odd form

$$\sigma = \frac{\rho}{(\omega, dx)}$$

as a relative density on the natural bundle $p : H \to U$. Its direct image $R(\rho) = p_*(\sigma)$ is by definition the Radon transform of the density ρ. We remark that the mapping p is a submersion, i.e., it has locally the structure of a projection $\mathbb{R}^{n-1} \times U \to U$. Therefore, according to the definitions of Sect. 2.1, $R(\rho)$ is a generalized function connected with the wave manifold W produced by the generating function $g = q^*(f)$, $q : H \to \mathbb{R}^n$, and the amplitude $b(x)a(t)$, with $N = n-1$. This means that $R(\rho)\,dS|dp|$ is a W-distribution.

We have

$$C_g = \{(x,\omega,p) \in H : \ f(x) = 0,\ t(f(x)) = 0\ ,\ \forall t \in T_x(H_{\omega,p})\}\ .$$

The first equation gives the hypersurface S and the equality $t(f(x)) = 0$ is possible only if t is tangent to S. Therefore C_g is the set of points on S together with the corresponding tangent hyperplanes. The mapping $g^* : C_g \to K(U)$ assigns to each such point the differential of the function g, which can be written in the form $dg \equiv dq^*(f) \bmod d((\omega,x) - p)$. We have further

$$dq^*(f) \equiv df - \lambda d((\omega,x) - p) = df - \lambda(\omega,dx) - \lambda(d\omega,x) + \lambda dp = \\ = \lambda(dp - (x,d\omega))\ ,$$

where λ is chosen so that $\lambda(\omega, dx) = df$. Thus $g^*(x, \omega, p)$ is a contact element at the point (ω, p) defined by the form $\pm(dp - (x, d\omega))$, and the sign is determined by whether the covectors df and ω point the same direction or not. This contact element is tangent at the point (ω, p) to the manifold Ω_x of hyperplanes passing through x. Consider the diagram

$$\begin{array}{ccc} K(\mathbb{R}^n) & \xrightarrow{\varkappa} & K(U) \\ \cup & & \cup \\ N^+(S) & \longrightarrow & W \end{array},$$

in which the mapping $\varkappa$ maps a contact element (x, K) to the element $(\omega, p, \varepsilon L)$, where $H_{\omega,p}$ is the hyperplane passing through x tangent to K and L is the tangent hyperplane to Ω_x, $\varepsilon = 1$ if K and $H_{\omega,p}$ are cooriented alike, and $\varepsilon = -1$ otherwise. The required wave manifold W is the image of the manifold $N^+(S)$ of the tangent hyperplanes to S cooriented by the form df. We remark that $\varkappa$ is consistent with the contact structures and consequently maps any wave manifold into a wave manifold.

We arrive at the same conclusion if we replace the Radon transform by integration over a family of cooriented hypersurfaces H_u, $u \in U$, in a smooth manifold X possessing the property that for every contact element K at a point $x \in X$ there is exactly one point u of the manifold U such that H_u is tangent to K and has the same coorientation.

2.3. The Stationary Phase Method

Theorem 4. *Let $f = f(x, \theta)$ be a generating function defined in the domain U and $g(x, \theta, \eta) = f(x, \theta) + h(\eta)$ an inflation with a form h of rank n and signature (p, q). Let $b = b(x, \theta, \eta, t)$ be an amplitude of class $\sum_{\rho,\delta}^m$, where $\delta < \frac{1}{2}$ and $\rho > 0$, such that the projection $p : \operatorname{supp} b \to U \times \mathbb{R}$ is proper. Then there exists an amplitude $a = a(x, \theta, t)$ of class $\sum_{\rho,\delta}^m$ such that*

$$a(x, \theta, f(x, \theta)) = \int_p b(x, \theta, \eta, g(x, \theta, \eta));, \tag{5.2.5}$$

and the Fourier transforms of these amplitudes are connected by the asymptotic relation

$$\begin{aligned} \tilde{a}(x, \theta, \tau) &= \frac{e_{p,q}(\tau)}{\sqrt{|\Delta|}} |\tau|^{-\frac{n}{2}} \exp\left(i \frac{\Box_0}{2\pi\tau}\right) \tilde{b}(x, \theta, \eta, \tau) \equiv \\ &\equiv \frac{e_{p,q}(1)}{\sqrt{|\Delta|}} (\tau + i0)^{-\frac{p}{2}} (\tau - i0)^{-\frac{q}{2}} \exp\left(i \frac{\Box_0}{2\pi\tau}\right) \tilde{b}(x, \theta, \eta, \tau), \end{aligned} \tag{5.2.6}$$

in which

$$e_{p,q}(\tau) = \exp\left(\frac{i\pi}{4}(p - q) \operatorname{sgn} \tau\right).$$

This relation means that for any natural number k we have $a - b_k \in \sum_{\rho,\delta}^{m-k(1-2\delta)}$, where

$$\tilde{b}_k = \frac{e_{p,q}(\tau)}{\sqrt{|\Delta|}} |\tau|^{-\frac{n}{2}} \sum_{j=0}^{k-1} \frac{1}{j!} \Big(\frac{i\Box_0}{2\pi\tau} \Big)^j \tilde{b} \,, \tag{5.2.7}$$

and $\Box_0^k$ is the functional $c(\cdot, \eta) \mapsto \Box^k c(\cdot, 0)$, where

$$\Box = \frac{1}{2} \sum h^{ij} \frac{\partial^2}{\partial\eta_i \partial\eta_j}$$

is the differential operator dual to the form $h(\eta) = \dfrac{1}{2} \sum h_{ij} \eta_i \eta_j$, and $\Delta =$ $\det \left(\frac{1}{2} h_{ij} \right)$.

Proof. Since $\Box$ is a second-order differential operator, the jth term of the series (5.2.7) belongs to the space $\sum_{\rho,\delta}^{m-j(1-2\delta)}$. Since $\delta < \frac{1}{2}$, the order $m - j(1 - 2\delta)$ tends to $-\infty$ as $j \to \infty$. Therefore one can construct a function $a_1(x, \theta, t)$ of class $\sum_{\rho,\delta}^{m}$ whose Fourier transform is equivalent to the right-hand side of (5.2.6). It remains to verify that this amplitude also satisfies (5.2.5), where both sides of the relation are interpreted using Proposition 1. Let us compute the integral on the right-hand side of this relation. We denote by H_s the hypersurface $h(\eta) = s$ in $\mathbb{R}^n$ oriented as the boundary of the set $h < s$, and we set

$$a_2(x, \theta, t) = \int b(x, \theta, \eta, t + h(\eta))\, d\eta = \int \int_{H_s} b(x, \theta, \eta, t + s) \frac{d\eta}{dh}\, ds \,.$$

To compute the inner integral we use Theorem 5 of Sect. 1.3 of Chapt. 4, and in doing this we need the following addendum to this theorem.

Addendum. Denote by $\sum^*$ the first sum on the right-hand sides of (4.1.2) and (4.1.3), and by $\sum_k^*$ a partial sum of it. For an integer $k \geq 0$ we consider the difference

$$R_k(s) = I(s) - \sum\nolimits_k^* - \sum_{j=0}^{\left[\frac{n}{2}\right]+k} c_j s^j \,,$$

where $I(s)$ is the integral on the left-hand sides of these relations. We claim that when $|s| \leq 1$ this quantity satisfies the inequalities

$$\left| \frac{d^i R_k(s)}{ds^i} \right| \leq C_k l(s) |s|^{\frac{n}{2}+k-i}, \quad i = 0, 1, \dots, \left[\frac{n}{2} \right] + k \,,$$

where $l(s) = 1 - \ln|s|$ in the case when p and q are odd and $l(s) = 1$ in all other cases.

The verification of this assertion is not difficult.

Corollary 5. *For every density $\varphi = a\, d\eta \in \mathcal{K}(\mathbb{R}^n)$ the following asymptotic expansion holds as $|\tau| \to \infty$*

$$\int e^{2\pi i\tau h(\eta)}\varphi \sim \frac{e_{p,q}(\tau)}{\sqrt{|\Delta|}}|\tau|^{-\frac{n}{2}} \exp\left(\frac{i\Box_0}{2\pi\tau}\right)a\,. \tag{5.2.8}$$

This expansion means that for any $k \geq 1$ the quantity

$$E_k(\tau) = \int e^{2\pi i\tau k}\varphi - \frac{e_{p,q}(\tau)}{\sqrt{|\Delta|}}\sum_{j=0}^{k-1}\frac{1}{j!}\left(\frac{i\Box_0}{2\pi\tau}\right)^j a$$

satisfies the inequality

$$|E_k(\tau)| \leq C_k|\tau|^{\frac{n}{2}-k}\,. \tag{5.2.9}$$

Proof of the Corollary. We write the integral (5.2.8) in the form

$$\int e^{2\pi i\tau s}I(s)\,ds$$

and instead of I we substitute the corresponding asymptotic expansion, which gives Theorem 5 of Sect. 1.3 of Chapt. 4. As a result we obtain the right-hand side of (5.2.8). The remainder term R_k is the Fourier transform of the function E_k, which in view of the addendum has at least $\left[\frac{n+1}{2}\right]+k$ integrable derivatives. Hence (5.2.9) follows.

We now continue the proof of Theorem 4. Using Corollary 5 we compute the Fourier transform of a_2:

$$\begin{aligned}
\tilde{a}_2(x,\theta,\tau) &= \int e^{-2\pi i\tau t}\int b(x,\theta,\eta,t+h(\eta))\,d\eta\,dt = \\
&= \int e^{-2\pi i\tau(t+s)}\int e^{2\pi i\tau s}\int\limits_{H_s} b(x,\theta,\eta,t+s)\frac{d\eta}{dh}\,ds\,dt = \\
&= \iint\limits_{H_s}\left(\int e^{-2\pi i\tau t'}b(x,\theta,\eta,t')\,dt'\right)\frac{d\eta}{dh}e^{2\pi i\tau s}\,ds = \\
&= \iint\limits_{H_s}\tilde{b}(x,\theta,\eta,\tau)\frac{d\eta}{dh}e^{2\pi i\tau s}\,ds = \int\tilde{b}(x,\theta,\eta,\tau)e^{2\pi i\tau h(\eta)}d\eta \sim \\
&\sim \frac{e_{p,q}(\tau)}{\sqrt{|\Delta|}}|\tau|^{-\frac{n}{2}}\exp\left(\frac{i\Box_0}{2\pi\tau}\right)\tilde{b}\,.
\end{aligned}$$

Thus the amplitudes a_1 and a_2, after their Fourier transforms are taken, have the same asymptotic expansions in decreasing powers of τ. It follows from this that the amplitudes themselves differ by an infinitely differentiable function. Under the substitution $t = f(x,\theta)$ they will differ by a function of class $C^\infty(U)$, which completes our reasoning.

§3. Versal Wave Manifolds

3.1. Let $p : W \to X$ be a wave manifold over X and x a point of its front, over which there is exactly one element K of this manifold. The *tangent space* to the front $F(W)$ at the point x is the space $T_x(F(W)) := dp(T_K(W))$. In general we assume that the front at the point x has as many tangent spaces $T_{x,K}(F(W))$, as contact elements $K \in W_x = W \cap K_x(X)$.

Example 1. Let W be the union of the trajectories of some Hamiltonian flow. In this case every *ray*, i.e., projection of a trajectory, is tangent to the tangent space (or one of them) to the front at every point.

Proposition 1. *Let W be the germ of a wave manifold at a point K. The tangent space $T_{x,K}(F(W))$ ($x = p(K)$) is contained in K. If $f(x, \theta)$ is a generating function of this germ, the codimension of $T_{x,K}(F(W))$ at K is equal to the corank of the quadratic form $d^2_\theta f$ taken at the point corresponding to K, i.e.,*

$$\dim T_x(X) - \dim T_{x,K}(F(W)) = 1 + N - \operatorname{rank} d^2_\theta f .$$

Proof. The first assertion follows from the fact that W is an integral manifold of canonical contact structure. Since the germ W is diffeomorphic to C_f over X, it follows that $T_{x,K}(F(W)) = dq(T_{x,\theta}(C_f))$, where q is the projection of the domain of definition of f on X. Therefore the dimension of this space is the difference between $\dim T_{x,\theta}(C_f) = \dim T_x(X) - 1$ and the number of independent vectors $t \in T_{x,\theta}(C_f)$ such that $dq(t) = 0$. This number equals the corank $d^2_\theta f$.

We shall say that a mapping of smooth manifolds is *transversal* to the wave manifold W defined over X if it is transversal to each tangent space to its front. If the condition of transversality is met only at the point x and W is closed, then the condition holds in a neighborhood of x.

Proposition 2. *If φ is transversal to a wave manifold W, the set of contact elements $L = d\varphi^{-1}(K)$, $K \in W$, is a wave manifold V over Y, and for any point $(y, L) \in V$ the following relation holds:*

$$T_{y,L}(F(V)) = d\varphi^{-1}(T_{x,K}(F(W))) .$$

Proof. If $f(x, \theta)$ is a generating function of a germ of W at some point K, the function $g(y, \theta) = f(\varphi(y), \theta)$ can serve as a generator for the germ (V, L), since condition (5.1.2) holds by virtue of the assumption of transversality. Indeed by virtue of this condition the product bundle $C_g = C_f \times_X Y$ is a germ of a smooth manifold. Hence follows the relation

$$T(C_g) = T(C_f) \times_{T(X)} T(Y) ,$$

which proves the second assertion.

Definition. The wave manifold V constructed over Y will be called the *pullback* of W and denoted $\varphi^*(W)$. Its front is the pullback of the front W under the mapping φ. For every point $x \in X$ we call the set $W_x = W \cap K_x(X)$ the *fiber* of W over this point. For any point $y \in Y$ there is a natural homeomorphism $V_y \approx W_{\varphi(y)}$ and an isomorphism of sheaves of $\mathbb{R}$-algebras $\mathcal{O}(W_{\varphi(y)}) \cong \mathcal{O}(V_y)$, i.e., an isomorphism of $\mathbb{R}$-ringed spaces. We define the sheaf $\mathcal{O}(W_x)$ as the quotient of the sheaf $\mathcal{O}(K(X))$ of germs of smooth functions on $K(X)$ over the sum of the sheaf I_W of ideals of functions equal to zero on W and the sheaf I_x of ideals of functions equal to zero on the fiber $K_x(X)$. Note that the ringed space $(V_y, \mathcal{O}(V_y))$ depends only on the value of the mapping φ at the point y and the value of its first differential.

3.2.

Definition. We call the wave manifold W over X *versal* over the point x if it is closed over some neighborhood of x and for any smooth transversal mapping $\varphi : Y \to X$ that maps a distinguished point y to x and any wave manifold V over Y that is closed over a neighborhood of y such that the fibers of $\varphi^*(W)$ and V over the point y coincide as ringed spaces there exists a mapping $\psi : Y \to X$ such that $\psi(y) = x$, $d\psi(y) = d\varphi(y)$, and $\psi^*(W) = V$ over a neighborhood of y.

This definition is analogous to the concept of a versal object in the theory of deformations of complex spaces. The description and construction of versal wave manifolds, which we shall now discuss, depends on this analogy.

In what follows we shall consider only closed wave manifolds. If W is such a manifold over X and x is a point, we shall define a sheaf on $K_x = K_x(X)$:

$$\mathcal{T}^1(W_x) = \mathcal{O}(W_x) \otimes_{\mathcal{O}(K_x)} \mathcal{O}_{K_x}(1) ,$$

where $\mathcal{O}_{K_x}(1)$ is the canonical bundle whose sections are the traces of linear functions on a fiber of the cotangent bundle.[3] We also define a linear mapping

$$D_x W : T_x(X) \to T^1(W_x) \equiv \Gamma(K_x, \mathcal{T}^1(W_x)) ,$$

for which we lift the vector $v \in T_x(X)$ to a vector field $\tilde{v}$ on the domain of definition U of the generating function f of the germ (W, w) and then apply this field to f. We regard the result as a section of the sheaf $\mathcal{O}(C_f)$. The latter is equal to the quotient of the sheaf $\mathcal{O}(U)$ of smooth functions on U over the sheaf of ideals $\mathcal{T}(U)$ generated in it by the functions f and $\dfrac{\partial f}{\partial \theta_j}$, $j = 1, \ldots, N$. In view of the fact that there is a homomorphism of the sheaves $\mathcal{O}(C_f) \cong \mathcal{O}(W) \to \mathcal{O}(W_x)$, and the latter sheaf is locally isomorphic to $\mathcal{T}^1(W_x)$, we can regard the result as a local section of the sheaf $\mathcal{T}^1(W_x)$. This section is independent of the choice of f and the extension of the vector v and depends

[3] We can omit the factor $\mathcal{O}_{K_x}(1)$ in the case of smooth manifolds since this sheaf has the non-vanishing section $T_x^*(X) \ni V \mapsto \|V\|$.

linearly on the vector itself. Therefore these local sections define a section of $\mathcal{T}^1(W_x)$ on K_x, which we denote $D_xW(v)$. The mapping D_xW thus defined will be called the *characteristic mapping* of the manifold W at the point x. For any mapping φ transversal to W the relation $D_y\varphi^*(W) = D_{\varphi(y)}W \cdot d\varphi$ holds (the chain rule).

For every point $w \in K_x$ we denote by $\mathfrak{m}_w$ the maximal ideal in the fiber $\mathcal{O}_w$ of the sheaf $\mathcal{O}(K_x)$, i.e., in the algebra of germs of smooth functions over K_x at the point w. This ideal consists of germs of smooth functions on the sphere K_x equal to zero at the point w. Since $\mathcal{T}^1 = \mathcal{T}^1(W_x)$ is a sheaf of modules over the sheaf $\mathcal{O}(K_x)$ the submodules of it $\mathfrak{m}_w^i\mathcal{T}^1$, $i = 1, 2, \ldots$, are defined.

Corollary 3. *The contact element w of the wave manifold W equals the pullback of $\mathfrak{m}_w\mathcal{T}^1$ under the characteristic mapping, and the tangent space to the front corresponding to this element is the pullback of $\mathfrak{m}_w^2\mathcal{T}^1$.*

Example 2. Let the wave manifold W be defined over $X = \mathbb{R}^k$ by the generating function

$$f(x,\theta) = \theta^{k+1} + x_{k-1}\theta^{k-1} + \cdots + x_2\theta^2 + x_1\theta + x_0$$

(versal deformation of a singular point of type A_k). The front of W is the discriminant set of the polynomial $f(x,\cdot)$, i.e., the set of points x for which $f(x,\cdot)$ has at least one multiple real root. The fiber W_0 consists of a single point w, equal to the contact element defined by the equation $dx_0 = 0$, and the tangent space to $F(W)$ at the point $x = 0$ is defined by the equations $dx_0 = dx_1 = 0$. In the case $k = 2$ it is the origin, and in the case $k = 3$ it is the line tangent to the x_2-axis—the "spine" of the "swallow's tail." Moreover we have $\mathcal{T}_0^1 = \mathcal{O}_0(\mathbb{R})/(\theta^k)$, and the characteristic mapping at the point $x = 0$ maps the vector $\partial/\partial x_i$, $i = 0, \ldots, k-1$ to $\theta^i \bmod (\theta^k)$ and consequently is an isomorphism.

Proposition 4. *A necessary condition for a wave manifold W to be versal over the point x is that the characteristic mapping D_xW be an epimorphism.*

We give a brief description of the proof. Assuming that D_xW is not epimorphic, we choose a section τ of the sheaf $\mathcal{T}^1$ that is not contained in the image of the characteristic mapping. Using the local generating functions f of the manifold W, we consider new functions $g(x,y,\theta) = f(x,\theta) + yt(\theta)$, depending on an additional variable $y \in \mathbb{R}$, where the function t generates τ locally. We then obtain a family of functions that generate a certain closed wave manifold over $X \times \mathbb{R}$. It follows from the chain rule for the characteristic mapping that it cannot be even a local inverse image of W and consequently W is not versal.

It follows from what has just been said that a necessary condition for W to be versal at x is that the space $T^1(W_x)$ be finite-dimensional. We shall say that W is *finite* over x if this space is finite-dimensional.

Proposition 5. *A necessary and sufficient condition for W to be finite over x is that the set W_x be finite and at each point w of it the algebra $\mathcal{O}_w(W_x)$ be finite-dimensional.*

3.3.

Theorem 6. *A necessary and sufficient condition for the wave manifold W over X to be versal over the point x is that the characteristic mapping be epimorphic at this point.*

The necessity is contained in Proposition 4 and the sufficiency will follow from Theorem 8.

Remark. The germ of the wave manifold W is Legendre stable in the sense of (Arnol'd, Varchenko, Gusejn-Zade 1982) if and only if the characteristic mapping of this germ is epimorphic. Thus Legendre stability turns out to be equivalent to versality. The proof follows from §2 of (Arnol'd, Varchenko, Gusejn-Zade 1982) and the criterion for versality of a deformation of a germ of a function (cf., for example, Theorem 8).

Corollary 7. *The set of points X over which a given closed wave manifold is versal is open.*

We call a wave manifold *miniversal* over the point x if the characteristic mapping is an isomorphism at that point. In contrast to versality this property is not open.

In Example 2 the manifold W is versal over each point but is miniversal only over the point $x = 0$.

Example 3. The wave manifold $W = W_0 = K_0(\mathbb{R}^n)$ over $\mathbb{R}^n$ is not versal, since the space $T^1(W_0) \cong \Gamma(K_0, \mathcal{O}_{K_0}(1))$ is infinite-dimensional. However, in the complex-analytic version the space $T^1(W_0) \cong \Gamma(CP_{n-1}, \mathcal{O}(1))$ is isomorphic to $T_0(\mathbb{C}^n)$, and the characteristic mapping is an isomorphism. Therefore a complex-analytic wave manifold $W = PT_0^*(\mathbb{C}^n)$ in the projectivized cotangent bundle $PT^*(\mathbb{C}^n)$ is miniversal.

Theorem 8. *If the wave manifold W over X is finite over a certain point q, there exists a wave manifold V over Y that is miniversal over a point r, and also a smooth mapping $\varphi : X' \to Y$ of some neighborhood $X' \ni q$ such that $\varphi(q) = r$ and $W\big|X' = \varphi^*(V)$.*

Proof. According to Proposition 5 the fiber W_q consists of a finite number of points, and at each of these points W the algebra $\mathcal{O}_w(W_q)$ is finite-dimensional. We choose a generating function $f = f(x, \theta)$ for the germ (W, w); let $f^*(q, 0) = w$. According to what was said in Sects. 3.1 and 3.2 this algebra can be represented as the quotient of the algebra $\mathcal{O}_{(q,0)}(U)$ over the ideal $J(f) + I_q$, where $J(f)$ is the ideal generated by the function f and its partial derivatives of first order with respect to θ. Let $e_1(\theta), \dots e_k(\theta)$ be any functions whose images in this quotient algebra form a basis of the algebra as a vector space. Consider the function

$$F(s,\theta)=f(\theta)+\sum_1^k s_j e_j(\theta),\quad f(\theta)=f(q,\theta)\,, \tag{5.3.1}$$

defined in $S\times U_q$, $S=\mathbb{R}^k$. The set $U_q=U\cap\{x=q\}$ is an open subset of $\mathbb{R}^N$. It gives a deformation of the singular germ of the hypersurface $\{f(\theta)=0\}$ at the point $\theta=0$. This means that there exist:

1) a smooth function $h(x,\theta)>0$ in a neighborhood of the point $(q,0)$,
2) a smooth change of variables $\theta'=\theta'(x,\theta)$ in a neighborhood of the point $(q,0)$ depending on x,
3) a smooth mapping $s=s(x):X'\to S$ such that $S(q)=0$ and

$$f(x,\theta)=h(x,\theta)F(s(x),\theta'(x,\theta))\,. \tag{5.3.2}$$

This fact is a corollary of the Malgrange preparation theorem (Malgrange 1966). We show below that the function F generates a certain germ of the wave manifold $V(w)$ over S, and consequently the mapping 3) leads via (5.3.2) to the equality $s^*(V(w))=(W,w)$. Further we denote by Y the direct product of the bases S of such wave manifolds constructed for all points $w\in W_q$, and we use r to denote the origin. Since there is a projection $Y\to S$, we can regard (5.3.1) as a function on $Y\times U_q$, and we can use it to generate the germ of a wave manifold over Y. The union of such germs over all $w\in W_q$ is the required wave manifold V over Y (these germs, as can easily be seen, are pairwise disjoint). We further remark that the mapping D_rV takes the vector $\partial/\partial s_j$ to the image of the function e_j in

$$T^1(V_r)\cong T^1(W_q)\cong \oplus_w\mathcal{O}_w(W_q)\,.$$

By the choice of these functions their images constitute a basis of $T^1(V_q)$. This shows that D_rV is an isomorphism, and consequently the manifold V is miniversal. A direct product of mappings like 3) gives the required mapping $\varphi: X'\to Y$.

It remains to check that the function F satisfies (5.1.2) at the point $(0,0)$. Suppose the contrary. Then there is a constant c and a vector field u on U_q such that the function $cF+u(F)$ vanishes at the point $(0,0)$ together with its first differential and either $c\neq 0$ or the value of the field u at the point $(0,0)$ is nonzero. We write the following relation:

$$1=\sum_1^k a_je_j+b_0F+\sum b_j\frac{\partial F}{\partial\theta_j}\,, \tag{5.3.3}$$

where a_j are constants and b_j are germs of smooth functions in U_q. This relation must hold since $e_1,\dots,e_k$ constitute a basis. We apply to it the differential operator $l=u+c$

$$c=\sum a_jl(e_j)+u(b_0)F+b_0l(F)+\sum\Big(u(b_j)\frac{\partial F}{\partial\theta_j}+b_jl\Big(\frac{\partial F}{\partial\theta_j}\Big)\Big)\,.$$

The quantity $l(e_j)$ is equal to the derivative of the function $l(F)$ with respect to s_j and consequently vanishes at the point $(0,0)$. We also have $F = l(F) = \frac{\partial F}{\partial \theta_j} = 0$ at this point, since it is a critical point. The function $l\left(\frac{\partial F}{\partial \theta_j}\right)$ is equal to the derivative of $l(F)$ with respect to θ_j and therefore also vanishes at the point $(0,0)$. From this we conclude that $c = 0$. We further write the representation analogous to (5.3.3) replacing the left-hand side by the coordinate functions $\theta_1, \ldots, \theta_N$ and verify similarly that all the coefficients of the field u at the point $\theta = 0$ are equal to zero. This contradiction proves (5.1.2) and completes the verification of the theorem.

Let us verify the sufficiency in Theorem 6. Since $D_x W$ is an epimorphism, it follows from the chain rule that $d\varphi$ is also an epimorphism. Since the germ (X, x) is smooth, it follows that φ is a submersion and there exists a retraction, i.e., a mapping of germs $\psi : (Y, y) \to (X, x)$ such that $\varphi \cdot \psi = \mathrm{id}\,(Y)$. Hence it follows that W is versal. In fact, in order to construct the mapping $\alpha : Z \to X$ so that the inverse image of W is equal to a given wave manifold over Z, it suffices to construct a similar mapping $\beta : Z \to Y$ and take its composition with ψ.

Proposition 9. *Let $V \to Y$ and $V' \to Y'$ be wave manifolds that are miniversal over points y and y' satisfying the hypotheses of Theorem 8. There is a diffeomorphism of germs $\psi : (Y', y') \to (Y, y)$ such that $\psi^*(V) = V'$.*

Proof. The mapping of germs ψ exists by the hypothesis that V is versal. According to the chain rule $DV' = DV \cdot d\psi$, from which we conclude that the mapping $d\psi$ is invertible. Hence we arrive at the relation $\psi^*(V) = V'$.

§4. The Geometry of Wave Fronts and Singularities of Distributions

4.1. The strong Huyghens principle says that the motion of a medium ceases after a wave passes. For the wave equation in $\mathbb{R}^n$ with constant coefficients it holds if and only if n is even and is equivalent to the assertion that the fundamental solution of the Cauchy equation for this equation vanishes not only outside, but also inside the cone of propagation (cf. Sect. 3.2.3) (when $n = 2$, it is constant inside this cone). I.G. Petrovskij (Petrovskij 1945) (cf. also (Atiyah, Bott, Gårding 1973)) considered the problem of lacunae for a fundamental solution of a hyperbolic equation. The name *lacuna* is given to each connected component of the complement to the cone of rays such that this solution vanishes in the component. A *weak lacuna* is a component in which the solution is a polynomial. Petrovskij found homological criteria for a lacuna and a weak lacuna for equations with constant coefficients (here we are speaking of stable lacunae): some cycle in the complement to the characteristic manifold must be homologous to zero. Gårding (1977) extended this criterion to the case of hyperbolic equations with variable (smooth) coefficients. In Gårding's criterion a certain local homological group is used and it is a question of sharpness of the front rather than lacunae.

Definition. Let $u \in \mathcal{K}'(U)$, and let $V \subset U$ be an open set in which u is infinitely differentiable. The generalized function u is *sharp* (or has a sharp front) at a point $x \in \partial V$ from the V-side if there exists a neighborhood U' of this point and an infinitely differentiable function u' in U' such that $u = u'$ in $V \cap U'$. Otherwise the front u is *nonsharp* at the point x from the V-side.

Later results in this direction can be found in (Atiyah, Bott, Gårding 1970; Gårding 1977; Tvorogov 1979; Vasil'ev 1986).

Incidentally a purely geometric condition of Davydova-Borovikov is known for sharpness of the front at a simple point x_θ of the cone of rays K for the case of equations with constant coefficients. The intersection K_0 of the cone K with the hyperplane $t = t(x_0)$ (where t is the time coordinate) has nonzero principle curvatures at such a point. Let p be the number of principal directions on K_0 along which this hypersurface is convex with respect to a given side V of the cone K. If p is even, the fundamental solution has a sharp front at x_0 from the V-side; when p is odd, the front is nonsharp since the solution can be expanded in a convergent series in half-integer powers of the coordinate φ, which vanishes on K and integer powers of the other coordinates (Borovikov 1959). For example, the exterior of the cone of propagation is always a lacuna, and for it $p = 0$, since it is always convex.

This criterion was extended to equations with variable coefficients by Hirschowitz and Piriou (1979), but only for a neighborhood of zero and in terms of the bicharacteristic conoid. The extension of the Davydova-Borovikov criterion in its initial form is impossible, since for equations with variable coefficients the signs of the principal curvatures of the ray conoid may vary without any change in the sharpness of the fundamental solution. Therefore, far away from the vertex of the conoid there is no rigid connection between the curvatures of the conoid and the sharpness.

Nevertheless, as we shall show below there is a local connection between the curvatures of the front and the sharpness, and not only for fundamental solutions, but for a larger class of distributions connected with wave manifolds. Roughly speaking locally for every nonsingular point of the wave front the sharp side for such a distribution is connected with the same sign of the Gaussian curvature. But if we compare two distant points of the front, one must still account for a certain index of the curve that joins them similar to the Morse index. This index is the number of critical points (taken with the appropriate multiplicities) of the projection of this curve on the fiber $K(X)$. The projection is determined by the connection chosen on X by means of which the sign of the Gaussian curvature is computed. For details and generalizations see (Tvorogov 1979).

4.2. The Local Geometry of a Front

Proposition 1. *If W is a closed finite wave manifold over X having a unique element w over a given point q, then for any germ of the hypersurface H that is tangent to w at the point q we have the inequality*

$$x \in F(W) \Rightarrow \operatorname{dist}(x, H) \leq C|x-q|^{1+\varepsilon}$$

with some positive C and ε.

Thus the distance from a point $x \in F(W)$ near q to H is infinitesimal in comparison with its distance from q. This means w is the hyperplane tangent to $F(W)$ in the metric sense of the word. If the front is nonsingular at the point q, then of course w coincides with the usual tangent.

Proof. Assume that the germ (W, w) is defined by a generating function of the form (5.3.1). Here we can assume that $d_\theta^2 f(0) = 0$, since otherwise f may be represented as an inflation of a function having this property and the latter can be used in formula (5.3.1). Since $\theta = 0$ is a critical point of the function $f(\theta)$ of finite multiplicity, we can convert it into a polynomial using a local change of variables. Further we may set $e_1 = 1$, $e_j = \theta_{j-1}$, $j = 2, \ldots, N+1$, and assume that $e_{N+2}, \ldots, e_k$ contain no linear terms. Then from the equations that define the set C_F, we find

$$|s_1| \leq c|\theta| \cdot |s'| + |f|, \quad s' = (s_2, \ldots, s_k)\,,$$

$$\left|\frac{\partial f}{\partial \theta_{j-1}}\right| \leq |s_j| + c|\theta| \cdot |s'|, \quad j = 2, \ldots, N+1\,.$$

Further, using the inequality $|f| \leq c|\theta| \cdot |\operatorname{grad} f|$, (Lojasiewicz 1959), we obtain

$$|s_1| \leq c|\theta| \cdot |s'| = o(|s'|)\,,$$

since $|\theta| \to 0$ if $(s, \theta) \in C_F$ and $|s| \to 0$. Since C_F is a real algebraic set and $F(W)$ is a projection of it, it follows from the Tarski-Seidenberg principle that this relation can be strengthened to the estimate $|s_1| \leq c|s'|^{1+\varepsilon}$. The hyperplane $s_1 = 0$ is tangent to w and consequently this estimate implies Proposition 1 for the case of the versal wave manifold W constructed in Theorem 8 of §3. In the general case $W = \varphi^*(V)$, $v \in V$, where V is a versal wave manifold and φ is a smooth mapping that is transversal to the contact element $v \in V$ defined by the equation $ds_1 = 0$ and w is the pullback of v. It follows from this that $\operatorname{dist}(x, H) \leq C(|s_1| + |s'|^2)$, where $s = s(\varphi(x))$, and it follows from the smoothness of φ that

$$|x-q| \geq C|\varphi(x) - \varphi(q)| \geq C|s'|\,.$$

As a result

$$\operatorname{dist}(x, H) \leq C'|s'|^{1+\varepsilon} \leq C''|x-q|^{1+\varepsilon}\,.$$

We note one more property of the metric tangent. We define an m-*dimensional foliation* in a domain $U \subset X$ as a representation of U in the form of the union of a smooth family of smooth m-dimensional nonintersecting closed submanifolds.

Proposition 2. *Again let W be a closed finite wave manifold over X having a unique element over q and $\mathcal{F}$ a one-dimensional foliation in a neighborhood of q that is transversal to w. There exists a neighborhood U of the point q and a number μ independent of $\mathcal{F}$ such that every curve of the foliation $\mathcal{F} \cap U$ intersects $F(W)$ in at most μ points.*

As Example 3 shows, without the condition of transversality this assertion is not true. The proof of Proposition 2, like the preceding proposition, reduces to the case when F is a polynomial of the form (5.3.1). In this case C_F is a part of the complex-analytic critical set of the projection of the hypersurface $\{F = 0\}$ onto the complexification of the base S. It is known that the projection of this critical set is an analytic set that can be defined by the equation $D(s_1, s') = 0$, where D is a pseudopolynomial in s_1 with leading coefficient 1 and degree equal to the Milnor number μ of the critical point $\theta = 0$ of the function $f(\theta)$. Therefore every line $s' =$ const intersects this analytic set in at most μ points. The same is true for the intersections of this line with the front $F(W)$, which is contained in this set. Moreover, on the zero set D we have the estimate $|s_1| \leq c|s'|^{1+\varepsilon}$ (cf. the proof of Proposition 1). Therefore this estimate of the number of points of intersection with $F(W)$ holds locally for any foliation transversal to w.

The front points of the wave manifold can be classified according to number and multiplicity of the points of the manifold itself that lie above them. The classification of such points according to multiplicity, modality, and other parameters was carried out by V. I. Arnol'd (Arnol'd, Varchenko, Gusejn-Zade 1982).

A front point is called *simple* if there is only one point of W over it and its multiplicity is 1, i.e., the rank of the mapping $p : W \to X$ at this point is $n - 1$. The set of nonsimple points of the front form, by Sard's theorem, a closed set of $n - 1$-dimensional measure zero.

Proposition 3. *The set of simple points of a front of any wave manifold is an open smooth $n - 1$-dimensional manifold. We shall call it the simple part of the front and denote it $F^0(W)$.*

Proof. Suppose there is only one point $w \in W$ over the point $x \in F(W)$, and that f is the generating function of the germ (W, w). A necessary and sufficient condition for a point x to be simple is that the differential $d_\theta^2 f$ be nondegenerate at the point corresponding to w. This condition holds in a neighborhood of this point, and consequently the simple part of the front is open. The system of equations $\dfrac{\partial f}{\partial \theta_j} = 0$, $j = 1, \dots, N$, by the implicit function theorem, has a smooth solution $\theta = \theta(x)$. Consider the function

$\varphi(x) = f(x, \theta(x))$. In order for the point x to belong to $F(W)$ it is necessary and sufficient that this function vanish. At the same time $d\varphi \neq 0$ by (5.1.2), which proves that the simple part of the front is smooth.

Remark. The simple part of a front has a canonical coorientation, since at each of its points the corresponding cooriented contact element $w \in W$ is tangent to the front. this coorientation can be extended continuously to every nonsimple point q of the front for which the hypotheses of Proposition 1 hold.

4.3.

Example 1. Let W have a unique point over q whose multiplicity is 2. If W is a minimal versal manifold over this point, then one can choose a generating function for it of the form $f(s, \theta) = \theta^3 + s_1\theta + s_0$. Consequently the front of W is the semicubical parabola defined by the equation $27s_0^2 + 4s_1^3 = 0$, and its coorientation is shown in Fig. 2. The s_1-axis is a metric tangent at the point $(0, 0)$, and the coorientation can be extended by continuity to that point.

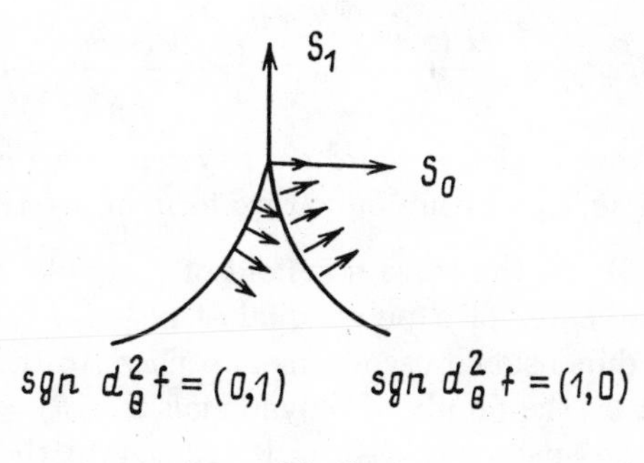

Fig. 2

In the general case $F(W)$ is the pullback of the front depicted on Fig. 2 for some smooth mapping $\varphi : X \to \mathbb{R}^2$, which must be transversal to the tangent space to the semicubical parabola at the origin. Since this space is zero, the condition means that $d\varphi$ is epimorphic at every point x that maps to the origin. These points form a submanifold $Y \subset X$ of codimension 2 called the *cuspidal edge* of $F(W)$, and the front $F(W)$ is itself locally diffeomorphic to the direct product of Y by the semicubical parabola.

Example 2. If the germ of W is defined by the generating function $f(s, \theta) = \theta^4 + s_2\theta^2 + s_1\theta + s_0$, its front is the "swallow's tail" shown in Fig. 3. It consists of a simple part, a cuspidal edge and a line of self-intersection (half of a parabola) formed by the points over which there are exactly two points of W. If we complete this curve to a whole parabola, we obtain an algebraic set that can be defined by the equation $D_3(s_0, s_1, s_2) = 0$, where D_3 is the discriminant of the polynomial $f(s,\cdot)$. The s_2-axis is the tangent space to this front at the origin.

Example 3. Suppose W has a unique point above q, and the multiplicity of this point is 3. The versal for the fiber W is the wave manifold shown in Example 2, and consequently the front of W is the pullback of the "swallow's tail" under some smooth mapping $\varphi : X \to \mathbb{R}^3$, which is transversal to the s_2-axis. Thus $d\varphi$ is not necessarily an epimorphism. In particular one can set

$$X = \mathbb{R}^3, \quad s_0(x) = x_1\,, \quad s_1(x) = x_2\,, \quad s_2(x) = \exp\Big(-\frac{1}{x_3^2}\Big)\sin\frac{1}{x_3}\,.$$

In this case the front is an infinite sequence of "paired" swallow's tails accumulating around the point $x = 0$.

4.4. The Curvature of the Simple Part of the Front. Since we have no preferred coordinate system, the curvature form cannot be a global invariant of the wave manifold. Moreover if we concentrate on the front of the wave manifold generated by the function (5.3.1), we notice that it is the union of $k - N - 1$-dimensional affine subspaces defined by the equations

$$F(s,\theta_0) = \frac{\partial F(s,\theta_0)}{\partial\theta_1} = \cdots = \frac{\partial F(s,\theta_0)}{\partial\theta_N} = 0\,.$$

By a small perturbation of the coordinate system s we can convert any of these into a curved surface with a curvature form of prescribed signature.

Example 4. Let W be the wave manifold of Example 2 of §3. Its front is the discriminant set D_{k+1} of a polynomial of degree $k+1$. This set is the union of the $k-2$-dimensional osculating subspaces to the curve S_{k+1} given in parametric form by the family of polynomials $F(s,\theta) = (\theta-\lambda)^k(\theta+k\lambda)$, $\lambda \in \mathbb{R}$, and also the space $s_0 = s_1 = 0$ tangent to the front at zero. In particular D_4 is a ruled surface.

Definition. Suppose the point $q \in X$ is the image of a unique point of a closed wave manifold W, and f is a generating function of a germ of W at this point, while $d_\theta^2 f = 0$ at the point corresponding to q. Let x be a simple point of the front of W and (x,θ) the corresponding point of C_F. We call the signature (P,Q) of the quadratic form $d_\theta^2 f(x,\theta)$ the *signature of the point x with respect to q*. It is clear that $P + Q = N$, since the form $d_\theta^2 f(x,\theta)$ is nondegenerate. It is not difficult to show that there is a neighborhood U of the point q at which the signature of every simple point of the front with respect to q is independent of the choice of the generating function. The signature is constant on every connected component of $U \cap F^0(W)$. The following result gives a geometric meaning to the signature.

Theorem 4. *Let W be a closed wave manifold over a Riemannian manifold X having a single element over the point q, and let p be the codimension in $T_q(X)$ of the tangent space to the front $T_q(F(W))$. Consider the bundle*

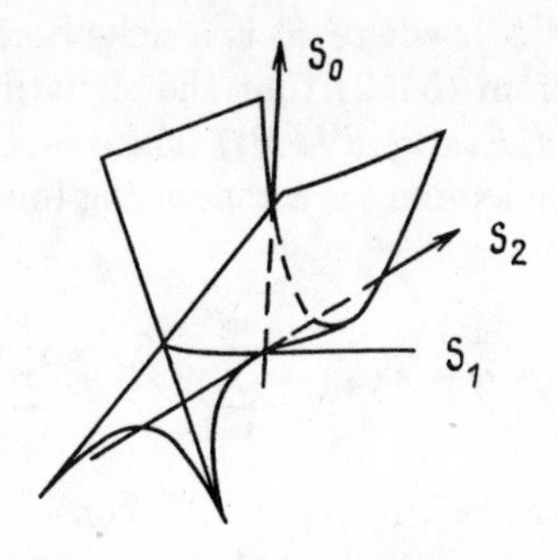

Fig. 3

$\mathrm{Gr}_p T(X)$ *of p-dimensional subspaces in $T(X)$, and assume that a closed set $G = \cup G_x \subset \mathrm{Gr}_p T(X)$ has been chosen so that every space $L \in G_q$ is transversal to $T_q(F(W))$. Then there exists a neighborhood U of the point q such that for every point $x \in U \cap F^0(W)$ and every space $L_x \in G_x$ the curvature form of the cooriented hypersurface $F^0(W)$ at the point x restricted to L_x has signature equal to the signature of the point x with respect to q. Moreover all the curvatures $F^0(W) \cap L_x$ tend to infinity as $x \to q$.*

Proof. The curvature form $F^0(W)$ has the same signature as the form $-d^2\varphi$, which restricted to the subspace $d\varphi = 0$, where $\varphi(x) = f(x, \theta)$, and the function $\theta = \theta(x)$ is defined by the equation $d_\theta f(x, \theta) = 0$. Differentiating this equation, we obtain

$$d_x d_\theta f(x, \theta) + d_\theta^2 f(x, \theta)\theta'_x = 0 . \tag{5.4.1}$$

Since $d\varphi = df(x, \theta)$, taking account of (5.4.1), we find

$$-d^2\varphi = -d_x^2 f - 2d_x d_\theta f \cdot \theta'_x - d_\theta^2 f \cdot \theta'_x \cdot \theta'_x = d_\theta^2 f \cdot \theta'_x \cdot \theta'_x - d_x^2 f . \tag{5.4.2}$$

Despite the fact that the form $d_\theta^2 f$ is degenerate at every nonsimple point of the front the following assertion holds.

Lemma. *The eigenvalues of the form $d_\theta^2 f(\theta'_x, \theta'_x)$, which is bounded on $L_x \cap T_x(F^0(W))$, where $L_x \in G_x$, tend to infinity as $x \to q$.*

Since the norm $\|d_x^2 f\|$ is bounded in a neighborhood of q, the assertion of the theorem will follow from (5.4.2).

Proof of the Lemma. According to Theorem 8 of §3, the germ of W is the pullback of a versal wave manifold V with a generating function of the form (5.3.1) under some mapping of bases $\sigma : X' \to S$, where X' is a neighborhood of the point q, and $\sigma(q) = 0$. Suppose the lemma is proved for V. In this case it is valid for W also. Indeed, since σ is transversal to the tangent space $T_0(F(V))$ whose pullback is equal to $T_q(F(W))$, it follows that for every $x \in X'$ and $L_x \in G_x$ the image $d\sigma(L_x)$ has dimension p, and the set of these images is a closed subset $\widetilde{G} \subset \mathrm{Gr}_p T(S')$, where S' is a neighborhood of zero in the base S. Moreover it follows from (5.3.2) that the signature of the form $d_\theta^2 f(x,\theta)$ equals that of the form $d_\theta^2 F(s(x), \theta'(s,\theta))$ (since $h > 0$).

It remains to verify the lemma for a generating function of the form (5.3.1). We rewrite it in the form

$$F(s,\theta) = f(\theta) + s_{N+1} + \sum_{j=1}^{N} s_j\theta_j + \sum_{N+2}^{k} s_j e_j(\theta)\,, \tag{5.4.3}$$

where all $e_{N+2}, \ldots, e_k$ are polynomials with constant term and linear term equal to zero (we assume without loss of generality that $d_\theta^2 F(0,0) = 0$). It is clear that $T_0(F(V))$ is given by the equations $s_1 = \cdots = s_N = s_{N+1} = 0$. We first verify the assertion of the lemma by choosing L_s to be the subspace in $T_s(S)$ spanned by the vectors $\dfrac{\partial}{\partial s_1}, \ldots, \dfrac{\partial}{\partial s_{N+1}}$. It contains the fields

$$v_i = \frac{\partial}{\partial s_i} - \theta_i(s)\frac{\partial}{\partial s_{N+1}}, \quad i = 1, \ldots, N\,,$$

which are tangent to the front $F(V)$, where we assume that $d_\theta F(s, \theta(s)) = 0$. Since $\theta(s)$ is independent of s_{N+1}, we have $v_i(\theta) = \theta_i' \equiv \dfrac{\partial\theta}{\partial s_i}$, and consequently

$$d^2F(v_i(\theta), v_j(\theta)) = d^2F(\theta_i', \theta_j')\,. \tag{5.4.4}$$

the right-hand side can be found by applying the identity (5.4.1) to F:

$$d^2F(\theta_i', \theta_j') = -\theta_{i,j}'\,.$$

It remains to be shown that the spectrum of the matrix $d^2F(\theta_i', \theta_j')$ tends to infinity as $s \to 0$. This assertion is equivalent to the assertion that the matrix inverse to $\theta_{i,j}'$ tends to zero. But from (5.4.1) we find

$$(\theta_{i,j}')^{-1} = -\frac{\partial^2 F}{\partial\theta_i \partial\theta_j}\,. \tag{5.4.5}$$

Our assertion follows from this since $d^2F(s,\theta) \to d^2F(0,0) = 0$.

Let us now turn to the general case. Let $\pi : C_F \to S$ be the natural projection; there is a distribution $d\pi(T(C_F))$ on $F(V)$ of dimension $k - p$, which is generated at every point by the images of some smooth fields $u_1, \ldots, u_{k-p}$ tangent to C_F. Since $p = N + 1$, for any point $s \in F^0(V) \cap S'$ and any space $L_s \in \widetilde{\mathrm{G}}_s$ the subspace $L_s \cap T_s(F^0(V))$ has dimension N. By what has been said any orthonormal frame $t_1, \ldots, t_N$ in this subspace can be expanded in terms of these tangent vectors

$$t_i = \sum_1^N a_{ij} v_j + \sum_1^{k-p} b_{ij} u_j, \quad i = 1, \ldots, N\,.$$

Here, by the condition of closedness and transversality, the coefficients a_{ij} and b_{ij}, and also the elements of the inverse matrix $(a_{ij})^{-1}$ are bounded by a constant depending only on the set $\widetilde{G}$. Using this expansion we write

$$d_\theta^2 F(t_i(\theta), t_j(\theta)) = \sum_{\alpha,\beta} a_{i\alpha} a_{j\beta} d_\theta^2 F(\theta'_\alpha, \theta'_\beta) + \\ + 2 \sum_{\alpha,\beta} a_{i\alpha} b_{j\beta} d_\theta^2 F(\theta'_\alpha, u_\beta(\theta)) + \sum b_{i\alpha} b_{j\beta} d_\theta^2 F(u_\alpha(\theta), u_\beta(\theta)) \,. \quad (5.4.6)$$

We need to show that the spectrum of the matrix formed by the left-hand sides with $i, j = 1, \ldots, N$ tends to infinity as $s \to 0$. This is true for the matrix formed by the first terms of the right-hand side, since by what has been proved the spectrum of $d_\theta^2 F(\theta'_\alpha, \theta'_\beta)$ tends to infinity, while the matrix $(a_{ij})^{-1}$ is bounded. It remains to be verified that the second and third terms of the right-hand side of (5.4.6) are bounded. From (5.4.1) we find that $d_\theta^2 F \cdot \theta'_\alpha = -\dfrac{\partial}{\partial s_\alpha} d_\theta F$ is a column of the identity matrix. The functions $u_\alpha(\theta)$ are bounded, since u_α is a field on C_F.

Example 5. The function $f(\theta_1, \theta_2) = 3\theta_1^2\theta_2 - \theta_2^3$ defines a germ of a particular curve of type D_4^-, and the function

$$F(s, \theta) = f(\theta) = s_3 + s_1\theta_1 + s_2\theta_2 + s_4(\theta_1^2 + \theta_2^2)$$

defines a versal deformation of this germ and consequently generates a wave manifold V over $\mathbb{R}^4$ that is versal over the point $s = 0$. The front $F(V)$ represents the elliptic umbilic catastrophe (Poston, Stewart 1978); its intersection with the subspace $s_4 = \text{const} > 0$ has the form shown in Fig. 4, where the plane of the drawing coincides with the s_1s_2-plane and the s_3-axis is directed toward the reader and defines the coorientation. The curvilinear triangle bounded by the cuspidal edges has a curvature of signature $(2, 0)$, and the other components of the nonsingular part of the front have signature $(1, 1)$. By Theorem 4 the relative signature on the components of the simple part of the front is similar, since the s_4-axis is its tangent space.

4.5. We now find a relation between the kind of discontinuity of a W-distribution at a simple point of the front, in particular the presence of sharpness, and the geometry of the front, to be specific, its curvature form. The discontinuity of w can be described by the profile $a(x, t)$ that is the germ of a distribution that is smooth for $t \neq 0$ and such that $w(x) \equiv a(x, \varphi(x)) \mod C^\infty$, where $\varphi(x) = 0$ is a local equation of $F^0(W)$ and $d\varphi$ defines the coorientation (compare §2). The sharpness of w ahead of (resp. behind) the front in a neighborhood of a given point is equivalent to the sharpness of the profile of a when $t > 0$ (resp. $t < 0$). On the other hand, if the amplitude of w is known, its profile can be found using formula (5.2.6), which contains the signature (p, q) of the generating function f for the germ of W over some point $x_0 \in X$

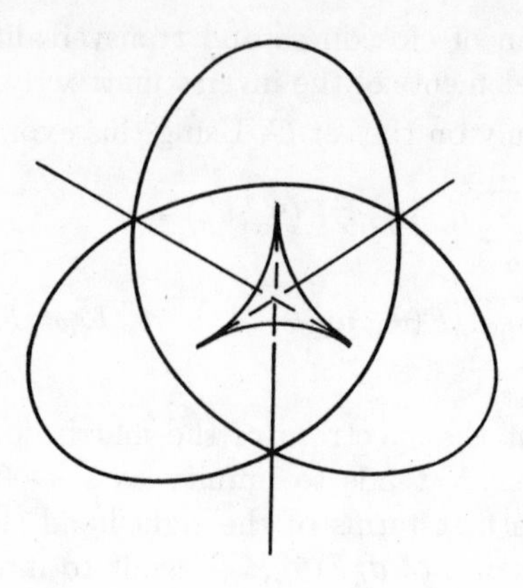

Fig. 4

(we assume that there is only one element of W over x_0). If we write f as an inflation of a function g using a quadratic form h, and $d_\theta^2 g(x_0, \cdot) = 0$, we find that

$$(p, q) = (P, Q) + (p_0, q_0) ,$$

where (P, Q) is the signature of the point x with respect to x_0 and (p_0, q_0) is the signature of the form h. According to Theorem 4 the signature (P, Q) can be computed in terms of the curvature of the front, but the signature (p_0, q_0) is independent of the front and of the wave manifold itself. However, this last quantity occurs as a term that is independent of the point x (near x_0). Therefore the curvature of the front makes it possible to determine the signature in (5.2.6) for all simple points of the front near x_0 up to an unknown additive constant. From this one can obtain the rule for transforming the profile of a distribution under transition from one local component of $F^0(W)$ to another. In particular one can judge the sharpness of the front; however, in order to obtain the law of transformation of such information in closed form, one must widen its scope to some extent.

Definition. We shall say that a W-distribution u is *cosharp* from ahead of (resp. behind) the front $F(W)$ at a simple point x_1 if it admits a profile $b(x, t)\, dt$ at this point which becomes sharp from the side $t > 0$ (resp. $t < 0$) after the Hilbert transform on t, i.e., the function

$$Hb(x, t) = \frac{1}{\pi} \int \frac{b(x, t')\, dt'}{t' - t}$$

is sharp, where the integral is interpreted as the Cauchy principal value. The function b, which was previously defined only in a neighborhood of the hyperplane $t = 0$ for x near x_1, we now extend as a function of compact support that is smooth for $t \neq 0$. Thus the function b in this integral is defined up to a smooth term of compact support in t. The Hilbert transform of such a term is infinitely differentiable; consequently the choice of the extension b has no influence on the sharpness property of Hb.

Note that the condition of cosharpness is preserved under transition from φ to another coorienting function ψ. We have $\varphi(x) = \alpha(x)\psi(x)$, where α is a smooth positive function. Hence $u(x) \equiv b(x, \varphi(x)) = c(x, \psi(x))$, where $c(x,t) = b(x, \alpha(x)t)$. It is easy to see that sharpness of Hb implies sharpness of Hc.

Since $H^2 = -\mathrm{id}$, cosharpness of b implies sharpness of Hb.

Proposition 5. *Let W be a germ of a wave manifold over a neighborhood of the point x_0 and f a generating function of it. Let u be a distribution of the form (5.2.3) with phase function f having amplitude a. Let x_1 be a simple point of the front sufficiently close to x_0 and let the form $d_\theta^2 f(x_1, \cdot)$ have signature (p,q). Then:*

I) *if the number q is even, sharpness of a from the side $t > 0$ implies the sharpness of u at the point x_1 ahead of the front, and cosharpness of a implies cosharpness of u;*

II) *if the number q is odd, sharpness of a for $t > 0$ implies cosharpness of u ahead and cosharpness of a implies sharpness of u.*

Similar assertions hold with q replaced by p and the side of the front and the half-line $t > 0$ by their opposites.

Proof. A function $a(x,t)$ that is sharp from the side $t > 0$ can be written as the sum of a smooth function of compact support with respect to t and a function $a_-(x,t)$ with support in the half-space $t \leq 0$. The converse is obviously also true. The Fourier transform $\tilde{a}(x,\tau) = F_{t\to\tau}a(x,t)$ is the sum of a function that is rapidly decreasing on τ and a function $\tilde{a}_-$ that is the boundary value on the real axis of some holomorphic function in the half-plane $\operatorname{Im}\tau > 0$, which has at most power growth as $\tau \to \infty$ and as $\operatorname{Im}\tau \to 0$. This characteristic of sharp functions a can be inverted if we add the condition that $\tilde{a}$ depends smoothly on $x \in U$ and that all estimates are uniform over $x \in K$, where K is any compact set in a neighborhood U of the point x_1. According to §2 the profile of b is computed from the amplitude $a(x,\theta,t)$ using formula (5.2.6). Therefore to preserve the sharpness it suffices that the coefficient in this formula be extended holomorphically into the upper half-plane of the variable τ. The exponential is holomorphic for $\tau \neq 0$, and the first factor $\mathrm{const} \cdot e_{p,q}(\tau)|\tau|^{-\frac{n}{2}}$ has an extension equal to $\mathrm{const} \cdot (\tau + i0)^{-\frac{n}{2}}$ only when q is even. This proves assertion I). To prove II) we remark that $F(Ha) = i \cdot \operatorname{sgn}\tau \cdot F(a)$. Therefore a necessary and sufficient condition for cosharpness of a is that the function $\operatorname{sgn}\tau \cdot F(a)$ be extendible up to a decreasing term to the upper half-plane as a holomorphic function with the estimates just shown. The subsequent reasoning is carried out just as in the first case.

Example 6. The profiles

$$\pi_{0,0}^k(t) = \begin{cases} \delta^{(-k-1)}(t), & k = -1, -2, \dots, \\ \operatorname{sgn} t \cdot t^k, & k = 0, 1, 2, \dots, \end{cases}$$

are sharp from both sides; the profiles

$$\pi_{1,1}^k(t) = \begin{cases} t^k, & k = -1, -2, \dots, \\ t^k \cdot \ln |t|, & k = 0, 1, 2, \dots, \end{cases}$$

are cosharp from both sides; and the profiles

$$\left.\begin{array}{l} \pi_{0,1}^{k+\frac{1}{2}}(t) = t_+^{k+\frac{1}{2}} \\ \pi_{1,0}^{k+\frac{1}{2}}(t) = t_-^{k+\frac{1}{2}} \end{array}\right\} \quad k = \dots - 2, -1, 0, 1, 2, \dots$$

are sharp from one side and cosharp from the other. This family of profiles is closed in the following sense: if the amplitude of a certain generalized function has the form $a(x, \theta)\pi(t)$, where a is a smooth function and π is one of the profiles of this family, then at any simple point of the front the profile of the same generalized function has the form $b(x)\pi'(t)$, up to smoother terms, where b is again a smooth function and π' is another element of this family. This follows from (5.2.6). The formula

$$a \cdot \pi_{\alpha,\beta}^m \mapsto b \cdot \pi_{\{\alpha+p\},\{\beta+q\}}^{m+\frac{N}{2}} \tag{5.4.7}$$

gives a more precise description of this correspondence of profiles. Here m is any integer or half-integer, $\{k\} = 0, 1$ denotes the residue modulo 2 of the number k, (p, q) is the signature of the corresponding quadratic form $d_\theta^2 f$, and N is its rank.

Example 7. The parametrix of a strictly hyperbolic operator of order m in $\mathbb{R}^n$ in a neighborhood of any simple point of a wave front can be represented as an asymptotically smooth series (Borovikov 1959; Babich 1961b; Gårding 1977)

$$E_y \sim b_0(x)\pi_{\alpha,\beta}^k(\varphi(x)) + b_1(x)\pi_{\alpha,\beta}^{k-1}(\varphi(x)) + \cdots,$$

where $\varphi(x) = 0$ is the local equation of the wave front and $\dfrac{\partial\varphi}{\partial t} > 0$, $b_j \in C^\infty$, $k = m - \dfrac{n}{2} - 1$, $\alpha = \{n - i\}$, $\beta = \{i\}$, and i is the Morse index of the ray from y to the given point x, i.e., the number of points of intersection of the ray or a nearby ray with the cuspidal edge of of the front. This formula is not difficult to obtain from what was said in Example 3 of §2 and from (5.4.7).

Remark. In a small neighborhood of the cuspidal edge the simple part of the front has two connected components, which can be called the posterior and anterior components with respect to the coorientation of the front (cf. Fig. 2). If the amplitude of the distribution u varies only "slightly" in a neighborhood of the corresponding point of the wave manifold (like, for example, the amplitude of the parametrix), the profile of u can be compared on these components of the front. It can be seen from the proof of Theorem 4 that the profile on the leading component is connected up to a scale factor with the profile on the posterior component by the transformation $-H$, where H is the Hilbert transform, up to smoother terms (Maslov 1972).

4.6. Let Γ be a continuous p-dimensional distribution defined in a neighborhood of a singular point of the front q which at that point is transversal to the tangent space to the front (cf. Theorem 4). According to what was said in Sect. 4.5 the qualitative properties of the profile of the generalized function connected with W depend on the signature of the curvature form along Γ of a simple point of the front near q. Here we shall describe the connection between the absolute growth of the profile as $x \to q$ and the growth of the absolute value of the Gaussian curvature of the front along Γ. Let Y be a connected component of the front $F^0(W) \cap X'$, where X' is a small neighborhood of q and φ is a smooth function defined in some neighborhood of Y; moreover $\varphi(x) = 0$ on Y and $|\mathrm{grad}\,\varphi| = 1$. We interpret the Gaussian curvature of Y along Γ as the determinant of the quadratic form $d^2_\Gamma \varphi$ that equals the restriction of $d^2\varphi$ to an orthonormal frame that generates the intersection $T(F^0(W)) \cap \Gamma$. We remark that near Y one can introduce local coordinates $(y, \varphi(x))$ in such a way that

$$x = y + \varphi(x)\mathrm{grad}\varphi\,(x)\,, \quad y \in Y\,.$$

Theorem 5. *Suppose the hypotheses of Theorem 4 are satisfied and that the W-distribution u has an amplitude of the form*

$$a(x, \theta, t) \equiv A(t) a_0(x, \theta) \quad \mathrm{mod}\, \Sigma^{m-1}_{\rho,\delta}\,, \quad \rho > 0\,, \quad \delta < \frac{1}{2}\,,$$

where $A \in \Sigma^m$ and a_0 is a smooth distribution. We claim that in a neighborhood of Y this function has a representation

$$u(x) \equiv (c(y) + o(1))|\det d^2_\Gamma \varphi(y)|^{\frac{1}{2}} B(\varphi(x)) a_0(y, \theta(y)) \quad \mathrm{mod}\, \Sigma^{m-1}_{\rho,\delta}\,, \tag{5.4.8}$$

where

$$\tilde{B}(\tau) = e_{p,q}(\tau)|\tau|^{-\frac{N}{2}} \tilde{A}(\tau)\,,$$

the function c is continuous on the closure of Y and nonzero on that set (it depends only on the phase function f and Γ), the quantity $o(1)$ tends to zero as $x \to q$, and the function $\theta(y)$ is determined from the equation $d_\theta f(y, \theta) = 0$.

To prove this theorem it suffices to make use of formula (5.2.6) taking account only of the first term of the expansion in powers of τ in this formula, then express the coefficient $|\Delta|$ in terms of the Gaussian curvature. To find this expression we pass from W to the versal manifold V just as in Theorem 4 and from the function φ to the function $\Phi(s) = F(s, \theta(s))$. The Gaussian curvature of $F^0(V)$ along Γ equals $b_1(s) \det d^2_\Gamma \Phi$, the functions b_1, b_2, b_3, being continuous and nonzero on the closure of any connected component of $F^0(V)$ or $F^0(W)$. Further, from (5.4.2), (5.4.4), and (5.4.6) we have

$$\det d_\Gamma^2 \Phi = (b_2 + o(1)) \det d^2 F(\theta_i', \theta_j') .$$

Using (5.4.5) and (5.3.2), we find

$$\left| \det d^2 F \cdot \theta_i' \cdot \theta_j' \right| = \left| \det \theta_{i,j}' \right| =$$

$$= \left| \det \frac{\partial^2 F}{\partial \theta_i \partial \theta_j} \right|^{-1} = (b_3 + o(1)) \cdot \left| \det \frac{\partial^2 f}{\partial \theta_i \partial \theta_j} \right|^{-1} = (b_3 + o(1)) |\Delta|^{-1} ,$$

which completes the proof.

Example 8. If the point q lies on a cuspidal edge, the absolute value of the Gaussian curvature of the front at the point x along every distribution transversal to the edge is equivalent to $d^{-\frac{1}{2}}$, where d is the distance from x to the edge. Therefore the second factor in (5.4.8) is equivalent to $d^{-\frac{1}{4}}$.

§5. Versal Integrals

5.1. If the wave manifold W is finite over a point $x_0 \in X$, then in some neighborhood of this point every W-distribution can be represented using a finite number of such functions, which we call *versal integrals.* To do this we consider a wave manifold V over $S = \mathbb{R}^d$ that is miniversal over the point $s = 0$ such that its pullback under some mapping $\varphi : X \to S$, $\varphi(x_0) = 0$, coincides with W over some neighborhood of the point x_0. Versal integrals are certain generalized functions defined in a neighborhood of zero in S and connected with V. Every W-distribution can be represented using two operations:

1) taking the inverse image of versal integral functions under the mapping φ, and
2) onefold convolution of the resulting functions with some distributions.

In such a representation exactly $d = \dim T^1(W_{x_0})$ versal integrals are needed. And if we allow the operation

3) differentiation of versal integrals,

a smaller number of such functions, namely the number of points of the fiber W_{x_0}, will suffice. Let us consider first the inverse image operation.

Proposition 1. *If a wave manifold W over X is the inverse image of the wave manifold V over Y under the mapping $\psi : X \to Y$, then for every V-function v the pullback $w = \psi^*(v)$ is defined and is a W-function. Here if $v \in \sum_{\rho,\delta}^m(V)$, then $w \in \sum_{\rho,\delta}^m(W)$.*

Proof. By hypothesis ψ is transversal to the wave manifold V and consequently to each element of it. Since the wave front v is contained in V, it follows that the inverse image $w = \psi^*(v)$ is defined as a generalized function on X. Let us find an explicit form for it. Using a partition of unity in a neighborhood of the fiber $K_{y_0}(Y)$, we reduce the problem to the case when v has the form

$$v(y) = \int b(y, \theta, g(y, \theta))\, d\theta \,,$$

where g is a generating function for the germ of V at some point of this fiber and $b \in \sum_{\rho,\delta}^{m}$ is an amplitude function. Making the substitution $y = \psi(x)$, we obtain

$$w(x) = \int b(\psi(x), \theta, g(\psi(x), \theta))\, d\theta \,,$$

and consequently w has amplitude $b(\psi(x), \theta, t)$ and phase function $f(x, \theta) = g(\psi(x), \theta)$ which generates W.

5.2. Construction of Versal Integrals. In view of the finiteness of V over the distinguished point $s = 0$, one may assume that there is only one element $K \in V$ over this point. In the general case one must combine the versal integrals corresponding to different points of the fiber V_0. According to Theorem 8 of §3, we may assume that V has a generating function F of the form (5.3.1), and $F^*(0, 0) = K$; this function is defined in some neighborhood $S' \times U$ of the origin and has no critical points there. In such a case the inverse image $\delta_F \in \mathcal{D}'(S' \times U)$ of the delta-function δ_0 of one variable under the mapping F is defined. For any density $\rho \in \mathcal{K}(U)$ the direct image $v_\rho = p_*(\rho\delta_F)$ under the mapping $p : S' \times U \to S'$ is defined. The generalized function v_ρ can be obtained using the construction of §2, applied to the amplitude $a(s, \theta, t) = \rho\delta_0(t)$, and at the same time, according to Sect. 3.3 of §3 of Chapt. 2, it can be written in the form

$$v_\rho(s) = \int\limits_{Z_s} \frac{\rho}{dF} \,, \tag{5.5.1}$$

where $Z_s \subset U$ is the set of zeros of the function $F(s, \cdot)$ with coorientation given by the form $d_\theta F$. This formula is demonstrably true in the domain $S' \setminus p(C_F)$ of noncritical values of the projection $Z \to S'$, where Z is the set of zeros of F, and in this domain v_ρ is a smooth function.

Proposition 2. *The function v_ρ defined by formula* (5.5.1) *outside the set $p(C_F)$, which is of measure zero, is locally integrable on S'. The generalized function corresponding to it coincides with $p_*(\rho\delta_F)$ and belongs to $\Sigma_{1,0}^{-\frac{N}{2}}(V)$.*

Proof. We choose an arbitrary compact set $G \subset S'$ and density $\sigma \in \mathcal{K}(S')$ with support contained in G. According to (2.3.3) we have

$$p_*(\rho\delta_F)(\sigma) = \delta_F(\sigma\rho) = \int\limits_Z \frac{\sigma\rho}{|\text{grad}\, F|} \,.$$

Let C_ε be the ε-neighborhood of $p(C_F)$. As $\varepsilon \to 0$

$$\int_{G\setminus C_\epsilon} v_\rho \cdot \sigma = \int_{Z\setminus p^{-1}(C_\epsilon)} \frac{\sigma\rho}{|\mathrm{grad}\, F|} \to \int_{Z\setminus Z^*} \frac{\sigma\rho}{|\mathrm{grad}\, F|}, \qquad (5.5.2)$$

where Z^* is the preimage of $p(C_F)$ under the mapping $Z \to S'$. This set has measure zero as a subset of the hypersurface Z endowed with Lebesgue measure. Therefore the right-hand side of (5.5.2) equals $p_*(\rho\delta_F)$. Assume that the density ρ is of constant sign, for example $\rho \geq 0$. In such a case $v_\rho \geq 0$, and it follows from the existence of the limit (5.5.2) that the function v_ρ is integrable on G and coincides with $p_*(\rho\delta_F)$ on that set. Since every density ρ can be represented as a sum of two densities of constant sign belonging to $\mathcal{K}(U)$, this inference is valid in the general case.

Definition. The generalized functions $v_j = v_{\rho_j}$, $j = 1, \ldots, k$ are the *versal integrals* connected with V if $\rho_j = r_j\, d\theta$, and the functions $r_1, \ldots, r_k$ generate a basis of the quotient space of the algebra $\mathcal{O}_0(U)$ of germs of smooth functions at the origin in $U \subset \mathbb{R}^N$ over the ideal $J(f)$ generated by the functions f, $\partial f/\partial\theta_1, \ldots, \partial f/\partial\theta_N$.

Theorem 3. *Let W be a closed wave manifold over X that is finite over the point x_0, and $\varphi : (X, x_0) \to (S, s_0)$ a smooth mapping of germs such that $\varphi^*(V) = W$, where V is a versal wave manifold over (S, s_0) and $s_1, \ldots, s_k$ are local coordinates on S equal to zero at the point s_0. For every distribution $w \in \Sigma_0^m(W)$ there exist distributions $b_j\, dt \in \Sigma_{1,0}^{m+\frac{N}{2}}$, $j = 1, \ldots, k$, such that in some neighborhood of the point x_0 the following relation holds:*

$$w(x) = \sum_{j=1}^{k} \int b_j(x,t) v_j(s_1(x) - t, s_2(x), \ldots, s_k(x))\, dt \pmod{C^\infty}, \qquad (5.5.3)$$

where $\varphi(x) = (s_1(x), \ldots, s_n(x))$, and we set $r_1(\theta) \equiv 1$.

We remark that the integrals are actually taken between finite limits, since all the functions v_j are of compact support with respect to s_1 due to the compactness of the support of all the densities ρ_j.

Proof. By hypothesis there are a finite number of germs of the manifold W over the point x_0, and each of them is finite over that point. It is clear that the general case reduces to the case in which there is exactly one germ of W over x_0. Consequently we can assume this and use the constructions of §2.

Consider the algebra A of germs at the point $(x_0, 0, 0)$ of infinitely smooth functions on the manifold $X \times S \times U$. It is a module over the algebra B of smooth functions at the point $x_0 \in X$. Let H be the ideal in A generated by the functions

$$F,\ \partial F/\partial\theta_1, \ldots, \partial F/\partial\theta_N;\ s_1 - s_1(x), \ldots, s_k - s_k(x)\,, \qquad (5.5.4)$$

and let $\mathfrak{m}_B$ be a maximal ideal in B. The sum $H + \mathfrak{m}_B A$ is also an ideal in A; taking into account the relation $s_j(x_0) = 0$, we conclude that this sum contains the germs of the coordinate functions $s_1, \ldots, s_k$ on S and consequently the whole ideal $J(f)$. It follows from this that

$$A/H + \mathfrak{m}_B A \cong \mathcal{O}_0(U)/J(f) ,$$

where $\mathcal{O}(U)$ is the sheaf of germs of smooth functions on U. In accordance with the condition that the germ of W is finite the right-hand side of this isomorphism is finite-dimensional. Therefore the B-module A/H is quasi-finite, and hence by a theorem of Malgrange (1966) it is finite, and the images of the functions $r_1 \ldots, r_k$ are generators of this module. Therefore every germ $a \in A$ can be written as

$$a(x,s,\theta) = \sum b_j(x) r_j(\theta) + a_0(x,s,\theta)F(s,\theta) + \\ + \sum a_j(x,s,\theta)\partial F(s,\theta)/\partial\theta_j + \sum c_j(x,s,\theta)(s_j - s_j(x)) , \quad (5.5.5)$$

where $b_j \in B$ and a_j and c_j belong to A.

We need the following semilocal version of this assertion. We set $I = (-1, 1)$.

Lemma. *There exists a neighborhood X' of the point x_0 and a neighborhood $P \subset X' \times S \times U$ of the point $(x_0, 0, 0)$ such that for every smooth function a defined on $P \times I$ there is a representation* (5.5.5) *in which a_j and c_j are smooth functions in $P \times I$ and b_j are smooth functions in $X' \times I$.*

Proof. Let $\mathcal{B}$ be the sheaf of germs of smooth functions on $X \times I$, $\mathcal{A}$ the sheaf of germs of smooth functions on $X \times S \times U \times I$, and $\mathcal{H}$ the sheaf of ideals in $\mathcal{A}$ generated by the functions (5.5.4), while $\pi : X \times S \times U \times I \to X \times I$ is projection on a direct factor. The sheaf $\mathcal{M} = \pi_*(\mathcal{A}/\mathcal{H}|P \times I)$ is a sheaf of $\mathcal{B}$-modules, and for any $\eta \in I$ the fiber of it over the point (x_0, η) is isomorphic, modulo the maximal ideal $\mathfrak{m}_\eta$ in the algebra $\mathcal{O}_\eta(I)$, to the $\mathcal{B}$-module $\mathcal{A}/\mathcal{H}$. It follows from this that this fiber is generated as a $\mathcal{B}$-module by the images of the functions $r_1, \ldots, r_k$. We shall show that for suitable X' and P the sheaf $\mathcal{M}$ is generated as a B-module over $X' \times I$ by the images of these functions. To do this using Tougeron's theorem (Tougeron 1968) we choose a local coordinate system $y = y(x)$ such that the function f becomes a polynomial and we assume that the basis functions $r_1, \ldots, r_k$ are polynomials in the variables θ (this is possible since the assertion of the lemma is independent of the choice of these functions). In such a case F is a polynomial; consequently we can consider the following analytic sheaves: the sheaf B_C of germs of holomorphic functions in some small ball $X_C \subset \mathbb{C}^k$ with center at the point 0, the sheaf A_C of germs of holomorphic functions in some complex neighborhood P_C of the point $(0,0,0)$, and the sheaf of ideals H_C in A_C generated by the functions (5.5.4), where $s_i(x)$ is replaced by x_i, $i = 1, \ldots, k$. The coherent sheaf A_C/H_C is finite over a sufficiently small ball X_C in view of the finiteness of the module A/H. Therefore its direct image $M_C = \pi_*(A_C/H_C|P_C)$ is also coherent (Grauert's theorem) and therefore is generated by the images of the polynomials $r_1, \dots, r_k$ over a sufficiently small ball X_C. This means that for any point $x \in X_C$ the images of these polynomials generate over the fiber of the sheaf B_C the direct sum of all fibers of the sheaf A_C/H_C at points of the form $(x, s, \theta) \in P_C$: It follows from this that if $P = \operatorname{Re} P_C$, then the direct sum of the fibers of the sheaf A/H at points of the form $(x, s, \theta, \eta) \in P \times I$, where the points $x \in X' = \operatorname{Re} X_C$ and $\eta \in I$ are fixed, is generated modulo the maximal ideal of this point by the images of the same polynomials. Therefore at each such point the theorem of Malgrange is applicable, from which there follows a local representation of the form (5.5.5) for every function $a = a(x, s, \theta, \eta)$ defined in $P \times I$. Using a suitable partition of unity we combine these local representations into a single representation with smooth coefficients defined on $X' \times I$ or $P \times I$ respectively. We then change x_i back to $s_i(x)$, $x \in X'$.

We now return to the proof of the theorem. Consider the following mapping of nuclear Fréchet spaces:

$$\Gamma(P \times I, \mathcal{A})^{k+N+1} \oplus \Gamma(X' \times I, \mathcal{B})^k \xrightarrow{\sigma} \Gamma(P \times I, \mathcal{A}) ,$$

in which the mapping σ takes a vector-valued function $\{c_j, a_j; b_j\}$ to the right-hand side of (5.5.5). By the lemma σ is an epimorphism. By virtue of a result of Grothendieck (1954) for every bounded set Σ in $\Gamma(P \times I, \mathcal{A})$ there exists a bounded set Λ in the first space such that $\sigma(\Lambda) \supset \Sigma$.

Let $w \in \Sigma_0^m(W)$, and let $a = a(x, \theta, t)$ be its amplitude of class $\Sigma_{1,0}^m$. Set $a^\lambda(x, s, \theta, \eta) := \exp(-(N/2+m)(\lambda+\eta))\tilde{a}(x, \theta, \exp(\lambda+\eta))$, $-1 < \eta < 1$, $\lambda \in \mathbb{R}$. It follows from (5.2.1) that the family $\{a^\lambda\}$ is a bounded set of sections of the sheaf $\mathcal{A}$ over $X \times S \times U \times I$ and consequently over $P \times I$. Therefore by what was said above, there exist bounded sets $\{a_j^\lambda\}$, $j = 0, \dots, N$; $\{c_j^\lambda\}$, $j = 1, \dots, k$, in the space $\Gamma(P \times I, \mathcal{A})$ and $\{b_j^\lambda, \lambda \in \mathbb{R}\}$, $= 1, \dots, k$, in $\Gamma(X' \times I, \mathcal{B})$ that define the representation (5.5.5) for the functions a^λ. Let the parameter λ in these representations range over the set of integers, and let $\{h^\lambda\}$ be a smooth partition of unity on the real axis such that $h^\lambda(\eta) = h^0(\eta - \lambda) \operatorname{supp} h^s \subset I$. We set $a_j' := \sum\limits_\lambda h^\lambda(\eta) a^\lambda(x, s, \theta, \eta)$, $j = 0, \dots, N$, and we define similarly the functions b_j' and c_j', $j = 1, \dots, k$. As a result we obtain a representation of the form (5.5.5) for the functions

$$\exp(-(N/2+m)\eta)\tilde{a}(x, \theta, \exp(\eta)) , \quad \eta \in \mathbb{R} ,$$

with coefficients a_j', c_j', b_j'. These coefficients are differentiable on $P \times \mathbb{R}$ any number of times and all of their derivatives are bounded on $P \times \mathbb{R}$. Setting $\eta = \log \tau$, $\tau > 1$, in this representation and multiplying by the function $\exp((N/2+m)\eta) = \tau^{N/2+m}$, we obtain a representation of the form (5.5.5) for the function $\tilde{a}(x, \theta, \tau)$ in which the coefficients satisfy the inequality (5.2.1) for $\tau > 1$, since $D_\tau = 1/\tau D_\eta$. By a similar device one can obtain a similar representation for $\tilde{a}(x, \theta, \tau)$ when $\tau < -1$, and also for $-2 < \tau < 2$. Gluing these representations together with a partition of unity, we finally arrive at a representation on the manifold $P \times \mathbb{R}$ in which the coefficients satisfy inequalities (5.2.1). Applying the inverse Fourier transform $\tau \to t$ to this representation, we obtain the equality

$$a(x,\theta,t) = \sum b_j^0(x,t)\, dt\rho_j(\theta) + a_0(x,s,\theta,t)F(s,\theta) + \\ + \sum a_j(x,s,\theta,t)\partial F(s,\theta)/\partial\theta_j + \sum c_j(x,s,\theta,t)(s_j - s_j(x)) \,, \quad (5.5.6)$$

in which a_j and c_j are amplitudes of class $\Sigma_{1,0}^m$, and $b_j\, dt \in \Sigma_{1,0}^{m+\frac{N}{2}}$.

We now set $t = F(s,\theta)$ in equality (5.5.6) and integrate it with respect to the variables t and θ.

$$\int\limits_P a_F = \sum_j \int b_j^0(x,t) \int\limits_{F=t} \rho_j/dF\, dt + \int\limits_P F(a_0)_F + \\ + \int\limits_P \partial F/\partial\theta_j (a_j)_F + \sum_j \int\limits_P (c_j)_F(s_j - s_j(x)) = \\ = \sum_j \int b_j^0(x,t)v_j(s_1 - t, s_2, \ldots, s_k)\, dt + \int\limits_P a_F^1 + \\ + \sum_j \int\limits_P (c_j)_F(s_j - s_j(x)),$$

where $P : X \times S \times U \times \mathbb{R} \to X \times S$ is the natural projection and we have used the fact that the equation $F(s,\theta) = t$ is equivalent to the equality $F(s_1 - t, s_2, \ldots, s_k, \theta) = 0$. Here

$$a^1(x,s,\theta,t) = t\{(a_0(x,s,\theta,t) - \sum_j \partial a_j(x,s,\theta,t)/\partial\theta_j\}$$

as a result of integration by parts. The amplitude a^1 belongs to the class $\Sigma_{1,0}^{m-1}$ and the same reasoning can be applied to it, after which we find

$$\int\limits_P a_F^1 = \sum \int b_j^1(x,t)v_j(s_1 - t, s_2, \ldots, s_k)\, dt + \int\limits_P a_F^2 + \sum_j \int\limits_P (c_j^1)_F \,,$$

where b_j^1 belongs to the class $\Sigma_{1,0}^{m-1+\frac{N}{2}}$. We then apply a similar operation to a^2, etc. It is easy to construct amplitudes $b_j \in \Sigma_{1,0}^{m+\frac{N}{2}}$ for which the series

$$b_j^0 + b_j^1 + b_j^2 + \cdots, \quad j = 1, \ldots, k$$

are asymptotic with respect to smoothness. As a result we obtain

$$\int\limits_q a_F \equiv \sum \int b_j(x,t)v_j(s_1 - t, s_2, \ldots, s_k)\, dt + \cdots \pmod{C^\infty} .$$

Setting $s_j = s_j(x)$ here, we arrive at (5.5.3). Theorem 3 is now proved.

Remark 1. The converse assertion is also true: for any $b_j \in \Sigma_{1,0}^{m+\frac{N}{2}}$ the right-hand side of (5.5.3) defines a generalized function of class $\Sigma_0^m(W)$. Indeed, we can transform each term of (5.5.3) to the form

$$\int b_j \int_{F=t} \rho_j/dF\,dt = \int \tilde{b}_j(x, F(s(x),\theta))\rho_j(\theta) = p_*((b_j\rho_j)_F)\,,$$

from which our assertion follows immediately.

Remark 2. Instead of versal integrals one can use integrals analogous to those considered by V.M. Babich (1961a)

$$I_\rho^{\lambda\pm}(s) = \int (F(s,\theta) \pm i0)^\lambda \rho\,, \quad \rho \in K(u)\,.$$

They are connected with versal integrals by the relations

$$I_\rho^{\lambda\pm}(s) = \int v_\rho(s_1 - t, s_2, \ldots, s_k)(t \pm i0)^\lambda\,dt\,,$$

which are a version of the Mellin transform. These relations make it possible to express versal integrals in terms of $I_\rho^{\lambda\pm}$.

A similar approach to the study of oscillatory integrals was developed by Duistermaat (1974), cf. also (Pearcey 1946; Guillemin, Sternberg 1977).

Commentary on the References

The reference list contains several textbooks that provide a more or less elementary introduction to the theory of generalized functions: (Jones 1982; Shilov 1968; Antosik, Mikusinski, Sikorski 1973; Arsac 1961; Wloka 1969). The book of L. Schwartz (1950/51), which was written forty years ago and has played such a large role in the theory, can still serve as an excellent introduction to the subject. Further study of the theory and its applications to differential equations is provided by the monographs (Vladimirov 1981; Gel'fand and Shilov 1958ac; Friedman 1963; Trèves 1967). The book (Hörmander 1983a) contains deeper harmonic analysis, and the geometric problems of the theory are studied in (Guillemin, Sternberg 1977). The books (Gel'fand, Graev, Vilenkin 1962) and (Helgason 1984) are devoted to harmonic analysis and integral geometry on symmetric spaces. An exposition of the theory of hyperfunctions can be found in (Sato, Kawai, Kashiwara 1973; Hörmander 1983a; Kashiwara, Schapira 1990).

References*

Antosik, P., Mikusinski, J., Sikorski, R. (1973): Theory of Distributions. The Sequential Approach. PWN-Polish Scientific Publishing, Warszawa, Zbl. 267.46028

Akhiezer, A.I., Berestetskij, V.B. (1969): Quantum Electrodynamics. Nauka, Moscow (Russian)

Arnol'd, V.I., Varchenko, A.N., Gusejn-Zade, S.M. (1982): Singularities of Differentiable mappings. The Classification of Critical Points of Caustics and Wave Fronts. Nauka, Moscow. English transl.: Birkhäuser, Basel-Boston-Stuttgart, Zbl. 513.58001

Arsac, J. (1961): Transformation de Fourier et théorie des distributions. Dunod, Paris, Zbl. 108,110

Atiyah, M.F. (1970): Resolution of singularities and division of distributions. Commun. Pure Appl. Math. *23*(2), 145–150, Zbl. 188,194

Atiyah, M.F., Bott, R., Gårding, L. (1970): Lacunas for hyperbolic differential operators with constant coefficients. 1. Acta Math. *124*, 109–189, Zbl. 191,112; II. Acta Math. *131*, 145–206, Zbl. 266.35045

Babenko, K.I. (1956): On a new problem of quasianalyticity and the Fourier transform of entire functions. Tr. Mosk. Mat. O.-va *5*, 523–542, Zbl. 72,51

Babich, V.M. (1961a): The analytic character of the field of a nonstationary wave in a neighborhood of a caustic. (Problems of the Dynamic Theory of Propagation of Seismic Waves, Vol. 5) Leningrad University Press, 115–144

Babich, V.M. (1961b): A fundamental solution of the dynamic equations of the theory of elasticity for an inhomogeneous medium. Prikl. Mat. Mekh. *25*(1), 38–45. English transl.: J. Appl. Math. Mech. *25*, 49–60 (1961), Zbl. 102,172

Berenstein, C.A., Dostal, M.A. (1972): Analytically uniform spaces and their applications to convolution equations. Lect. Notes Math. *256*, 1–130, Zbl. 237.47025

Berenstein, C.A., Struppa, D. (1989): Complex analysis and convolution equations. Itogi Nauki Tekh., Ser. Sovr. Probl. Mat., Fundam. Napravleniya *54*, 5–111, Zbl. 706.46031

Berenstein, C.A., Taylor, B.A. (1980): Interpolation problems in $\mathbb{C}^n$ with applications to harmonic analysis. J. Anal. Math. *38*, 188–254, Zbl. 464.42003

Berenstein, C.A., Taylor, B.A. (1983): Mean-periodic functions. Int. J. Math. Math. Sci. *3*, 199–236, Zbl. 438.42012

Bernstein, I.N. (1972): Analytic continuation of generalized functions with respect to a parameter. Funkts. Anal. Prilozh. *6*(4), 26–40. English transl.: Funct. Anal. Appl. *6*, 273–285 (1973), Zbl. 282.46038

Bernstein, I.N., Gel'fand, S.I. (1969): The meromorphic function p^λ. Funkts. Anal. Prilozh. *3*(1), 84–86. English transl.: Funct. Anal. Appl. *3*, 68–69 (1969), Zbl. 208,152

Björk, J.E. (1979): Rings of Differential Operators. North-Holland Publ. Co., Math. Lib. Ser. *21*, Amsterdam-Oxford-New York, Zbl. 499.13009

Bochner, S. (1932): Vorlesungen über Fouriersche Integrale. Leipzig, Zbl. 6,110

Bogoliubov, N.N. (=Bogolyubov, N.N.), Parasiuk, O.S. (1957): Über die Multiplikation der Kausalfunktionen in der Quantentheorie der Felder. Acta Math. *97*, 227–266, Zbl. 81,433

Bogolyubov, N.N., Shirkov, D.V. (1976): Introduction to Quantum Field Theory. Nauka, Moscow (Russian)

* For the convenience of the reader, references to reviews in *Zentralblatt für Mathematik* (Zbl.), compiled using the MATH database, have, as far as possible, been included in this bibliography.

Bony, J.M. (1976): Propagation des singularités différentiables pour une classe d'opérateurs différentiels à coefficients analytiques. Astérisque *34/35*, 43–91, Zbl. 344.35075

Borovikov, V.A. (1959): Fundamental solutions of linear partial differential equations with constant coefficients. Tr. Mosk. Mat. O.-va *8*, 199–257. English transl.: Transl., II. Ser., Am. Math. Soc. *25*, 11–76 (1963), Zbl. 90,312

Bros, J., Iagolnitzer, D. (1975): Tuboids et structure analytique des distributions. I, II. Sémin. Goulaouic-Lions-Schwartz, 1974/1975, Exp. No. 16, 18, Zbl. 333.46028–29

Browder, F.E. (1963): On the "edge of the wedge" theorem. Can. J. Math. *15*, 125–131, Zbl. 128,304

Brychkov, Yu.A., Prudnikov, A.P. (1977): Integral Transforms of Generalized Functions. Nauka, Moscow. English transl.: Gordon & Breach, New York, 1989, Zbl. 464.46039

Colombeau, J.F. (1984): New Generalized Functions and Multplication of Distributions. North-Holland, Amsterdam-Oxford-New York, Zbl. 532.46019

Courant, R., Hilbert, D. (1962): Methods of Mathematical Physics, II: Partial Differential Equations. Interscience, New York-London, Zbl. 99,295

Dirac, P. (1926–27): The physical interpretation of the quantum dynamics. Proc. Roy. Soc. Lond., Ser. A *113*(62), 1–641

Duistermaat, J.J. (1974): Oscillatory integrals, Lagrange immersions and unfolding of singularities. Commun. Pure Appl. Math. *27*(2), 207–281, Zbl. 285.35010

Duistermaat, J.J., Hörmander, L. (1972): Fourier integral operators. II. Acta Math. *128*, 183–269, Zbl. 232.47055

Dyson, F.J. (1958a): The connection between local commutativity and regularity of Wightman functions. Phys. Rev., II. Ser. *110*, 579–581, Zbl. 82,423

Dyson, F.J. (1958b): Integral representation of causal commutators. Phys. Rev., II. Ser. *110*, 1460–1464, Zbl. 85,454

Ehrenpreis, L. (1954): Solution of some problems of division. I. Am. J. Math. *78*(4), 883–903, Zbl. 56,106

Ehrenpreis, L. (1955): Mean periodic functions. Am. J. Math. *77*(2), 293–328, Zbl. 68,317

Ehrenpreis, L. (1956a): On the theory of the kernels of Schwartz. Am. J. Math. *7*, 713–718, Zbl. 70,334

Ehrenpreis, L. (1956b): Solution of some problems of division. III. Am. J. Math. *78*(4), 685–715, Zbl. 72,328

Ehrenpreis, L. (1960): Solution of some problems of division. IV. Am. J. Math. *82*(3), 522–588, Zbl. 98,84

Ehrenpreis, L. (1970): Fourier Analysis in Several Complex Variables. Wiley Interscience, New York, Zbl. 195,104

Ehskin, G.I. (1961): Generalized Paley-Wiener-Schwartz theorems. Usp. Mat. Nauk *16*(1), 185–191, Zbl. 109,306

Erdélyi, A. (1975): Fractional integrals of generalized functions. Lect. Notes Math. *457*, 151–170, Zbl. 309.26008

Euler, L. (1743): De integratione æquationum differentialium altiorum gradum. Miscell. Berd. *7*, 193–242

Fedoryuk, M.V. (1959): Inhomogeneous generalized functions of two variables. Mat. Sb., Nov. Ser. *49(91)*(4), 431–446. English transl.: Transl., II. Ser., Am. Math. Soc. *34*, 223–240 (1963), Zbl. 91,113

Friedman, A. (1963): Generalized Functions and Partial Differential Equations. Prentice-Hall, Englewood Cliffs, N.J., Zbl. 116,70

Gårding, L. (1947): The solution of Cauchy's problem for two totally hyperbolic linear differential equations by means of Riesz integrals. Ann. Math., II. Ser. *48*, 785–826, Zbl. 29,216

Gårding, L. (1977): Sharp fronts of paired oscillatory integrals. Publ. Res. Inst. Math. Sci. *12*, Suppl., 53–68; Corrections *13*, 821, Zbl. 369.35062

Gel'fand, I.M., Shilov, G.E. (1958a): Generalized Functions and Operations on Them. Fizmatgiz, Moscow. English transl.: Acad. Press, London-New York, 1964, Zbl. 91,111

Gel'fand, I.M., Shilov, G.E. (1958b): Spaces of Fundamental and Generalized Functions. Fizmatgiz, Moscow. English transl.: Acad. Press, London-New York, 1968, Zbl. 91,111

Gel'fand, I.M., Shilov, G.E. (1958c): Some Problems of Differential Equations. Fizmatgiz, Moscow (Russian), Zbl. 91,111

Gel'fand, I.M., Vilenkin, N.Ya. (1961): Some Applications of Harmonic Analysis. Rigged Hilbert Spaces. Fizmatgiz, Moscow. English transl.: Acad. Press, London-New York, 1964, Zbl. 103,92

Gel'fand, I.M., Gindinkin, S.G., Shapiro, Z.Ya. (1979): A local problem of integral geometry in spaces of curves. Funkts. Anal. Prilozh. *13*(2), 11–31. English transl.: Funct. Anal. Appl. *13*, 87–102 (1979), Zbl. 415.53046

Gel'fand, I.M., Graev, M.I., Vilenkin, N.Ya. (1962): Integral Geometry and Related Questions of the Theory of Representations. Fizmatgiz, Moscow, Zbl. 115,167

Gindikin, S.G., Khenkin, G.M. (1986): Several Complex Variables. IV. Algebraic Aspects of Complex Analysis. Itogi Nauki Tekh., Ser. Sovr. Probl. Mat., Fundam. Napravleniya *10*. English transl.: Springer-Verlag, Berlin-Heidelberg-New York (1990)

Grothendieck, A. (1954): Sur les éspaces (F) et (DF). Summa Brasil. Math. *3*, 243–280, Zbl. 58,98

Grothendieck, A. (1955): Produits tensoriels topologiques et éspaces nucléaires. Mem. Am. Math. Soc. *16*, 1–191 (Résumé des resultats essentiels dans la théorie des produits tensoriels topologiques et des éspaces nucléaires (1954). Ann. Inst. Fourier *4*, 73–112), Zbl. 55,97

Guillemin, V., Sternberg, S. (1977): Geometric Asymptotics. American Mathematical Society, Providence, Rhode Island, Zbl. 364.53011

Gurevich, D.I. (1974): Closed ideals with exp-polynomial generators in rings of entire functions of two variables. Izv. Akad. Nauk Arm. SSR, Mat. *9*(6), 459–472, Zbl. 304.46032

Gurevich, D.I. (1975): A counterexample to a question of L. Schwartz. Funkts. Anal. Prilozh. *9*(2), 29–35. English transl.: Funct. Anal. Appl. *9*, 116–120 (1975), Zbl. 326.46020

Hadamard, J. (1932): Le problème de Cauchy et les équations aux dérivées partielles linéaires hyperboliques. Hermann, Paris, Zbl. 6,205

Helgason, S. (1980): The Radon Transform. Birkhäuser, Boston-Basel-Stuttgart, Zbl. 453.43011

Helgason, S. (1984): Groups and Geometric Analysis. Academic Press, Orlando, Zbl. 543.58001

Hirschowitz, A., Piriou, A. (1979): Propriétés de transmission pour les distributions intégrales de Fourier. Commun. Partial Diff. Equations *4*(2), 113–217, Zbl. 456.58028

Hörmander, L. (1955): La transformation de Legendre et le théorème de Paley-Wiener. C. R. Acad. Sci. Paris *240*(4), 392–395, Zbl. 64,103

Hörmander, L. (1958): On the division of distributions by polynomials. Ark. Mat. *3*, 555–568, Zbl. 131,119

Hörmander, L. (1971): Fourier integral operators. I. Acta Math. *127*(1–2), 79–183, Zbl. 212,466

Hörmander, L. (1974): Uniqueness theorem and wave front sets for solutions of linear differential equations with analytic coefficients. Commun. Pure Appl. Math. *24*, 671–704, Zbl. 226.35019

Hörmander, L. (1983a): The Analysis of Linear Partial Differential Operators. I. Distribution Theory and Fourier Analysis. Springer-Verlag, Berlin-Heidelberg-New York, Zbl. 521.35001

Hörmander, L. (1983b): The Analysis of Linear Partial Differential Operators. II. Differential Operators with Constant Coefficients. Springer-Verlag, Berlin-Heidelberg-New York, Zbl. 521.35002

Hörmander, L. (1985): The Analysis of Linear Partial Differential Operators. IV. Fourier Integral Operators. Springer-Verlag, Berlin-Heidelberg-New York, Zbl. 612.35001

Jeanquartier, D. (1970): Développement asymptotique de la distribution de Dirac. C. R. Acad. Sci. Paris, Ser. A *271*, 1159–1161, Zbl. 201,165

John, F. (1955): Plane Waves and Spherical Means. Interscience, New York-London, Zbl. 67,321

Jones, D.F. (1982): The Theory of Generalised Functions. Cambridge University Press, 2nd ed., Zbl. 477.46035

Jost, R, Lehmann, H. (1957): Integral-Darstellung kausaler Kommutatoren. Nuovo Cimento, X. Ser. *5*, 1598–1610, Zbl. 77,424

Kaneko, A. (1972): Representation of hyperfunctions by measures and some of its applications. J. Fac. Sci. Univ. Tokyo, Sect. 1A *19*, 321–352, Zbl. 247.35007

Kashiwara, M., Schapira, P. (1990): Sheaves on Manifolds. Springer-Verlag, Berlin-Heidelberg-New York, Zbl. 709.18001

Kawai, T. (1970): On the theory of Fourier hyperfunctions and its applications to partial differential equations with constant coefficients. J. Fac. Sci. Univ. Tokyo, Sect. 1A *17*(3) 467–517, Zbl. 212,461

Khenkin, G.M. (1985): Method of Integral Representations in Multidimensional Analysis. In: Itogi Nauki Tekh., Sovr. Probl. Mat. *65*, VINITI, Moscow. English transl. in Enc. Math. Sc. *7*, Springer-Verlag, Berlin Heidelberg New York 1988

Komatsu, H. (1973): Ultradistributions. I. Structure theorems and a characterization. J. Fac. Sci. Univ. Tokyo, Sect. 1A *20*, 25–105, Zbl. 258.46039

König, H. (1955): Multiplikation von Distributionen. Math. Ann. *128*, 420–452, Zbl. 64,113

Kuchment, P.A. (1981): Representation of the solutions of invariant differential equations on certain symmetric spaces. Dokl. Akad. Nauk SSSR *259*(3), 532–535. English transl.: Sov. Math., Dokl. *24*, 104–106 (1981), Zbl. 497.58025

Kuchment, P.A. (1982): Floquet theory for partial differential equations. Usp. Mat. Nauk *37*(4), 3–52. English transl.: Russ. Math. Surv. *37*, No. 4, 1–60 (1982), Zbl. 519.35003

Kushnirenko, A.G. (1983): Introduction to Quantum Field Theory. Vysshaya Shkola, Moscow (Russian)

Lax, P.D. (1957): Asymptotic solutions of oscillatory initial value problems. Duke Math. J. *24*, 627–646, Zbl. 83,318

Leont'ev, A.F. (1949): Differential-difference equations. Mat. Sb., Nov. Ser. *24*, 347–374, Zbl. 41,423

Lions, J.L. (1953): Supports dans la transformation de Laplace. J. Anal. Math. *2*, 369–380, Zbl. 51,335

Lions, J.L., Magenes, E. (1968/70): Problèmes aux limites non homogènes et applications. Vol. I–III. Dunod, Paris, Zbl. 165,108; Zbl. 197,67

Lojasiewicz, S. (1959): Sur le problème de la division. Stud. Math. *3*, 87–136, Zbl. 115,102

Lützen, J. (1982): The Prehistory of the Theory of Distributions. Springer-Verlag, Berlin-Heidelberg-New York, Zbl. 494.46038

Malgrange, B. (1956): Existence et approximation des solutions des équations aux dérivées partielles et des équations de convolution. Ann. Inst. Fourier *6*, 271–368, Zbl. 71,90

Malgrange, B. (1960): Sur la propagation de la régularité des solutions des équations à coefficients constants. Bull. Sci. Math. Phys. R.S.R. *3*(4), 433–440, Zbl. 109,320

Malgrange, B. (1964): Systèmes différentiels à coefficients constants. Sémin. Bourbaki *15* (1962/63), No. 246, 11pp., Zbl. 141,273

Malgrange, B. (1966): Ideals of Differentiable Functions. Oxford University Press, Zbl. 177,179

Malgrange, B. (1971): Le polynôme de Bernstein d'une singularité isolée. Invent. Math. *14*, 123–142; see also Lect. Notes Math. *459*, 98–119 (1975), Zbl. 308.32007

Malgrange, B. (1974/75): Intégrales asymptotiques et monodromie. Ann. Sci. Ec. Norm. Supér., IV. Sér. 7(3), 405–430, Zbl. 305.32008

Martineau, A. (1967): Equations différentielles d'ordre infini. Bull. Soc. Math. Fr. *95*(2), 109–154, Zbl. 167,442

Martineau, A. (1969): Théorème sur le prolongement analytique du type "Edge of the wedge theorem". Sémin. Bourbaki *68*(1967/1968), No. 340, 17 pp., Zbl. 209,148

Maslov, V.P. (1972): The Fourier *A*-transform. Tr. Mosk. Inst. Elektr. Mashinostr. *25*, 56–99

Maslov, V.P., Fedoryuk, M.V. (1976): The Quasiclassical Approximation for the Equations of Quantum Mechanics. Nauka, Moscow. English transl.: D. Reidel, Dordrecht-Boston-London (1981), Zbl. 449.58002

Methée, P.D. (1954): Sur les distributions invariantes dans la groupe des rotations de Lorentz. Commun. Math. Helv. *28*(3), 225–263, Zbl. 55,341

Methée, P.D. (1955): Transformation de Fourier des distributions invariantes. I. C.R. Acad. Sci. Paris *240*, 1179–1181, Zbl. 64,115; II. C.R. Acad. Sci. Paris *241*, 684–686, Zbl. 65,350

Mikusinski, J. (1960): Une simple démonstration du théorème de Titchmarsh sur la convolution. Bull. Acad. Pol. Sci. *7*, 715–718, Zbl. 94,260.

Natterer, F. (1986): The Mathematics of Computerized Tomographies. Wiley, New York and Teubner, Stuttgart, Zbl. 617.92001

Palamodov, V.P. (1962a): The general form of the solutions of linear differential equations with constant coefficients. Dokl. Akad. Nauk SSSR *143*(6), 1278–1281. English transl.: Sov. Math., Dokl. *3*, 595–598 (1962), Zbl. 168,449

Palamodov, V.P. (1962b): The Fourier transform of rapidly growing infinitely differentiable functions. Tr. Mosk. Mat. O.-va *11*, 309–350, Zbl. 161,325

Palamodov, V.P. (1967): Linear Differential Operators with Constant Coefficients. Nauka, Moscow. English transl.: Springer-Verlag, Berlin-Heidelberg-New York, 1970, Zbl. 191,434

Palamodov, V.P. (1968a): Remarks on the exponential representation of the solutions of differential equations with constant coefficients. Mat. Sb., Nov. Ser. *76*(5), 417–434. English transl.: Math. USSR, Sb. *5*, 401–416 (1968), Zbl. 162,409

Palamodov, V.P. (1968b): Differential operators with coherent analytic sheaves. Mat. Sb., Nov. Ser. *77*(3), 390–422. English transl.: Math. USSR, Sb. *6*, 365–391 (1968), Zbl. 177,190

Palamodov, V.P. (1975): The complex of holomorphic waves. Tr. Semin. Im. I. G. Petrovskogo, No. 1, 175–210. English transl.: Transl., II. Ser., Am. Math. Soc. *122*, 187–222 (1984), Zbl. 322.46048

Palamodov, V.P. (1977): From hyperfunctions to analytic functionals. Dokl. Akad. Nauk SSSR *235*(3), 534–537. English transl.: Sov. Math. Dokl. *18*, 975–979 (1978), Zbl. 377.46033

Palamodov, V.P. (1985): Asymptotic expansions of integrals in the complex and real domains. Mat. Sb., Nov. Ser. *127*(2), 209–238; English transl.: Math. USSR, Sb. *55*(1) (1986), 207–236, Zbl. 596.41046

Palamodov, V.P. (1993): Harmonic synthesis of solutions of elliptic equations with periodic coefficients. Ann. Inst. Fourier, *43*, 751–768, Zbl. 784.35023

Paley, R., Wiener, N. (1934). The Fourier Transform in the Complex Domain. New York, Zbl. 11,16
Pearcey, T. (1946): The structure of an electromagnetic field in the neighbourhood of a cusp of a caustic. Phil. Mag. *37*, 311–317
Petrovskij, I.G. (= Petrovsky, I.G.) (1945): On the diffusion of waves and the lacunas for hyperbolic equations. Mat. Sb., Nov. Ser. *17*(3), 289–370, Zbl. 61,213
Polyakov, P.L., Khenkin, G.M. (1986): Homotopy formulas for the $\bar{\partial}$-operator on $\mathbf{CP}^n$ and the Radon-Penrose transform. Izv. Akad. Nauk SSSR, Ser. Mat. *50*(3), 566–597. English transl.: Math. USSR, Izv. *28*, 555–587 (1987), Zbl. 607.32003
Poston, T., Stewart, I. (1978): Catastrophe Theory and its Applications. Pitman, London, Zbl. 382.58006
Rajkov, D.A. (1967): Some linear-topological properties of the spaces $\mathcal{D}$ and $\mathcal{D}'$. (Appendix to the book *Topological Vector Spaces* by A.P. and W. Robertson.) Mir, Moscow, Zbl. 153,162
Ramanujan, S. (1914): Some definite integrals connected with Gauss' sums. Messeng. Math. *44*, 75–85
de Rham, G. (1955): Variétés différentiables. Hermann, Paris, Zbl. 65,324
Riesz, M. (1949): L'intégrale de Riemann-Liouville et le problème de Cauchy. Acta Math. *81*, 1–223, Zbl. 33,276
de Roever, J.W. (1977): Complex Fourier Transformation and Analytic Functionals with Unbounded Carriers. Mathematish Centrum, Amsterdam, Zbl. 406.46032
Roumieu, M.C. (1960): Sur quelques extensions de la notion de distribution. Ann. Sci. Ec. Norm. Supér., II. Sér. *77*(1), 41–121, Zbl. 104,334
Sato, M., Kawai, T., Kashiwara, M. (1973): Microfunctions and pseudodifferential equations. Lect. Notes Math. *287*, 264–529, Zbl. 277.46039
Schaefer, H.H. (1966): Topological Vector Spaces. Macmillan, New York, Zbl. 141,305
Schapira, P. (1970): Théorie des hyperfonctions. Lect. Notes Math. *126*, Zbl. 192,473
Schwartz, L. (1947): Théorie générale des fonctions moyennes-périodiques. Ann. Math., II. Ser. *48*(4), 857–929, Zbl. 50,150
Schwartz, L. (1950/51): Théorie des distributions. Vol. I, II. Hermann, Paris, Zbl. 37,73; Zbl. 42,115
Schwartz, L. (1951): Analyse et synthèse harmonique dans les éspaces de distributions. Can. J. Math. *3*, 503–512, Zbl. 43,330
Schwartz, L. (1952): Théorie des noyaux. Proc. Int. Congr. Math. 1952 *1*, 220–230, Zbl. 48,351
Schwartz, L. (1957): Distributions semi-regulières et changement de coordonnées. J. Math. Pures Appl. *36*(6), 109–127, Zbl. 77,316
Sebastião a Silva, J. (1958): Les fonctions analytiques comme ultradistributions dans le calcul opérationnel. Math. Ann. *136*(1), 58–96, Zbl. 195,413
Shilov, G.E. (1968): Generalized Functions and Partial Differential Equations. Gordon and Breach, London, Zbl. 177,363
Shirokov, Yu.M. (1979): The algebra of three-dimensional generalized functions. Teor. Mat. Fiz. *40*(3), 348–354. English transl.: Theor. Math. Phys. *40*, 790–794 (1980), Zbl. 425.46029
Shubin, M.A. (1978): Pseudodifferential Operators and Spectral Theory. Nauka, Moscow. English transl.: Springer-Verlag, Berlin-Heidelberg-New York, 1987, Zbl. 451.47064
Sobolev, S.L. (1936): Méthode nouvelle à resoudre le problème de Cauchy pour les équations linéaires hyperboliques normales. Mat. Sb., Nov. Ser. 1, 39–72
Sobolev, S.L. (1988): Some Applications of Functional Analysis in Mathematical Physics, 3rd ed.. Nauka, Moscow. English transl.: Transl. Math. Monogr. *90*, Providence, Zbl. 662.46001, Zbl. 123,90

Stein, E., Weiss, G. (1971): Introduction to Fourier Analysis on Euclidean Spaces. Princeton University Press, Zbl. 232.42007

Struppa, D.C. (1983): The fundamental principle for systems of convolution equations. Mem. Am. Math. Soc. *273*, 167 pp., Zbl. 503.46027

Tillmann, H.G. (1961): Darstellung der Schwartzschen Distributionen durch analytische Funktionen. Math. Z. *77*(2), 106–124, Zbl. 99,97

Tougeron, J.C. (1968): Ideaux de fonctions différentiables. Ann. Inst. Fourier *18*(1), 177–240, Zbl. 188,451

Trèves, F. (1961): Lecture on linear partial differential equations with constant coefficients. Inst. de Math. Pura Appl., Rio de Janeiro, Zbl. 129,69

Trèves, F. (1967): Topological Vector Spaces, Distributions, and Kernels. Academic Press, New York-London, Zbl. 171,104

Trèves, F. (1980): Introduction to Pseudodifferential and Fourier Integral Operators. II. Plenum Press, New York, Zbl. 453.47027

Tvorogov, V.B. (1979): The sharp front and singularities of solutions of a class of nonhyperbolic equations. Dokl. Akad. Nauk SSSR *244*(6), 1327–1331. English transl.: Sov. Math., Dokl. *20*, 240–244 (1979), Zbl. 415.35012

Varchenko, A.N. (1980): The Gauss-Manin connection of isolated singular points and Bernstein polynomial. Bull. Sci. Math., II. Ser. *104*(2), 205–223, Zbl. 434.32008

Varchenko, A.N. (1987): On normal forms of nonsmoothness of solutions of hyperbolic equations. Izv. Akad. Nauk SSSR, Ser. Mat. *51*(3), 652–665. English transl.: Math. USSR, Izv. *30*, 615–628 (1988), Zbl. 646.35054

Vasil'ev, V.A. (1986): Sharpness and the local Petrovskij condition for strictly hyperbolic operators with constant coefficients. Izv. Akad. Nauk SSSR, Ser. Mat. *50*(2), 243–284. English transl.: Math. USSR, Izv. *28*, 233–273 (1987), Zbl. 615.35012

Vladimirov, V.S. (1964): Methods of the Theory of Functions of Several Complex Variables, Nauka, Moscow. English transl.: MIT Press, Boston 1966, Zbl. 125,319

Vladimirov, V.S. (1981): The Equations of Mathematical Physics, 4th ed. Nauka, Moscow. 5th, expanded edition: Nauka, Moscow 1988, Zbl. 652.35002. English transl. of 1967 edition: Dekker, New York 1971, Zbl. 207,91

Whitney, H. (1948): On ideals of differentiable functions. Am. J. Math. *70*, 635–658, Zbl. 37,355

Wloka, J. (1969): Grundräume und verallgemeinerte Funktionen. Lect. Notes Math. *82*, Zbl. 169,158

Yano, T. (1978): On the theory of *b*-functions. Publ. Res. Inst. Math. Sci., Kyoto Univ. *14*(1), 111–202, Zbl. 389.32005

Zemanian, A.H. (1968). Generalized Integral Transformations. Interscience, New York, Zbl. 181,127

II. Optical and Acoustic Fourier Processors

V. S. Buslaev

Translated from the Russian
by Roger Cooke

Contents

§1. Introduction 131
§2. The Optical Fourier Transform 136
2.1. Light Beams 136
2.2. Filters 139
2.3. Formation of an Image Using a Lens 141
2.4. The Optical Fourier Transform 142
2.5. Two Remarks 143
2.6. Holographic Measurement of a Field and Holographic Filters 145
§3. Notes and Comments 147
3.1. The Influence of the Lens Size 147
3.2. Spatial Filtering 148
3.3. Complex Amplitude Filters 151
3.4. Phase Gratings 153
3.5. Improving the Resolution 157
3.6. Incoherent Signal Processing 158
§4. Acoustic and Acousto-Optical Fourier Processors 160
4.1. The Electro-Acoustic Transducer 160
4.2. Determination of the Parameters of a Rayleigh-Wave Transducer 161
4.3. Convolvers Based on Surface Acoustic Waves 164
4.4. Acoustic Fourier Processors 166

4.5. The Acousto-Optical Cell 167
4.6. Acousto-Optical Processors 168
Comments on the References 173
References 173

§1. Introduction

A *Fourier processor* is a physical device inside which the Fourier image of a signal falling on the entrance to the device is formed. Signals processed in such devices are usually either *radio-electronic signals* propagating in electronic (radio) circuits, or *light beams* arriving through optical devices. Along with the electronic and optical elements in Fourier processors widespread use is made of acoustic components. In many Fourier processors repeated transformation of electronic, optical and acoustic signals into one another occurs. The central elements by means of which the Fourier transform is realized are usually the optical and acoustic components of the processors. The use of such devices has become widespread. In this connection a special area in optics has even arisen, known as *Fourier optics.* Optical and acoustic processors are used as computer parts or specialized devices for complex signal processing. Like the majority of computing devices of this kind, they have both natural advantages and natural disadvantages in comparison with purely digital computers. The sphere of application of Fourier processors in which the decisive value lies with their advantages is rapidly spreading.

The purpose of the present survey is to give a popular description of the principle on which devices of this kind operate. The author is not an expert in this circle of questions and became familiar with it as a specialist in mathematical physics. In the choice of material the guiding purpose, however, was to satisfy the interests of a wider circle of mathematicians to whom this topic is not a priori alien.

In what follows we shall touch on practical applications only briefly. In certain cases these applications have a very delicate and even exotic character, though in the majority of cases they are based on the most general properties of the Fourier transform, which is one of the basic elements of the mathematical machinery of the leading physical disciplines. Fourier processors can be used, for example, to solve differential equations with constant coefficients. In many situations the applications are connected with the fact that the Fourier image of a physical signal characterizes it more expressively than a direct representation. Certain important operations on physical signals (of filter discrimination type) can be performed immediately in terms of the Fourier transform.

We shall investigate in most detail the principles of operation of *optical Fourier processors*, designed for processing optical signals. Optical Fourier processors have a very simple and efficient structure. For example, to obtain the Fourier transformation of a picture imprinted on a photographic slide, only the commonest glass lens and light source forming a monochromatic *plane light wave* are required. The structure of such a source has no significance. At present it is simplest to use a *laser* as such a source. The elements just listed are arranged as shown in Fig. 1. Here Δ is the *aperture* that restricts the plane wave, D is the slide, L the *lens* with focal length F, and S the screen.

In this situation the square of the absolute value of the Fourier transform of the translucent picture fixed on the slide can be observed on the screen.

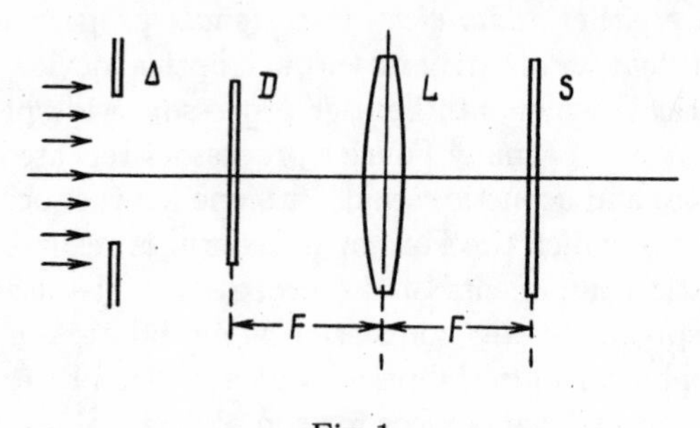

Fig. 1

What does this rather exotic-sounding assertion mean in the language of formulas? It means that the illumination intensity $I(x)$ of the screen as a function of the point x of the screen $x \in \mathbb{R}^2$ is given by the formula

$$I(x) = I_0 |\widetilde{\psi}(2\pi\lambda^{-1}F^{-1}x)|^2 ,$$

in which $\widetilde{\psi}$ is the Fourier transform of the function $\psi(y)$,

$$\widetilde{\psi}(x) = \int\limits_{\mathbb{R}^2} e^{-i\langle x,y\rangle} \psi(y)\, dy ,$$

that describes the transparency of the slide as a function of the point y of the plane of the slide, λ is the wave length of the incident light, F is the focal length, and I_0 is a certain constant depending on λ and F.

In all Fourier processors, including optical processors, the Fourier transform can be realized as the composition of the operation of multiplication by a function and the operation of *convolution.* This possibility is based on one of the two elementary formulas that connect a function $\psi : \mathbb{R}^d \to C$ and its Fourier transform $\widetilde{\psi} = \mathcal{F}[\psi]$:

$$\widetilde{\psi}(y) = \int\limits_{\mathbb{R}^d} e^{-i\langle y,x\rangle} \psi(y)\, dy .$$

The formulas just mentioned have the following form:

$$\widetilde{\psi}(y) = e^{-\frac{i}{2}y^2} \int\limits_{\mathbb{R}^d} e^{\frac{i}{2}(y-x)^2} e^{-\frac{i}{2}x^2} \psi(x)\, dx , \tag{1.1}$$

$$\widetilde{\psi}(z) = c_d \int\limits_{\mathbb{R}^d \times \mathbb{R}^d} e^{\frac{i}{2}(z-y)^2} e^{-\frac{i}{2}y^2} e^{\frac{i}{2}(y-x)^2} \psi(x)\, dx\, dy , \tag{1.2}$$

where $c_d = (2\pi i)^{d/2}$. The first of these formulas is obvious and the second is dual to it with respect to the Fourier transform, which changes multiplication into convolution. The second formula can also be verified in an elementary manner by explicitly carrying out the integration with respect to y:

$$\int_{\mathbb{R}^d} e^{\frac{i}{2}(z-y)^2} e^{-\frac{i}{2}y^2} e^{\frac{i}{2}(y-x)^2}\, dy = e^{-i\langle z,x\rangle} \int_{\mathbb{R}^d} e^{\frac{i}{2}(y-x-z)^2}\, dy = c_d e^{-i\langle z,x\rangle} .$$

We shall use the notation

$$\mathcal{F}_h[\psi](x) = \mathcal{F}[\psi](h^{-1}x), \quad K_h(x) = e^{\frac{i}{2}h^{-1}x^2} ,$$

and then the two formulas exhibited above assume the following form:

$$\mathcal{F}_h[\psi] = K_{-h} \cdot (K_h * (K_{-h} \cdot (\psi))) \tag{1.3}$$

and

$$\mathcal{F}_h[\psi] = c_d h^{d/2} K_h * (K_{-h} \cdot (K_h * \psi)) , \tag{1.4}$$

where

$$(\psi * \varphi)(y) = \int_{\mathbb{R}^d} \psi(x)\varphi(y-x)\, dx$$

is the *convolution* of ψ and φ.

Thus, having at our disposal devices that perform multiplication of a given function by K_{-h} and convolution of it with K_h, we can construct a Fourier processor from a circuit of three such devices. Naturally an actual Fourier processor must also contain suitable input and output elements.

In view of the simple two-way connection between the problems of forming the Fourier transform and the convolution, in each text devoted to Fourier processors there is inevitably a discussion of so-called *convolvers*—processors that form the convolution of two signals—and *correlators*—processors that form the *cross-correlation* $\psi \star \varphi$ of two signals ψ and φ:

$$(\psi \star \varphi)(y) = \int_{\mathbb{R}^d} \psi^*(x)\varphi(x+y)\, dx .$$

When $\psi = \varphi$, the cross-correlation is called the *autocorrelation*. The cross-correlation is widely used to separate weak signals from background noise in image-recognition devices and many others. The following obvious formula holds.

$$\mathcal{F}[\psi \star \varphi] = \left(\mathcal{F}[\psi]\right)^* \cdot \mathcal{F}[\varphi] .$$

It is now necessary to go into more detail as to which specific optical and electronic signals are the objects of processing in Fourier processors. In the optical case these are the relatively narrow rectilinear monochromatic *light beams* that arise, for example, in restricting monochromatic plane waves by an *aperture*, which are then transmitted through various regions of relatively smooth inhomogeneities. In analytic language the light field of such a beam can be described (neglecting polarization effects) by a complex-valued function of the form

$$u(x, y, z) = e^{ikz}\psi(x, y, z) , \tag{1.5}$$

where $k = 2\pi\lambda^{-1}$, λ is the wave length, and x, y, and z are Cartesian coordinates in three-dimensional space. The function ψ is assumed to vary with z and (as a consequence) with x and y rather slowly in comparison with the plane-wave type factor $\exp(ikz)$; in other words, the differentiation operators ∂_x, ∂_y, and ∂_z are considered to be small in comparison with the operation of multiplication by k with respect to suitable norms. Naturally the function ψ is assumed to be sufficiently smooth and rapidly decreasing as $(x, y) \to \infty$. In this case the function (1.5) forms a beam that propagates along the z-axis. An optical Fourier processor is constructed so that for some $z = z_2$ the field $\psi(x, y, z)$ is the Fourier transform on the variables (x, y) of the field ψ incident at the entrance to the processor, i.e., the field ψ for some $z = z_1$, $z_1 < z_2$.

In a similar manner a *radio-electronic signal* that is the object of processing in a Fourier processor is usually a modulated high-frequency signal propagating in electronic circuits. In other words, it is a signal that can be described by a complex-valued function of the form

$$u(t, z) = e^{-i\omega t}\psi(t, z) , \tag{1.6}$$

where ω is the *carrier frequency* and z is a coordinate that makes it possible to order the elements of the circuit in the entrance–exit direction. The function $\psi(t, z)$ is considered to vary rather slowly as a function of time t and to be smooth and rapidly decreasing with respect to t. There is a perfectly distinct and very far-reaching parallelism between the variables z in formulas (1.5) and (1.6) and between the variables (x, y) and t in these formulas. As a result of this parallelism many discoveries in the sphere of electronic signal processing were successfully used in developing optical Fourier processors and subsequently the opposite influence began to play a significant role.

However, we shall pay little attention to the history of the topic here, since it is closely connected with specific technical applications that it makes no sense to describe.

The parallelism just mentioned encompasses both the theoretical and technical sides of the matter. We shall say that the direction of an optical beam has as an analog the carrier frequency of an electronic signal. It is comparatively easy to fix both the direction of the beam and the carrier frequency of the signal: this can be done in the first case using a system of apertures, and in the second case by standard *electronic filters* using the traditional elements of radio hookups, i.e., concentrated inductance and capacitance. The rearranged electronic filters, which make it possible to conduct a complete Fourier analysis of an electronic signal, however, are extremely inefficient from the technical point of view. Similarly, a system of apertures by means of which it would be theoretically possible to obtain the Fourier transform of an optical signal can hardly be technically efficient.

There is one other essential point that must be noted. In practice a section of a light beam is inevitably bounded (just as the time during which one can observe the electronic signal is also bounded). As a consequence the Fourier transform of an optical signal cannot be of compact support, but in an experiment only its restriction to a bounded region can be measured. Depending on the nature of the function $\psi(x, y, z)$ different manipulations of the apertures that restrict the beam or the region in the plane where the Fourier transform is measured may be reasonable. The problems that arise here, which essentially relate to the *uncertainty principle*, have not been developed to a sufficient extent in the physical literature, and here, of course an interesting field of activity arises for specialists in analysis. This is by no means the only point that may attract the involvement of the mathematician. If one so desires, the whole topic can be interpreted as a collection of interesting original questions (problems, if you will) for the analyst and the specialist in mathematical physics in particular. In problems that relate to mathematical physics we shall adhere to the heuristic level throughout. The reason is not only the popular nature of the text, but also the fact that a large portion of these problems have not been developed to a sufficient extent. Almost all of the physical schemes considered here, and even their separate components, are in need of systematic investigation from the point of view of mathematical physics. It may be of interest, first of all, to have a more advanced description with formulas. In certain cases it seems to be possible to count on results raised to the status of theorems. Analogously we shall also ignore many questions connected with the boundedness of both a section of a light beam and the duration of an electronic signal, although from the point of view of technical characteristics of the devices under consideration these questions are extremely important. Specialists in analysis might be able to give recommendations making it possible to improve these characteristics—to do this one must be able to combine the restrictions introduced by the uncertainty principle with criteria of the theory of image recognition.

The structure of our exposition is as follows. Section 2 is devoted to the description of the fundamental principles on which the structure of optical Fourier processors is based. Section 3 exhibits certain selected applications and characterizes certain lines of further development of the ideas discussed in Sect. 2. In Sect. 4 we describe the decisive propositions on which the various structures of acoustic and acousto-optical Fourier processors are based.

§2. The Optical Fourier Transform

In an *optical Fourier processor* the operation of convolution is realized in the propagation of a rectilinear light beam in the homogeneous space between two planes orthogonal to the beam. The operation of multiplication by a suitable function is realized by interrupting the beam en route with a *thin lens*.

2.1. Light Beams. In the majority of optical devices one is dealing with narrow *light beams* propagating along some axis (or several axes). This peculiarity of the light field in optical devices makes it possible to simplify significantly the mathematical machinery used to describe the field. We shall neglect the effects of polarization and, instead of Maxwell's equations, we shall use the scalar *wave equation* to describe the light field $v(t,r) \in C$ (where t is time, and r is a point of physical space, $r = (x, y, z) \in \mathbb{R}^3$):

$$v_{tt} = c^2(v_{xx} + v_{yy} + v_{zz}) ,$$

in which $c = c(r)$ is the speed of light. For the stationary case

$$v = e^{-i\omega t}u(r)$$

the wave equation assumes the form

$$\omega^2 u + c^2(u_{xx} + u_{yy} + u_{zz}) = 0 . \tag{2.1}$$

Assuming that the medium is homogeneous, i.e., c is independent of r, we describe the transition to the equation that characterizes the light beam. We distinguish a line in space and make the z-axis coincide with it. The solution

$$u = e^{ikz}, \quad k = \omega c^{-1} > 0 ,$$

of Eq. (2.1), as is known, admits a simple ray interpretation; it corresponds to a family of rays parallel to the z-axis. To describe the light beam propagating along the z-axis, we set

$$u = e^{ikz}\psi(x, y, z) , \tag{2.2}$$

Substitution into Eq. (2.1) gives

$$\psi_{xx} + \psi_{yy} + \psi_{zz} + 2ik\psi_z = 0 , \tag{2.3}$$

We arrive at an effective equation for describing a beam propagating along the z-axis by rejecting the derivative ψ_{zz} in this equation:

$$\psi_{xx} + \psi_{yy} + 2ik\psi_z = 0 . \tag{2.4}$$

The transition from (2.3) to (2.4) is asymptotic in nature and relies on the assumption $k \gg 1$, i.e., $\omega \gg c$. This transition does not even preserve the type of the equation: if Eq. (2.3) is elliptic, then (2.4) is an *equation of Schrödinger type*. For Eq. (2.4), in contrast to (2.3), one can pose the *Cauchy problem* of seeking $\psi(x, y, z)$ for $z > z_0$ as a solution of (2.4) satisfying the initial condition

$$\psi(x, y, z)\big|_{z=z_0} = \psi_0(x, y) . \tag{2.5}$$

The meaning of the transition (2.3) → (2.4) is that in the set of solutions of Eq. (2.4) there exists a rich class of solutions ψ for which $|\psi_{zz}| \ll 2k|\psi_z|$.[4] Such a property, for example, is possessed by the solution of the Cauchy problem with initial condition independent of k. A narrow stationary monochromatic light beam propagating (or rather extending) along the z axis can usually be described with good precision by Eq. (2.4).

The solution of the Cauchy problem (2.4)–(2.5) can be represented in the form

$$\psi(z) = U_{z-z_0}\psi_0 \tag{2.6}$$

(x and y being omitted). Here U_z is the *resolvent operator* for the Cauchy problem. It is an integral operator whose kernel (the *Green's function for the Cauchy problem*)

$$U_z(x, y; x', y') = g_z(x - x', y - y') , \tag{2.7}$$

is given by the formula

$$g_z(\zeta) = \frac{k}{2\pi i z} e^{\frac{i}{2}\frac{k}{z}\zeta^2} , \quad \zeta = (x, y) . \tag{2.8}$$

Here

$$U_z\psi_0 = g_z * \psi_0 . \tag{2.9}$$

We note that the operator U_z is a unitary operator on $L_2(\mathbb{R}^2)$.

If $z/k \to 0$, then $g_z(\zeta) \to \delta(\zeta)$, and therefore for $(z - z_0)/k \ll 1$ we have

$$\psi(\zeta, z) \approx \psi_0(\zeta) . \tag{2.10}$$

For the solution of Eq. (2.1) this leads to the following expression

$$v(r) \approx e^{ikz}\psi_0(\zeta) . \tag{2.11}$$

Formula (2.11) is precisely the *ray approximation* for wave optics in the beam approximation. The ray approximation corresponds to rejecting all the second-order derivatives in Eq. (2.3); it is described by the equation

$$\psi_z = 0 . \tag{2.12}$$

Our computations show that this approximation can be used at distances

$$(z - z_0)/k \ll 1 . \tag{2.13}$$

[4] This relation may be violated on certain sparse sets; however, this does not interfere with the transition (2.3) → (2.4).

Since $k = 2\pi\lambda^{-1}$, where λ is the wave length, condition (2.13) has the form $(z - z_0)\lambda \ll 2\pi$.

In deriving formula (2.10) it was assumed that the function ψ_0 is essentially independent of all parameters. In practice this means that the unit of length l was chosen so that the derivatives of the function ψ_0 assume tempered values. This means that the tempered values have derivatives $x \to \psi_0(lx)$. We shall introduce this scale explicitly into condition (2.13) in order to make it dimensionless:

$$\frac{(z - z_0)\lambda}{2\pi l^2} \ll 1 . \tag{2.14}$$

For visible light $\lambda \approx 10^{-5}$ cm. If $l = (1 - 10)10^{-3}$ cm, then (2.14) assumes the form

$$(z - z_0) \ll (1 - 10^2)\text{cm} .$$

Beyond such distances the transition to the ray description becomes impossible.

Thus we have given the equation that describes the nature of the field in a light beam. Does this equation have the necessary self-consistency, in other words, does Eq. (2.4) have solutions that can be interpreted as narrow light beams? To understand this we estimate the behavior of a solution of (2.6) as $z - z_0 \to \infty$. To this end, we represent the solution ψ in terms of the Fourier transform $\widetilde{\psi}_0$ of the initial condition ψ_0:

$$\psi(\zeta, z) = (2\pi)^{-2} \int\limits_{\mathbb{R}^2} e^{i\langle\zeta,\zeta_1\rangle - \frac{i}{2}\frac{(z-z_0)}{k}\zeta_1^2} \widetilde{\psi}_0(\zeta_1)\, d\zeta_1 . \tag{2.15}$$

The asymptotic behavior of this solution as $(z - z_0)/k \to \infty$ can be described using the stationary phase method

$$\psi(\zeta, z) \sim \frac{k}{2\pi i(z - z_0)} e^{\frac{i}{2}\frac{k}{z-z_0}\zeta^2} \widetilde{\psi}_0\left(\frac{k}{z - z_0}\zeta\right) . \tag{2.16}$$

If the unit of length that characterizes the function ψ_0 is l, then the function $\zeta_1 \to \widetilde{\psi}_0(l^{-1}\zeta_1)$ also has derivatives that are uniformly bounded and rapidly decreasing. This means that the function $\psi(\zeta, z)$ is noticeably different from zero only for

$$|\zeta| \leq A\frac{|z - z_0|}{kl} , \tag{2.17}$$

where A is a number that characterizes the size of the region on which the function $\widetilde{\psi}_0(l^{-1}\zeta_1)$ (which contains no parameters) is noticeably different from zero. Condition (2.17) means that the beam has a final angle of spread α:

$$|\tan\alpha| \leq \frac{A}{kl} . \tag{2.18}$$

With the parameters shown above $\alpha \leq A(1 - 10) \cdot 10^{-4}$.

2.2. Filters. Optical devices that perform a Fourier transform also contain a certain number of *filters*. Each filter is located in a certain plane $z = z_0$. The propagation of the field between the filters is characterized by Eq. (2.4). The action of a filter on the field ψ is to multiply the field by the *transmittance coefficient* $s : \mathbb{R}^2 \to \mathbf{C}$ of the filter

$$\psi(\zeta, z_0 + 0) = s(\zeta)\psi(\zeta, z_0 - 0) \,. \tag{2.19}$$

Various filters are encountered in optical Fourier processors. Certain filters of comparatively complicated structure will be studied in the next section, but for now we shall limit ourselves to a very preliminary classification.

1) *Amplitude filter:* $s = s^*$. Such filters are often called *transparents.* They satisfy the condition $|s(\zeta)| \leq 1$. A typical example of a transparent is a (semi)transparent developed photographic plate.
2) *Phase filter:* $|s| = 1$. Such filters are absolutely transparent and thus, strictly speaking, invisible. In practice phase filters of course possess a certain absorption, so that $|s|$ is slightly less than one. Phase filters can be transparent crystals, liquids, biological preparations, or glass plates (sometimes having an inhomogeneous transparent coating).

As a particular class of phase filters we distinguish

3) *Thin lenses.* A thin lens can be regarded as a pure phase filter for which

$$s(\zeta) = s_F(\zeta) = K_{-F/k}(\zeta) = e^{-\frac{i}{2}\frac{k}{F}\zeta^2} \,. \tag{2.20}$$

Here F is the so-called *focal length* of the lens.

The expression "thin lens" is a precise term used in optics to denote an optical device familiar to everyone. For our purposes it is a three-dimensional body filled with a homogeneous optically transparent substance (glass) with a certain *refractive index* n larger than the refractive index of air, $n > 1$. This body is assumed to be bounded by two surfaces having a common axis of revolution. We shall regard these surfaces as spheres. If the lens is penetrated by a sufficiently narrow light beam, this assumption does not entail any additional restrictions. A lens is considered thin if its dimensions are much less than the radii R_1 and R_2 of these spheres.

Since the lens is assumed thin, the transmission of light through it can be described in terms of the *ray approximation.* For beams in a homogeneous medium the ray approximation actually reduces to the statement that the field $u_1(\zeta)$ in the section $z = z_1$ is connected with the field $u_2(\zeta)$ in the section $z = z_2$ by the following factor:

$$u_2(\zeta) = \exp[ik(z_2 - z_1)]u_1(\zeta) \,. \tag{2.21}$$

In an inhomogeneous medium characterized by a refractive index $n = n(\zeta, z)$ formula (2.21) must be replaced by the formula

$$u_2(\zeta) = \exp\left[ik \int_{z_1}^{z_2} n(\zeta, z)\, dz\right] \cdot u_1(\zeta) . \tag{2.22}$$

The validity of (2.22) assumes that we are dealing with short-wave beams of small wavelength. This time the restrictions on the wavelength cannot be reduced to a condition of type (2.13); it is necessary to require in addition that a light beam at the wavelength under consideration suffers no significant loss of rectilinearity. Although the introduction of a lens leads to refraction of rays and hence loss of rectilinearity, this effect is not able to manifest itself noticeably at distances of order equal to the thickness of the thin lens, so that formula (2.22) is applicable in this case.

Since the refractive index n is constant inside the lens, formula (2.22) implies

$$u_2(\zeta) = \exp[ik(\Delta_0 + (n-1)\Delta(\zeta))]u_1(\zeta) , \tag{2.23}$$

where Δ_0 is the maximal lens thickness and $\Delta(\zeta)$ is its thickness in the direction of the axis of the system at the point ζ (cf. Fig. 2). Passing to the functions

$$\psi(\zeta, z) = \exp\left[-ik \int^{z} n(0, z)\, dz\right] \cdot u(\zeta, z) \tag{2.24}$$

and referring the jump of ψ to some intermediate section of the lens $z = z_0$, we obtain

$$\psi(\zeta, z_0 + 0) = e^{ik(n-1)(\Delta(\zeta)-\Delta_0)}\psi(\zeta, z_0 - 0) . \tag{2.25}$$

Thus the lens must be associated with the transmittance coefficient

$$s(\zeta) = e^{ik(n-1)(\Delta(\zeta)-\Delta_0)} \tag{2.26}$$

To find $\Delta(\zeta)$ we take account of the specific geometry of the lens (cf. Fig. 2)

$$\Delta(\zeta) = \left(\sqrt{R_1^2 - \zeta^2} - r_1\right) + \left(\sqrt{R_2^2 - \zeta^2} - r_2\right) \approx \Delta_0 - \frac{1}{2}\left(\frac{1}{R_1} + \frac{1}{R_2}\right)\zeta^2 . \tag{2.27}$$

Finally, from (2.26)–(2.27) we have (2.20), where

$$F^{-1} = (n-1)\left(\frac{1}{R_1} + \frac{1}{R_2}\right) , \tag{2.28}$$

F has the physical dimensions of length.

Formula (2.20) does not take account of the boundedness of the dimensions of the lens. If a section of the light beam is much less than the transverse dimension of the lens, such a point of view does not introduce any additional errors. In practice, along with the lens one usually sets up an aperture covered by the lens. The influence of this aperture will be discussed separately at the beginning of the next section, and this effect will be ignored in the remainder of the discussion.

Along with amplitude and phase filters, of course, one uses filters that cannot be classified in either of these groups, their transmittance coefficient being a quite general complex number.

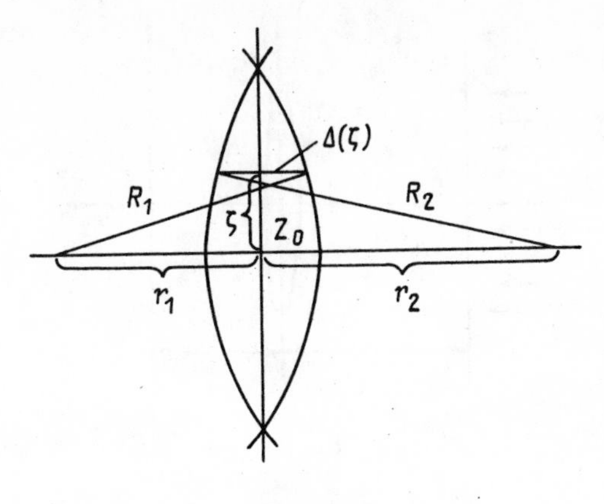

Fig. 2.

We must now discuss a fundamental but delicate question connected with the concept of a filter. A physical filter is a certain plate. Its influence on an incident beam shows up in more than just the transformation (2.19). A filter of course causes a rather complicated scattering of the light field incident on it. Can we neglect these complicated phenomena and limit ourselves to effects covered by the transmittance coefficient? At a minimum the answer to this question depends on the transmittance coefficient itself. The characteristic unit of length l introduced by the transmittance coefficient must not be too small, or else the beam will begin to spread rapidly in passing through the filter. We limit ourselves to these brief remarks for the time being.

2.3. Formation of an Image Using a Lens. It is well-known that a lens is most often used to *form an image.* To understand the way a lens works in this capacity we consider the transformation

$$U_b s_F U_a \tag{2.29}$$

of a freely propagating light beam which occurs on two abutting intervals of lengths a and b between which a lens is set up with focal length F, cf. Fig. 3. It is easy to show that the kernel of the operator (2.29) is given by the integral

$$A = \frac{k^2}{(2\pi i)^2 ab} \int\limits_{\mathbb{R}^2} e^{\frac{ik}{2b}(\zeta_3-\zeta_2)^2 - \frac{ik}{2F}\zeta_2^2 + \frac{ik}{2a}(\zeta_2-\zeta_1)^2} \, d\zeta_2 \,. \tag{2.30}$$

Here and in what follows it will be convenient for us to use the symbols z and ζ with subscripts to denote the variables z and ζ in various planes orthogonal to the axis of the beam. This subscript will coincide with the subscript of the planes P_i indicated in Fig. 3. We agree to endow the function ψ with the analogous subscripts.

If the *lens formula*

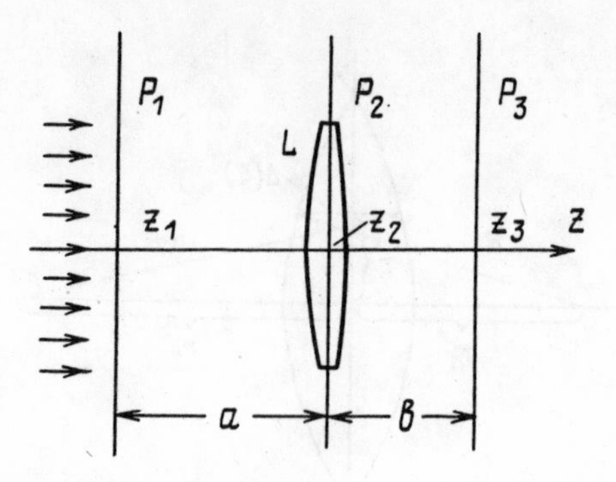

Fig. 3

$$\frac{1}{a}+\frac{1}{b}=\frac{1}{F}\,, \tag{2.31}$$

holds, then the integral assumes the form

$$A=\frac{k^2}{(2\pi i)^2 ab}e^{\frac{ik}{2b}\zeta_3^2+\frac{ik}{2a}\zeta_1^2}\int\limits_{\mathbb{R}^2} e^{-ik\left(\frac{\zeta_3}{b}+\frac{\zeta_1}{a},\zeta_2\right)}\,d\zeta_2=$$

$$=\frac{-1}{ab}e^{\frac{ik}{2b}\zeta_3^2+\frac{ik}{2a}\zeta_1^2}\delta\Big(\frac{\zeta_3}{b}+\frac{\zeta_1}{a}\Big)\,. \tag{2.32}$$

Thus when the lens formula holds, we have

$$\psi_3(\zeta_3)=(U_b s_F U_a\psi_1)(\zeta_3)=-\frac{1}{ab}e^{\frac{ik}{2b}\zeta_3^2+\frac{ik}{2a}\zeta_1^2}\psi_1\Big(-\frac{a}{b}\zeta_3\Big)\,. \tag{2.33}$$

Since only the *field intensity*

$$I=|\psi|^2\,, \tag{2.34}$$

is usually measured (in particular when examining light pictures visually), the phase factor can be omitted

$$I_3(\zeta_3)=|\psi_3(\zeta_3)|^2=\Big|\frac{1}{ab}\psi_1\Big(-\frac{a}{b}\zeta_3\Big)\Big|^2=\frac{1}{a^2b^2}I_1\Big(-\frac{a}{b}\zeta_3\Big)\,. \tag{2.35}$$

This result, in particular, means that under these conditions the lens forms the image up to reflection and change of scale. Setting up a slide in the plane P_1 and a screen in the plane P_3 and illuminatng the system with a plane wave (cf. Fig. 3), we will see on the screen an inverted image of the picture imprinted on the slide, with a change of scale.

2.4. The Optical Fourier Transform. We shall consider designs of two basic devices used to perform the Fourier transform.

Design 1 is depicted in Fig. 4. This design has already appeared in Sect. 1. In it the field ψ in the plane P_3 is compared with the field in the plane P_1. A thin lens L with focal length F is situated in a plane P_2 between these planes. It is assumed that $z_3 - z_2 = z_2 - z_1 = F$. It is easy to see that

$$\psi_3 = g_F*(K_{-F/k}\cdot(g_F*\psi_1)) = \left(\frac{k}{2\pi iF}\right)^2 K_{F/k}*(K_{-F/k}\cdot(K_{F/k}*\psi_1)) \,. \quad (2.36)$$

According to formula (1.4)

$$\psi_3 = \frac{k}{2\pi iF}\mathcal{F}_{F/k}[\psi_1] \,. \quad (2.37)$$

Thus ψ_1 and ψ_3 are connected essentially by the Fourier transform.

Design 2 is depicted in Fig. 5. It realizes formula (1.3) technically. Two lenses L_1 and L_2 having the same focal length F are situated at a distance F from each other. The field in plane P_1 immediately preceding lens L_1 is compared with the field in the plane P_2 that is located immediately behind lens L_2. The fields ψ_1 and ψ_2 in these planes are connected by the relation

$$\psi_2 = K_{-F/k}\cdot(g_F*(K_{-F/k}\cdot\psi_1)) = \frac{k}{2\pi iF}\mathcal{F}_{F/k}[\psi_1] \,. \quad (2.38)$$

As already noted, actual Fourier processors contain, in addition to the elements just indicated, certain input and output elements. If a filter with transmittance coefficient s is set up in the plane P_1 of Design 1 and illuminated with a planar wave, the Fourier transform of the function s will form in the plane P_3:

$$\psi_3 = \frac{k}{2\pi iF}\mathcal{F}_{F/k}[s] \,. \quad (2.39)$$

A measuring device is located in this plane, for example, a photographic film. After exposure and development of the film the blackness density B will be proportional to the intensity of the light field

$$B = \gamma I_3 \,. \quad (2.40)$$

We shall discuss in more detail the questions relating to the measurement of a field, including the question of how to measure the complete field ψ and not just its intensity I, in Sect. 2.6.

2.5. Two Remarks. All the designs just described make it possible to perform operations on functions of two variables $\psi = \psi(x,y)$. If, however, the transmittance coefficients of the filters used depend only on x, then beams of the form

$$u = u(x,z) = e^{ikz}\psi(x,z) \quad (2.41)$$

can propagate in such systems. To form the Fourier transform in the class of beams of this type, i.e., the Fourier transform of functions of one variable, one must use the so-called *cylindrical lenses* instead of ordinary lenses (cf. Fig. 6). Their transmittance coefficient has the form

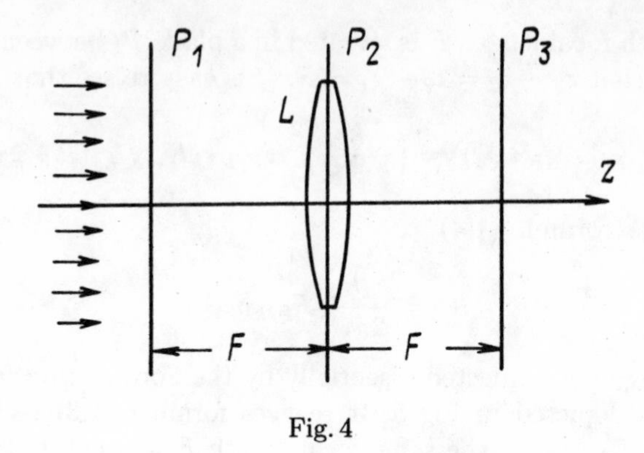

Fig. 4

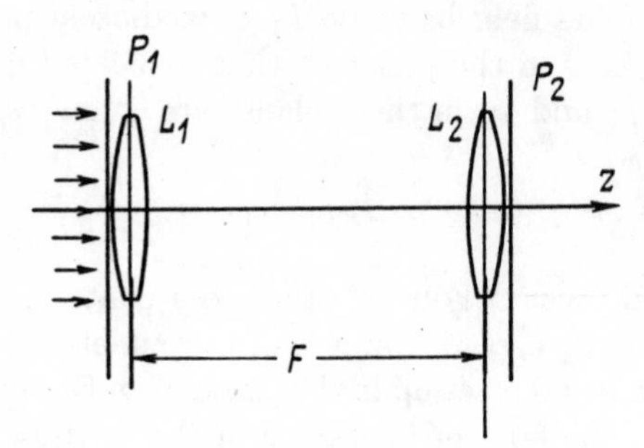

Fig. 5

$$s = s(x) = e^{-\frac{i}{2}\frac{k}{F}x^2} \tag{2.42}$$

In physical and engineering practice, as we have already said, it is not only the Fourier transform that has immediate interest, but also the convolution and cross-correlation of two functions of rather general form. Suppose that at the entrance to the system described by Design 1 (or 2), i.e., in the plane P_1, two filters Φ_1 and Φ_2 are set up in a row with transmittance coefficients s_1 and s_2. In this case a field arises in the plane P_3

$$\psi_3(\zeta_3) = \frac{k}{2\pi i F}\mathcal{F}_{F/k}[s_2 \cdot s_1](\zeta_3) = \frac{k}{2\pi i F}(2\pi)^{-2}(\tilde{s}_2 * \tilde{s}_1)\Big(\frac{k}{F}\zeta_3\Big) . \tag{2.43}$$

If we reverse the filter Φ_2, we obtain a filter Φ_2' with transmittance coefficient $s_2'(\zeta) = s_2(-\zeta)$. This time

$$\psi_3(\zeta_3) = \frac{k}{2\pi i F}\mathcal{F}_{F/k}[s_2' \cdot s_1](\zeta_3) = \frac{k}{2\pi i F}(2\pi)^{-2}(\tilde{s}_2^* * \tilde{s}_1)\Big(\frac{k}{F}\zeta_3\Big) . \tag{2.44}$$

Formulas (2.43) and (2.44) mean that Fourier processors can be used (with certain stipulations) to form the convolution (and cross-correlation) of the functions t_1 and t_2 if we construct filters whose transmittance coefficients are

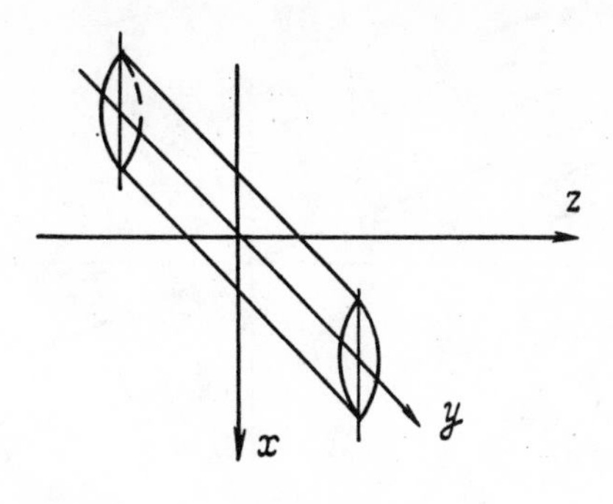

Fig. 6

the Fourier transforms $\tilde{t}_1$ and $\tilde{t}_2$ of these functions. Such filters can be formed using the same Fourier processors. We shall encounter other designs below at whose entrance one can input the functions t_1 and t_2 immediately.

2.6. Holographic Measurement of a Field and Holographic Filters. Both ordinary and exotic devices that make it possible to measure an optical field (the eye, a photographic plate, a photoelectric cell, photosensitive liquid layers) respond only to its intensity and ignore the phase of the field. This means in particular that only the square of the absolute value of the Fourier transform of the field input at the entrance of the Fourier processor can be measured directly. This last circumstance, of course, significantly decreases the possibilities for applying Fourier processors. The ability to measure a complex optical signal arose after the idea of holography was introduced into Fourier optics.

The ideas of holography in application to the problem of measuring a complex optical signal can be described as follows. Suppose a beam falls on a plane P, $z = 0$, forming an optical signal $\psi(\zeta)$ in the plane. Let us provide additional illumination to P by a so-called supporting beam—a plane wave $A \exp ik(x \sin\theta + z\cos\theta)$ with A constant—propagating at a certain angle θ to the z-axis (cf. Fig. 7). In the plane P there arises a field

$$u = \psi(\zeta) + Ae^{ik_1 x}, \tag{2.45}$$

where $k_1 = k \cdot \sin\theta$. We measure it with a photographic plate. After development the photographic plate can be regarded as an amplitude filter whose transmittance coefficient is proportional to

$$I = |u|^2 = I_0 + I_+ + I_- = \left(|\psi|^2 + |A|^2\right) + Ae^{ik_1 x}\psi^* + A^* e^{-ik_1 x}\psi \,. \tag{2.46}$$

From the filter obtained (which we shall call a *holographic* filter) it is theoretically possible to extract the original signal ψ. In fact the function ψ must be regarded as slowly varying compared to $\exp ik_1 x$, so that the coefficients of the different powers of $\exp ik_1 x$ are determined from the expression (2.46) with only an insignificant error. But how can one compute the signal ψ from

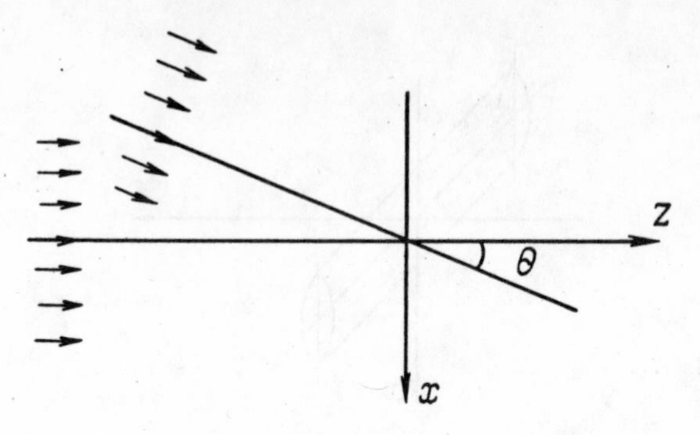

Fig. 7

the filter (2.46) in practice? To this end the filter (2.46) is set up in the entrance plane P_1 of a Fourier processor assembled, for example, according to Design 1. The Fourier image (2.46) of the function forms in the plane P_3 when this is done. The Fourier transform of the first term I_0 will essentially be concentrated near the origin. The rapidly varying exponentials in the second and third terms will lead to a noticeable displacement of the supports of their Fourier images in the neighborhood of the points $\left(-\frac{k_1}{k}F, 0\right)$ and $\left(\frac{k_1}{k}F, 0\right)$. If we set up an aperture in the plane P_3 that lets through only part of the beam intersecting P_3 in a neighborhood of the point $\left(\frac{k_1}{k}F, 0\right)$, then immediately behind this plane the optical signal will be equal to

$$\frac{k}{2\pi iF}\mathcal{F}_{F/k}[I_-](\zeta_3) = \frac{k}{2\pi iF}\widetilde{\psi}\left(\frac{k}{F}x_3 - k_1, \frac{k}{F}y_3\right) \tag{2.47}$$

(cf. Fig. 8). This signal carries complete information about the complex field ψ. If necessary, one can set up an analogous device behind the Fourier processor just considered and use it to recover the signal ψ in the original representation.

We conclude with a simple remark. Our description of a holographic filter was not completely accurate. The fact is that in the design depicted in Fig. 8 three beams instead of one form behind the holographic filter: the first of them is generated by the term I_0 of the transmittance coefficient (2.46) and propagates along the z-axis, while the other two are generated by the other terms in the expression (2.46) and propagate at angles $\pm\theta$ to the z-axis. This picture actually takes us beyond the scope of the single-beam approach to the propagation of optical signals. One could easily construct the necessary version of a multibeam theory; nevertheless for the sake of simplicity in exposition we have kept the single-beam approach, assuming that θ is small. An analogous simplification has been introduced into the following discussion as well.

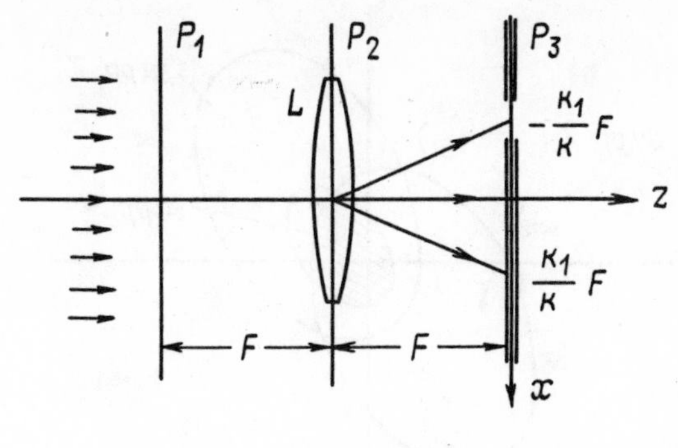

Fig. 8

§3. Notes and Comments

3.1. The Influence of the Lens Size. Of the parameters of the various components of an optical system, by means of which the Fourier transform is carried out, the one having most influence on the boundedness of the section of the beam is the size of the lens. Up to now we have neglected the effects caused by the bounded dimensions of the lens. Let us now consider the question of these effects in the context of Scheme 1 (Fig. 4). We assume that directly behind the lens L there is an *aperture* whose opening is completely covered by the lens. Let χ be the characteristic function of this opening. The operator $U_F s_F U_F$ that describes the transformation of the light signal in Scheme 1 taking account of the influence of the aperture acquires the form $U_F(s_F\chi)U_F$. Its kernel is given by the formula

$$U_F(s_F\chi)U_F \sim \left(\frac{k}{2\pi i F}\right)^2 e^{-i\frac{k}{F}\langle\zeta_3,\zeta_1\rangle} I_\chi(\zeta_1+\zeta_3)\,,$$
$$I_\chi(\zeta) = \int\limits_{\mathbb{R}^2} e^{\frac{i}{2}\frac{k}{F}(\zeta_1-\zeta)^2}\chi(\zeta_1)\,d\zeta_1\,. \tag{3.1}$$

The parameter kF^{-1} must be considered large, so that the considerations of the stationary phase method can be applied to the integral I_χ. If the point ζ is located outside a small neighborhood of the boundary of the aperture, then the integral I_χ is close to $2\pi i F/k \cdot \chi(\zeta)$. Neglecting transient effects, which arise only for ζ close to the boundary of the aperture, we can assume that

$$U_F(s_F\chi)U_F \sim \frac{k}{2\pi i F}\chi(\zeta_1+\zeta_3)e^{-i\frac{k}{F}\langle\zeta_3,\zeta_1\rangle}\,. \tag{3.2}$$

We shall explain this formula using Fig. 9. Applying the operator (3.1) to the function ψ, we must restrict the integration to the intersection $\operatorname{supp}\psi \cap \operatorname{supp}\chi(\cdot+\zeta_3)$. The result coincides with the Fourier transform of the function ψ only when $\operatorname{supp}\chi(\cdot+\zeta_3) \supset \operatorname{supp}\psi$. This relation is demonstrably violated for sufficiently large ζ_3.

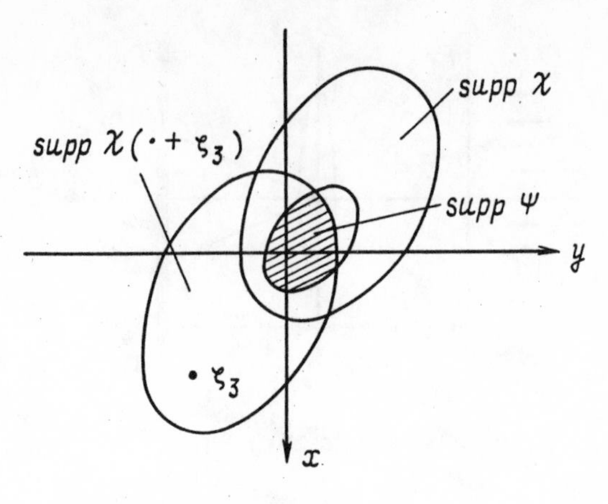

Fig. 9

3.2. Spatial Filtering. We shall touch briefly below on certain episodes in the history of the development of the ideas leading to the widespread propagation of Fourier processors. In this section and the one following we shall also exhibit certain simple applications of Fourier processors. One of the fundamental optical schemes constructed on the basis of the Fourier properties of a thin lens is represented in Fig. 10. It consists of two schemes of type 1 in a row. This means that the plane P_2 contains the Fourier transform $\widetilde{\psi}\left(\frac{k}{F}\zeta_2\right)$ of the field ψ formed in the plane P_1, i.e., at the entrance to the system, and the plane P_3 contains the Fourier transform of the field formed by the first part of the scheme in the plane P_2, in other words, the function ψ is recovered in the plane P_3. More precisely the function ψ is transformed by this scheme into the function $M\psi(M\zeta_3)$, $M = -F_1F_2^{-1}$.

The operation on the function ψ effected by introducing a filter Φ with suitable transmittance coefficient in the plane P_2 is called *spatial filtering.* We denote by the letter C the linear operator corresponding to this transform. We represent the transmittance coefficient of the filter Φ in the form $s(\zeta) = \tilde{h}(\frac{k}{F_1}\zeta)$. Then

$$(C\psi)(\zeta_3) = M(h * \psi)(M\zeta_3) \,. \tag{3.3}$$

Formula (3.3) shows that the processor depicted in Fig. 10 can be regarded as a convolver.

Among the first papers in which spatial filtering was carried out it is customary to mention the 1873 works of Abbe (1904). Abbe studied the formation of the image in microscopes, looking at periodic objects. By adapting the description of Abbe's experiments to the purposes of our present discussion, we can represent the situation as follows. If we put the filter shown in Fig. 11 in the plane P_1, a grating of luminous dots of various intensities arises in the plane P_2 (in order to describe these dots analytically one must take into account the finite size of the cross-section of the beam). Setting up a mask-filter in the plane P_2 that lets through only the central horizontal line of dots, we

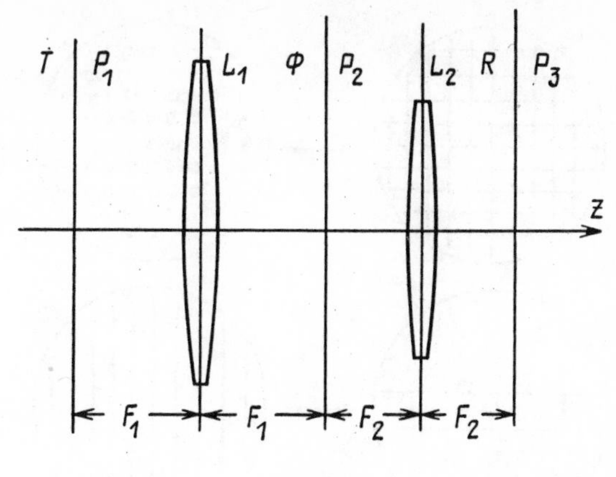

Fig. 10

obtain a system of bands in a plane P_3 instead of the grating. The interpretation given to these results by Abbe was incomplete from the point of view of the ideas now under consideration. It was significantly improved by Porter (1906), who identified the luminous dots in the plane P_2 with the terms of a double Fourier series. We shall not dwell on the elementary explanation of Abbe's results from the point of view of formula (3.3).

The next step in this topic is connected with the work of Zernike (1942). Zernike studied microscopic *phase-contrast objects*, i.e., transparent objects whose transmittance coefficient has a purely phase character

$$s(\zeta) = \exp i\varphi(\zeta) .$$

Such objects cannot be seen using ordinary optical systems that form an image. Zernike placed them in the plane P_1 and illuminated them with a plane wave, while in the plane P_2 he set up near the axis of the optical system a small transparent plate causing a phase shift of $\pi/2$ (we are simplifying the description of the original system slightly). Assuming that φ is small, we can represent the expression for s as

$$s(\zeta) \approx 1 + i\varphi(\zeta) .$$

It is clear that the first term, whose support in the plane P_2 coincides with the point 0, is multiplied by i under the action of the operator C. The second term, due to the smallness of the plate set up, undergoes almost no additional transformation:

$$(Cs)(\zeta_3) \approx M(i + i\varphi(M\zeta_3)) .$$

The intensity of the light beam in the plane P_3 thus turns out to be

$$I \approx M^2|1 + \varphi(M\zeta_3)|^2 \approx M^2(1 + 2\varphi(M\zeta_3)) .$$

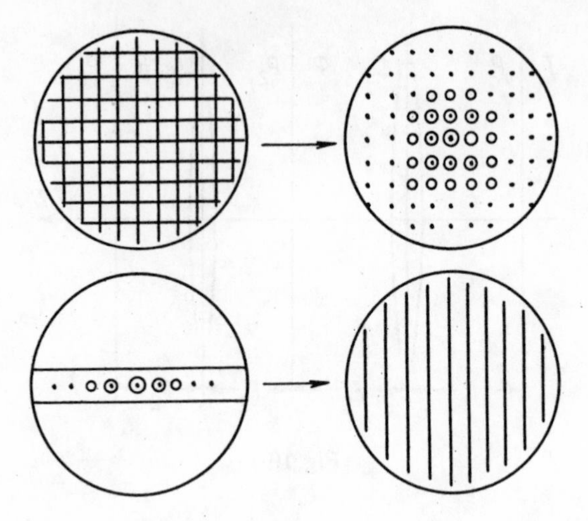

Fig. 11

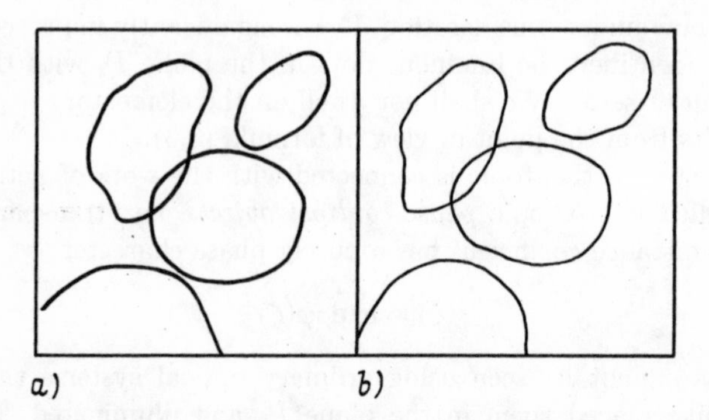

Fig. 12. Schematic reproduction of the microphotography of cells: a) phase-contrast; b) without contrast

This means that the phase-contrast object has become visible. The efficiency of this device is illustrated in Fig. 12, in which the phase-contrast object is photographed both with the phase-rotating plate and without it.

Many subsequent episodes of the story of the development of Fourier optics that we are interested in occurred in the late 1940's and early 1950's. Duffieux (1946) used the Fourier integral explicitly in analyzing signal transformation in optical devices. Elias et al. (1952, 1953) exhibited parallels with the theory of linear coupled systems, which up to that time had developed mainly under the influence of the problems of radio-electronics. These parallels were further developed by O'Neill (1956). A significant impetus to progress in Fourier optics was provided by the pioneering work in information theory of Shannon (1949) and Wiener (1949). All this made possible a quick evaluation of optical devices that perform a Fourier transform. It became clear that the processing of the Fourier spectrum of an optical signal opens up many possibilities that could not be achieved by operating directly on the original representation of the signal. An important role in the development of optical processors was played by the introduction into optics of powerful sources of high-quality light beams such as lasers. The importance of the ideas of holography in problems of Fourier optics has already been mentioned.

After the work of Maréchal (1953) the early 1950's brought broad investigations into the improvement of the quality of scientific photographs using spatial filtering. The idea of these investigations was very simple. A photographic plate was used as a transparent and set up in the plane P_1. If it became necessary to increase the contrast of the fine details, an opaque mask of suitable size was placed in the plane P_2 to block the low-frequency spatial components of the optical signal. An example of this operation is shown in Fig. 13. If, in contrast, the fine details arose as the result of interference, whose influence it would be desirable to remove, a very narrow aperture was set up in the plane P_2 to block the high-frequency components. Gradually more refined methods of controlling the photographic imaging using spatial filters came to be employed.

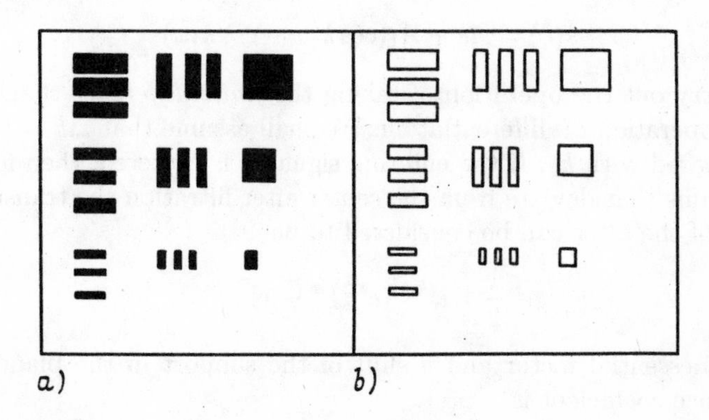

Fig. 13. Schematic reproduction of photographs: a) original, b) low frequencies blocked

3.3. Complex Amplitude Filters. Some important operations on signals are realized by using special classes of filters. We shall pause first to discuss amplitude filters.

Filters with transmittance coefficients of the form

$$s(\zeta) = A_0 + A_1 \cos(k_1 x + \varphi) ,$$

where A_0, A_1, k_1, and φ are certain constants, $|A_1| \leq |A_0|$, are called *grating* filters. We shall assume that a signal

$$\psi_1(\zeta_1) = f_1\Big(x_1 - F_1\frac{k_1}{k}, y_1\Big) + f_2\Big(x_1 + F_1\frac{k_1}{k}, y_1\Big)$$

formed by the superposition of two signals whose supports are concentrated near the points $(\frac{k_1}{k}F_1, 0)$ and $(-\frac{k_1}{k}F_1, 0)$ is input at the entrance to a processor into which such a filter has been introduced. A field

$$\psi_2(\zeta_2) = \frac{k}{2\pi i F_1}\Big[e^{ik_1x_2}\tilde{f}_1\Big(\frac{k}{F_1}\zeta_2\Big) + e^{-ik_1x_2}\tilde{f}_2\Big(\frac{k}{F_1}\zeta_2\Big)\Big]$$

arises in front of the filter in the plane P_2. Behind the filter a combination forms whose individual terms can be grouped in powers of the exponentials $\exp(ik_1x_2)$. In particular the expression

$$\frac{kA_1}{4\pi i F_1}e^{-i\varphi}\Big\{\tilde{f}_1\Big(\frac{k}{F_1}\zeta_2\Big) + e^{2i\varphi}\tilde{f}_2\Big(\frac{k}{F_1}\zeta_2\Big)\Big\}$$

arises at the zeroth power. After the Fourier transform the different powers of $\exp(ik_1x_2)$ will lead to the spread of echoes in the plane P_3. The Fourier transform of the expression given above will be measured around the optical axis, which actually generates the sum $f_1 + f_2$ in the case $\varphi = 0$ and the difference $f_1 - f_2$ in the case $\varphi = \pi/2$. In order to change φ it suffices to move the filter along the x-axis.

Using a *double grating filter* with transmittance coefficients

$$s(\zeta) = A_0 + A_1(\cos k_1' x - \cos k_1 x)$$

one can carry out the operation of taking the *finite difference* that approximates the operation of differentiation. We shall assume that $\Delta k = k_1' - k_1$ is small compared with k_1. If the entering signal ψ is centered, then for one of the two beams that deviate from the center after filtration the transmittance coefficient of the filter can be considered to be

$$\frac{1}{2}A_1 e^{ik_1x}(e^{i\Delta k\cdot x} - 1)\,.$$

Up to an inessential factor and a shift of the support in the plane P_3 this transmittance coefficient is

$$e^{i\Delta k\cdot x} - 1\,,$$

and it generates the transformation

$$\psi(\zeta_1) \mapsto M\Big[\psi\Big(M\Big(x_3 + \frac{\Delta k}{k}F_1\Big), My_3\Big) - \psi(M\zeta_3)\Big]\,.$$

And this says precisely that the double grating filter can be used to form the finite-difference operator.

To build a grating filter it is necessary to illuminate a photographic plate by two beams propagating at a certain angle to each other. Exponentiating the plate twice in different ways, one can obtain a double grating filter. By suitably illuminating the plate 5 times, one can obtain a filter that realizes the finite-difference approximation of the *Laplacian operator*. Since the symbol of the Laplacian is of constant sign, it is not difficult to construct amplitude filters that directly realize the Laplacian operator and its inverse. However, at the end of the preceding section, in discussing the problem of measuring a complex optical signal, we explained how to use the ideas of holography to construct the so-called *complex amplitude filters*, which are also called *Vander Lugt filters* (Vander Lugt 1963). Vander Lugt filters make it possible to construct an optical realization of differential operators with complex symbols. A grating filter can, if one wishes, be interpreted as a special case of a Vander Lugt filter. Computers are frequently used in constructing complex amplitude filters with a quite arbitrary transmittance coefficient to control the process of exponentiating the photographic film. Thus it is possible to form transparents containing as many as 10^6 small darkened areas.

3.4. Phase Gratings. The various complex phase filters find equally wide application. We shall give a relatively detailed discussion of these devices, since such filters with controllable characteristics are technically rather easy to build. Frequent use is made of *phase gratings* with transmittance coefficients of the form

$$s(\zeta) = Ae^{i\varphi \cos k_1 x}, \tag{3.4}$$

where A, k_1, and φ are certain constants. The formula

$$e^{i\varphi \cos k_1 x} = \sum_{n=-\infty}^{\infty} i^n J_n(\varphi) e^{ink_1 x},$$

in which J_n is a Bessel function, shows that phase gratings generate multiple shifts of the original signal. If $|\varphi| \ll 1$, the approximation

$$s(\zeta) \approx A(1 - i\varphi \cos k_1 x) \tag{3.5}$$

can be used for the transmittance coefficient.

A phase grating with the transmittance coefficient (3.4) can be obtained by setting up a crystal plate with refractive index

$$n = n_0 + \gamma \cos k_1 x \tag{3.6}$$

on the path of a ray of light (cf. Fig. 14). The applicability of the ray approximation to evaluating the influence of such a plate is characterized by the condition (2.14). In our case the scale l of the transverse inhomogeneity has order k_1^{-1}, so that condition (2.14) takes on the form

$$\Delta \cdot k_0^{-1} \cdot k_1^2 \ll 1,$$

where k_0 is the wave number of the incident light wave and Δ is the thickness of the crystal. When this condition holds, the crystal can be considered thin, and its transmittance coefficient, according to formula (2.22), is determined by the phase overrun

$$s(\zeta) = e^{ik_0(n_0-1)\Delta} e^{ik_0\gamma\Delta\cos k_1 x} .$$

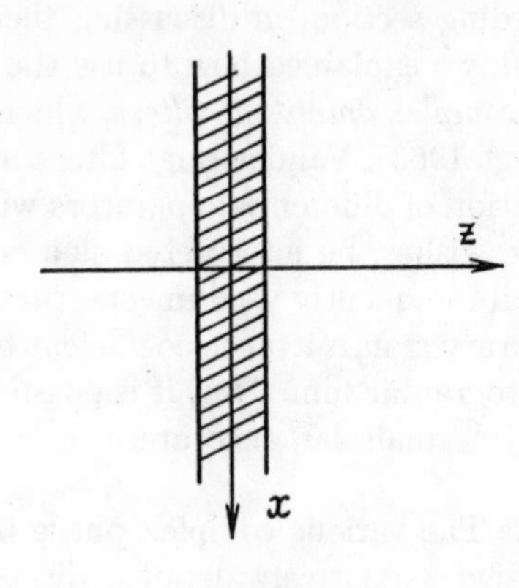

Fig. 14

A thick crystal, i.e., a crystal for which

$$\Delta \cdot k_0^{-1} \cdot k_1^2 \gg 1 ,$$

is a much more interesting object from the point of view of mathematical physics. Such a crystal is usually set up at a small angle θ to the direction of the incident wave (cf. Fig. 15). In this case the wave equation (2.1) in the crystal has the form

$$u_{x'x'} + u_{yy} + u_{z'z'} + k_0^2 n^2 u = 0 ,$$

where $n = n_0 + \gamma \cos k_1 x'$. For the function ψ,

$$u = e^{ik_0 n_0 (z'+\Delta/2)} \psi ,$$

by repeating verbatim the reasoning of Sect. 2.1, we can obtain an *equation of Schrödinger type*

$$\psi_{x'x'} + \psi_{yy} + k_0^2(n^2 - n_0^2)\psi + 2ik_0 n_0 \psi_{z'} = 0 .$$

The qualitative deductions at which we arrive will not change if, regarding γ as small, we slightly simplify the coefficient of the equation

$$n^2 - n_0^2 \approx 2n_0\gamma \cos k_1 x' .$$

In the case when a plane wave

$$u = e^{ik_0 z}$$

is incident on the crystal, a Cauchy problem arises with initial condition

$$\psi\big|_{z'=-\frac{\Delta}{2}} = e^{-ik_0\frac{\Delta}{2}\cos\theta}\, e^{-ik_0x'\sin\theta}\,.$$

Neither the initial condition nor the coefficient of the equation contains the variable y, so that ψ is independent of y. As a result the equation for ψ reduces to the following form:

$$\psi_{x'x'} + 2k_0^2 n_0\gamma\cos k_1x'\cdot\psi + 2ik_0n_0\psi_{z'} = 0\,.$$

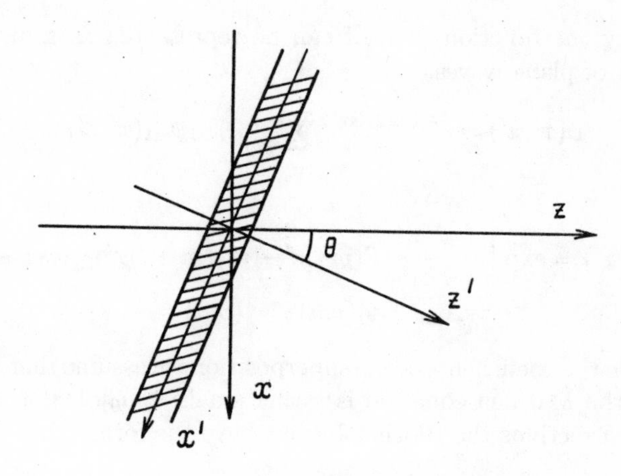

Fig. 15

The function ψ, a solution of the resulting Cauchy problem, can be expressed in terms of the Bloch solutions of the *Mathieu equation*

$$-\varphi_{x'x'} - 2k_0^2n_0\gamma\cos k_1x'\cdot\varphi = E\varphi\,.$$

As is known, the Bloch solutions $\varphi(x',p)$ are characterized up to normalization by the representation

$$\varphi(x',p) = e^{ipx'}\chi(x',p),\quad E = \mathcal{E}(p)\,,$$

in which χ is a periodic function of x' with period $2\pi k_1^{-1}$ and p is an arbitrary complex parameter. Under a suitable normalization the Bloch solutions generate an expansion theorem, which can be written in a manner completely analogous to the standard Fourier integral. As a corollary,

$$\psi(x',z') = e^{-\frac{i}{2}k_0\Delta\cos\theta}\frac{1}{2\pi}\int\limits_{\mathbb{R}} dp\, e^{-\frac{i}{2k_0n_0}\mathcal{E}(p)z'}\varphi(x',p)\psi_B(p)\,,$$

where

$$\psi_B(p) = \int\limits_{\mathbb{R}} d\tilde{x}\varphi^*(\tilde{x},p)e^{-ik_0\tilde{x}\cdot\sin\theta}\,.$$

It is reasonable to point out that the function $p \mapsto \varphi(x', p)$ undergoes discontinuities at the points $p = \frac{1}{2}mk_1$, $m \in \mathbf{Z}$. Recalling the periodicity of χ:

$$\chi(x', p) = \sum\nolimits_m \chi_m(p) e^{imk_1 x'},$$

it is not difficult to surmise that the support of the function $\psi_B(p)$ is the discrete grating

$$p = p_m, \quad p_m = mk_1 - k_0 \sin\theta, \quad m \in \mathbf{Z}.$$

As a corollary the function ψ itself can be represented as a discrete linear superposition of plane waves

$$\psi(x', z') = e^{-\frac{i}{2}k_0 \Delta \cos\theta} \sum\nolimits_{m,l} A_{ml} \psi_{ml}(x', z'),$$

where

$$\psi_{ml}(x', z') = \exp\left[-\frac{i}{2k_0 n_0} \mathcal{E}(pm) z' + ik_1(m+l)x' - ik_0 x' \sin\theta \right],$$
$$A_{ml} = \chi_l(pm) \chi^*_{-m}(pm).$$

To evaluate the coefficients of the superposition we assume that the periodic potential in the Mathieu equation is rather small: $k_0 n_0 |\gamma| \ll 1$. In this case up to small corrections the Bloch solutions have the form

$$\varphi(x', p) \approx e^{ipx'}, \quad \mathcal{E}(p) \approx p^2.$$

This holds everywhere except in small neighborhoods of the points $p = \pm\frac{1}{2}k_1$. In the neighborhood of the point $\frac{1}{2}k_1$, for example, the Bloch solutions are given by the following formulas (McLachlan 1953):

$$\varphi(x', p) \approx e^{ipx'}(1 - e^{-ik_1 x + 2i\tau}),$$
$$p \approx \frac{1}{2}k_1 - i\frac{2k_0^2 n_0 \gamma}{k_1} \sin 2\tau,$$
$$\mathcal{E} \approx \frac{1}{4}k_1^2 - 2k_0^2 n_0 |\gamma| \cos 2\tau,$$

for $p < \frac{1}{2}k_1$, we have $\tau = i\theta$, $\theta > 0$, and for $p > \frac{1}{2}k_1$, we have $\tau = \pi/2 + i\theta$, $\theta > 0$.

Moreover we must assume that $\varphi(x', -p) = \varphi(x', p)$ and $\mathcal{E}(-p) = \mathcal{E}(p)$. If the angle θ is such that $k_0 \sin\theta$ is not close to $\frac{1}{2}k_1 \pmod{k_1}$, then the function ψ reduces to a "straight" plane wave $\psi \approx \psi_{00}$. This mode is not of practical interest. Let us fix the angle $\theta = \theta_B$ by the condition

$$k_0 \sin\theta_B = \frac{1}{2}k_1,$$

and for definiteness assume that $k_0 \sin\theta_B = \frac{1}{2}k_1 - 0$. The angle θ_B is called the *Bragg angle*. When $\theta = \theta_B$, we have $p_0 = -\frac{1}{2}k_1 + 0$ and $p_1 = \frac{1}{2}k_1 + 0$.

In this approximation only the following coefficients are nonzero: $A_{00} = 1$, $A_{01} = -1$, $A_{10} = 1$, $A_{1,-1} = 1$. Thus when $\theta = \theta_B$,

$$\psi \approx e^{-\frac{i}{2}k_0\Delta\cos\theta_B}\left[(\psi_{00} + \psi_{1,-1}) + (-\psi_{01} + \psi_{10})\right] =$$

$$= 2e^{-\frac{i}{2}k_0\Delta\cos\theta_B} e^{-\frac{ik_1^2}{8k_0n_0}z'}\left[\cos qz'\, e^{-ik_0x'\sin\theta_B} + i\sin qz'\, e^{ik_0x'\sin\theta_B}\right],$$

where $q = k_0|\gamma|$.

It is now not difficult to verify that in leaving the crystal the solution U splits into two plane waves

$$U \approx A[\cos q\frac{\Delta}{2}\cdot e^{ik_0z} + i\sin q\frac{\Delta}{2}\cdot e^{ik_0(z\cos 2\theta_B + x\sin 2\theta_B)}], \qquad (3.7)$$

where A is a certain constant (cf. Fig. 16). Thus a thick crystal manifests itself very peculiarly, and its influence on a plane wave cannot be reduced to multiplication by a transmittance coefficient.

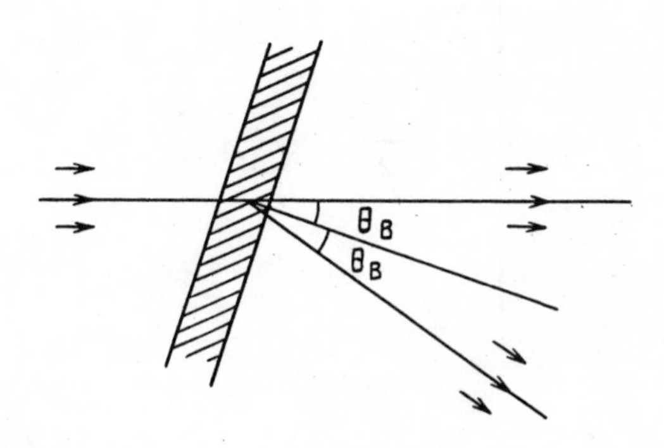

Fig. 16

3.5. Improving the Resolution. We shall now exhibit another application of Fourier processors. An ideal optical device intended for forming images must be described (up to a simple change of coordinates in the plane of the images) by the identity operator: $\psi_1(\zeta_1) \mapsto \psi_2(\zeta_2) = \psi_1(\zeta_2)$. Here ψ_1 is the optical signal at the entrance to the device and ψ_2 the signal at the exit. In practice, because of the boundedness of a section of a beam and the imperfection of the components of the device this operator usually has the form

$$\psi_1(\zeta_1) \mapsto (h * \psi_1)(\zeta_2) = \int_{\mathbb{R}^2} h(\zeta_2 - \zeta_1)\psi_1(\zeta_1)\, d\zeta_1\,. \qquad (3.8)$$

The function h as a rule has the property $h(\zeta) = h_0(|\zeta|)$. A typical graph of such a function is depicted in Fig. 17. The qualities of the device are determined by the function h. In particular the form of this function determines the possibility of distinguishing two nearby luminous points (the *resolution* of the device). It is possible to improve significantly the qualities of the device by placing at its entrance a processor designed for spatial filtering. If we set a filter having transmittance coefficient $\left[\tilde{h}(\frac{k}{F}\zeta) + i\varepsilon\right]^{-1}$, with ε sufficiently small, in the plane P_2 of this processor, the transformation of the signal in the combined device will be represented by the operator

$$\psi_1 \mapsto \psi_2 = \mathcal{F}[\tilde{h}(\tilde{h} + i\varepsilon)^{-1}\tilde{\psi}_1] , \tag{3.9}$$

which may turn out to be much closer to the identity than the operator (3.8). In some cases this trick makes it possible to improve significantly the capability of the device to distinguish luminous points that are close together (in this regard cf. (Vasilenko and Taratorin 1966), (Gurevich, Konstantinov, et al., 1978)).

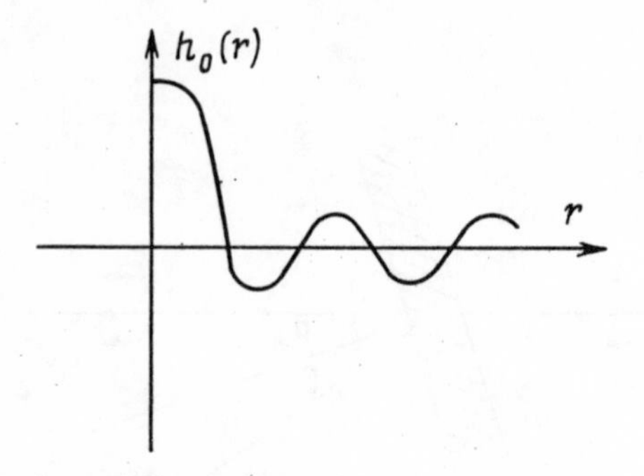

Fig. 17

3.6. Incoherent Signal Processing. In our discussion up to now we have assumed that the optical signal has a regular determinate structure. The signals generated by some *lasers* approximate this ideal very well, or, as it is customarily expressed, are *coherent*. However the optical signal from the self-luminous or diffusively reflecting bodies with which we usually have to deal are always essentially *incoherent*. The field corresponding to a self-luminous object is generated by the radiation of many atoms or molecules; the times of radiation, frequencies, and phases of the fields formed by the individual particles are to some degree independent. This causes the optical field $u(t, r)$ to become random. One of its characteristics may be the *correlation function*

$$\Gamma(r_1, r_2, \tau) = \langle u(r_1, t + \tau) u^*(r_2, t) \rangle ,$$

where the angle brackets on the right denote the probability mean. Totally spatial incoherent illumination is characterized by the relation

$$\Gamma(r_1, r_2, 0) = I(r_1)\delta(r_1 - r_2) .$$

It is natural to call the function I the *intensity* of the incoherent optical field.

If an optical system intended to form an image transforms a coherent signal in accordance with formula (3.8), then in the case of a incoherent signal, as one can easily verify, it will also transform the intensity linearly

$$I_1(\zeta_1) \mapsto I_2(\zeta_2) = \int G(\zeta_2 - \zeta_1) I_1(\zeta_1)\, d\zeta_1 \,, \tag{3.10}$$

where $G(\zeta) = |h(\zeta)|^2$. What is important for our purposes is that the convolution again appears in this circle of questions, as a result of which a natural return to the context of Fourier analysis occurs. It was through a formula of type (3.10) that the integral Fourier transform first penetrated the general theory of processing of optical signals.

Formula (3.10) finds numerous applications in the so-called *incoherent Fourier optics*. In particular it can also be used as the basis for fundamentally new designs of Fourier processors. Without developing this line of thought in detail, we limit ourselves to a description of one practical design for convolvers and correlators using formula (3.10). Incoherent processors have both advantages and disadvantages. In practice they are widely applied on an equal footing with coherent processors.

The design for a *correlator* that uses formula (3.10) is depicted in Fig. 18. According to this design, light passes through two photographic plates T_1 and T_2, the second of which can be moved in its plane, after which the light is focused on the photodetector by a lens in accordance with condition (2.31). The photodetector measures the full intensity of the light incident on it. If σ_1 and σ_2 are the squares of the transmittance coefficients of the plates, the current at the exit of the photodetector, taking account of the movement of the photographic plate T_2, is given by the formula

$$\int \sigma_1(\zeta')\sigma_2(\zeta + \zeta')\, d\zeta' \,.$$

This design will realize a convolution rather than a cross-correlator if one of the plates is reversed.

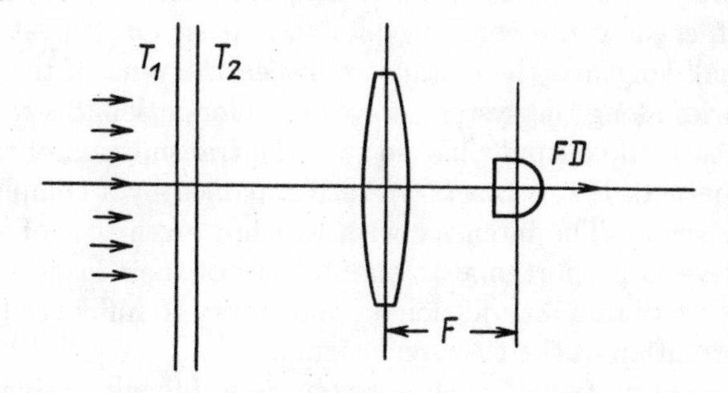

Fig. 18

§4. Acoustic and Acousto-Optical Fourier Processors

4.1. The Electro-Acoustic Transducer. A variety of different analog devices are used to obtain the Fourier image of a radio-electronic signal. Such devices, however, are almost never constructed of purely electronic elements. Acoustic components, specifically crystals in which elastic waves, i.e., (*ultra*)-*sonic waves*, can propagate at the frequency of the radio signals, are often introduced to build them into a system of electronic elements. In doing this one chooses crystals possessing piezoelectric properties. *Piezoelectric crystals* are characterized by the fact that the propagation of elastic waves in them is accompanied by the appearance of an electric field. The electric field in turn causes elastic deformations in piezoelectrics. The use of acoustic components relies mainly on the fact that the speed of sound is much less than the speed of propagation of an electronic signal. As a consequence radio-electronic components that would otherwise have to be tens of meters in size can be replaced by acousto-electronic elements only centimeters in size.

Various types of oscillations in crystals are used in practice and the necessary properties of the devices containing such crystals are sought in different ways. We shall describe briefly here a completely definite type of acousto-electronic device that can be regarded as an analog of optical lenses. These devices are called *Rayleigh-wave transducers* (RWTs) on *surface-acoustic Rayleigh waves* (SAWs). Rayleigh waves can propagate along the surface of an elastic crystal while dying out rapidly with depth. In a crystal occupying the half-space of (x, y, z)-space where $z > 0$ a monochromatic Rayleigh wave (ignoring its vector character) has the form

$$w = w_0 e^{-xz} e^{i\omega t - ik_1 x - ik_2 y} .$$

The base of the Rayleigh-wave transducer is a piezoelectric crystal cut in the form of a thin rectangular plate on whose longer edge a system of metal electrodes is attached by some means or other (cf. Fig. 19). The electronic signal is input at the contacts 1a and 1b. The stress on the pins, which can be considered equal along the whole system of pins attached to contacts 1a and 1b respectively, produces a Rayleigh wave that propagates along the system of pins in the direction of the contacts 2a and 2b. A given pair of neighboring input pins attached to the various contacts is most efficient at producing a wave whose half-length is the distance between the pins. If the distance between pins varies along the system, waves of different lengths travel different distances to reach the output pins, so that the transmission of the electronic signal from contacts 1 to contacts 2 is accompanied by a complicated deformation of the signal. The intensity with which a given pair of opposed pins produces a wave is proportional to the overlap of their lengths. By varying these overlaps we obtain an additional opportunity to influence the character of the transformation of the electronic signal.

The exact computation of such a system is a difficult problem of mathematical physics. There exists, however, a simple heuristic device for describing

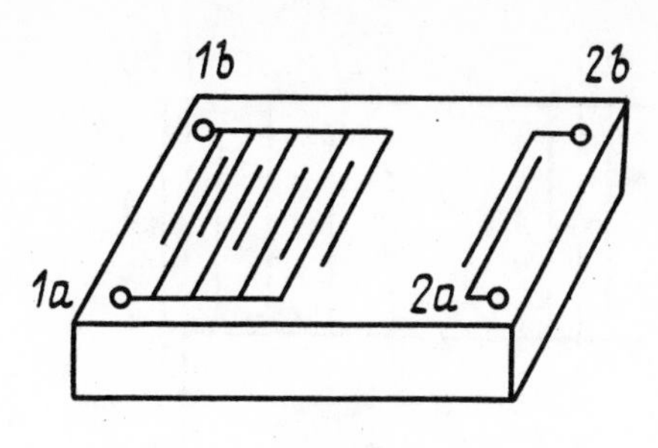

Fig. 19

the properties of this system. It is based on a principle according to which the input signal produces a system of linear sources of a Rayleigh wave situated midway between the opposite pins (cf. Fig. 20). The planar Rayleigh waves generated by these sources are assumed to propagate independently in the direction of the opposite pins. The time required for the plane wave produced by the pair of pins with numbers n and $n+1$ to reach the opposite pins is $t_n = v^{-1}(x_{\text{out}} - x_n)$, where v is the speed of propagation of the wave. According to the principle just described, a δ-shaped input electronic signal

$$u_1(t) = \delta(t)$$

is accompanied at the exit by a signal of the form

$$u_2(t) = h(t)\,, \quad h(t) = C \sum_{n=N_1}^{N_2} (-1)^n l_n \delta(t - t_n)\,. \tag{4.1}$$

Here C is a constant, and l_n is the length of the overlap of the pair of pins. The sign $(-1)^n$ reflects the fact that nearby linear sources produce fields with opposite phases. This is clear from the character of their mutual location. If the signal u_1 has general form, then the signal u_2 should be given by the formula

$$u_2(t) = \int_{-\infty}^{t} u_1(\tau) h(t-\tau)\, d\tau\,.$$

4.2. Determination of the Parameters of a Rayleigh-Wave Transducer. At this stage the purely analytic problem arises of finding numbers $N_2 - N_1$ and sequences $\{t_n\}_{n=N_1}^{N_2}$, $\{l_n\}_{n=N_1}^{N_2}$, $t_{n+1} < t_n$, $l_n > 0$, such that the function h has the necessary properties. What does it mean for h to have the necessary properties? We have said in the introduction that a typical electronic signal has the form

$$u(t) = e^{-i\omega t}\psi(t), \text{ more precisely}\,, \quad u(t) = \operatorname{Re}\left[e^{-i\omega t}\psi(t)\right],$$

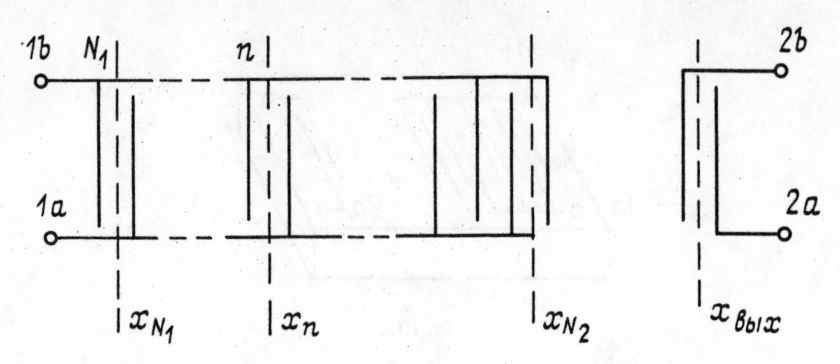

Fig. 20

where $\psi(t) = A(t)\exp[i\varphi(t)]$ is a function that varies slowly in comparison with $\exp(-i\omega t)$. In radio-electronics the typical function h should have the related structure

$$h(t) = \operatorname{Re}\left[A(t)e^{-i\omega t + i\varphi(t)}\right] . \tag{4.2}$$

It is clear that such a function h cannot be represented in the form (4.1) in the precise sense. We shall nevertheless trace through the computation proposed in the physical literature.

We choose a sequence t_n so that the equation

$$\chi(t_n) = n\pi \,, \quad \chi(t) = -\omega t + \varphi(t) \tag{4.3}$$

holds. We also assume that l_n can be described by the formula

$$l_n = L(t_n) \,, \tag{4.4}$$

where L is a certain smooth function. Then the function h will assume the form

$$h(t) = Ce^{i\chi(t)}L(t)|\chi'(t)| \sum\nolimits_n \delta(\chi(t) - n\pi) .$$

The summation extends over all integers, while the restriction $n \in [N_1, N_2]$ is taken care of by the condition

$$L(t) = 0, \quad \chi(t) \notin \pi[N_1, N_2] .$$

We use the Poisson summation formula

$$\sum\nolimits_n \delta(\chi - n\pi) = \frac{1}{\pi} \sum\nolimits_l e^{2il\chi} ;$$

$$h = \frac{C}{\pi} e^{i\chi} L|\chi'| \sum\nolimits_l e^{2il\chi} .$$

We now isolate the two terms in this expression with carrier frequency ω:

$$h(t) = \frac{2C}{\pi} L|\chi'| \cos\chi + \frac{C}{\pi} L|\chi'| \sum_{l \neq 0,1} e^{i(2l+1)\chi} . \tag{4.5}$$

It is proposed to identify the first term of formula (4.5) with the given function (4.2)

$$2CL|\chi'| = \pi A\,, \tag{4.6}$$

making it possible to determine the function L. The support of A (the function A must be of compact support) determines the support of L and, consequently the system of indices N_1, N_2.

The parasitic terms in formula (4.5) are eliminated technically by setting up the usual *electronic filter* at the exit from the Rayleigh-wave transducer to let through the electronic signal with *carrier frequency* ω and block signals with carrier frequencies 0 and $l\omega$, $l \geq 2$. Such a filter is a routine eletronic device.

A Rayleigh-wave transducer used as the analog of an optical lens is characterized by the function

$$h(t) = \zeta(\theta^{-1}t)\cos\left(\omega t - \frac{1}{2}\mu t^2\right)\,, \tag{4.7}$$

where μ is a certain constant and ζ is a smooth cutoff function whose form is reflected in Fig. 21. The parameter θ is analogous to the size of the lens. In what follows we shall neglect the finite sizes of Rayleigh-wave transducers, but just now, in discussing the technical parameters of Rayleigh-wave transducers, we shall take account of them.

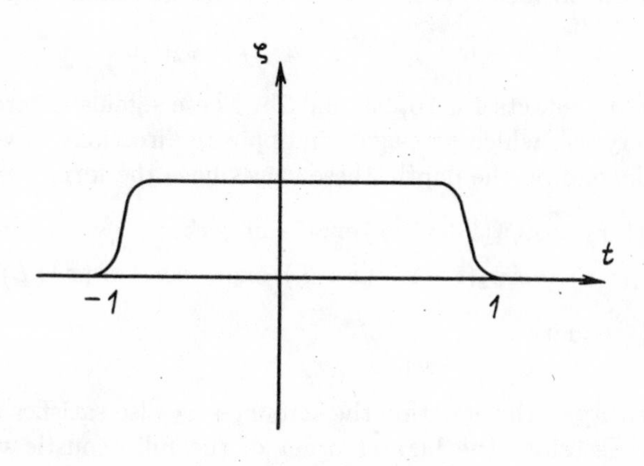

Fig. 21

According to (4.3)

$$\chi(t_n) = -\omega t_n + \frac{1}{2}\mu t_n^2 = n\pi\,; \tag{4.8}$$

according to (4.6)

$$L(t) = \frac{\pi}{2C}\zeta(\theta^{-1}t)|\omega - \mu t|^{-1}\,, \tag{4.9}$$

and the restrictions on the subscript n have the form $|t_n| \leq \theta$. It follows from (4.8) and (4.4) that

$$t_n = \frac{1}{\mu}(\omega - \sqrt{\omega^2 + 2\pi n\mu})\,, \tag{4.10}$$

$$Cl_n = \frac{\pi}{2}\zeta(\theta^{-1}t_n)|\omega^2 + 2\pi n\mu|^{-1/2}\,, \tag{4.11}$$

and the location of the pins is given by the coordinates

$$x_{\text{out}} - x_n = vt_n\,. \tag{4.12}$$

A real Rayleigh-wave transducer operating in a crystal of quartz ($v \approx 3 \cdot 10^3$m/sec) at frequency $f = \omega/2\pi = 14 \cdot 10^6$Hz with $\theta = 4 \cdot 10^{-6}$sec, $B = \frac{1}{2\pi}\mu\theta = 5.5 \cdot 10^6$Hz may have length of order 4cm and width around 1cm, and several hundred electrodes.

Thus we have described the structure of a device that outputs an electronic signal of the form (4.7) under the action of a pulsed input signal u_1. This device, a Rayleigh-wave acousto-electronic transducer, is the electronic analog of a lens.

4.3. Convolvers Based on Surface Acoustic Waves. The theoretical design of a typical *convolver*, i.e., a device designed to form a convolution, is depicted in Fig. 22. Two pairs of contacts with pins and two additional flat contacts are placed on the surface of a crystal plate. Radio-electronic signals

$$u_1(t) = \psi_1(t)e^{-i\omega t}\,, \quad u_2(t) = \psi_2(t)e^{-i\omega t}$$

are input at the contacts 1a, 1b, 2a, and 2b. These signals generate Rayleigh waves in the crystal, which propagate in opposite directions. If we ignore the factors that depend on the depth, these waves have the form

$$w_1(t,x) = A\psi_1(t - v^{-1}x)\exp(-i\omega t + ikx)\,, \quad k = v^{-1}\omega\,,$$
$$w_2(t,x) = A\psi_2(t + v^{-1}(x - L))\exp(-i\omega t + ik(x - L))$$

and satisfy the equation

$$w_{tt} - v^2 w_{xx} = 0\,.$$

By the linearity of the equation the sum $w_1 + w_2$ also satisfies the equation and correctly describes the highest order of the full acoustic wave field in the crystal. The convolver reacts, however, not to this sum, but to small corrections to it, which arise by virtue of the nonlinearity of the process of propagation of acoustic waves. The mechanism by which the corrections we are interested in arise can be understood starting from the equation

$$w_{tt} - v^2 w_{xx} = \alpha w^2\,, \quad |\alpha| \ll 1\,,$$

which has, of course, only a symbolic connection with the real nonlinear equation for the acoustic field. The small corrections w' to the sum $w_1 + w_2$ must be determined from the equation

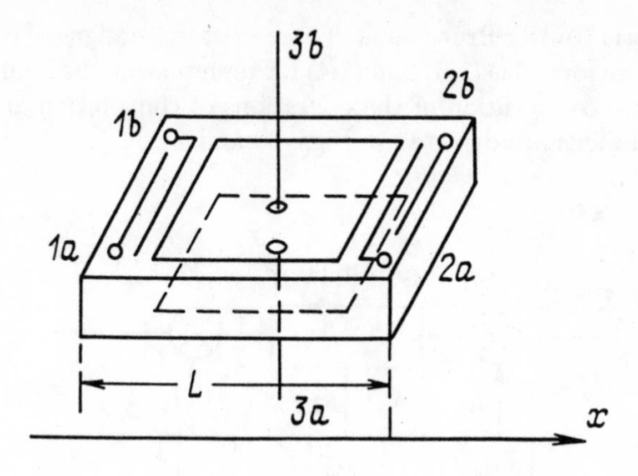

Fig. 22

$$w'_{tt} - v^2 w'_{xx} = \alpha(w_1 + w_2)^2 .$$

It is not difficult to isolate from this correction w' a term w'_0 that depends on t with double frequency 2ω and varies only slowly with x:

$$w'_0 \approx -\frac{\alpha A^2}{2\omega^2}\psi_1(t - v^{-1}x)\psi_2(t + v^{-1}(x - L))e^{-i(2\omega t + kL)} .$$

Since the flat contacts measure the integral

$$I = \gamma \int_0^L (w_1 + w_2 + w')\, dx ,$$

only the terms that vary slowly with x will make a significant contribution to the signal at the contacts 3. If in addition a filter is set up at the exit from the convolver which lets through only the carrier frequency 2ω, only the contribution made by w_0 to the integral I will be measured:

$$I \to -\frac{\alpha\gamma A^2}{2\omega^2} e^{-i(2\omega t + kL)} \int_0^L \psi_1(t - v^{-1}x)\psi_2(t + v^{-1}(x - L))\, dx .$$

If the supports of the signals u_1 and u_2 are sufficiently small, so that for some t the supports of the functions $\psi_1(t - v^{-1}x)$ and $\psi_2(t + v^{-1}(x - L))$ are bounded with respect to x by the interval $[0, L]$, then

$$I \to -\frac{\alpha v\gamma A^2}{2\omega^2} e^{-i(2\omega t + kL)} (\psi_1 * \psi_2)(2t - v^{-1}L) .$$

In what follows, in discussing convolvers, we shall ignore the inessential modification of the argument of the convolution that must be performed to bring the convolution into canonical form.

Thus, by means of devices that use surface acoustic waves, one can form the convolution of two signals.

4.4. Acoustic Fourier Processors. There exist two designs of Fourier processors that realize formulas (1.3) and (1.4) for representing the Fourier transform in terms of the composition of the operations of convolution and multiplication. These designs are depicted in Figs. 23 and 24.

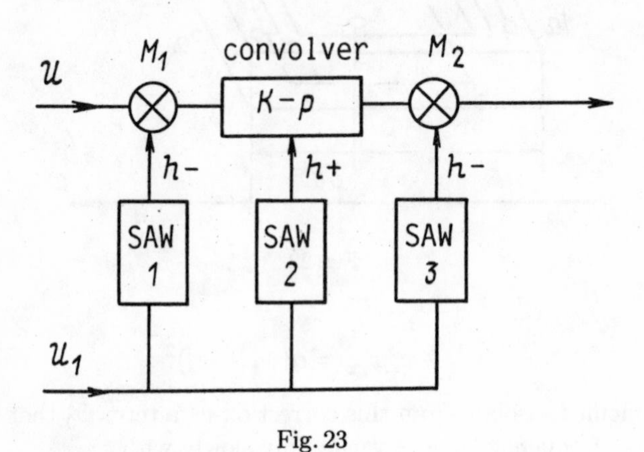

Fig. 23

A radio-electronic signal

$$u(t) = \psi(t)$$

and a pulsed signal $u_1(t) = \delta(t)$ are input at to the model depicted in Fig. 23. The latter signal is input directly to the Rayleigh-wave transducer 1, forming a response

$$h_-(t) = e^{-i\omega t - i\frac{\mu}{2}t^2} .$$

For the sake of a certain simplification in the exposition we have resorted to complex notation. The signals u and h_- meet in the *multiplier* M_1, a device that forms their product

$$u(t)h_-(t) = e^{-i\omega t - i\frac{\mu}{2}t^2}\psi(t) .$$

We say nothing about the technical structure of multipliers, since they are easily constructed using a variety of simple elements. As a multiplier one can use, for example, an ordinary triode vacuum tube or a transistor. The product $u \cdot h_-$ is then input to the convolver, at whose other entrance a signal arrives from the Rayleigh-wave transducer 2

$$h_+(t) = e^{-i\omega t + i\frac{\mu}{2}t^2} .$$

The convolution of the signals $u \cdot h_-$ and h_+ is formed in the convolver:

$$h_+ * (u \cdot h_-) = h_+ \cdot \mathcal{F}_{\mu^{-1}}[\psi] ,$$

It remains only to multiply the convolution and the signal h_- output from the Rayleigh-wave transducer 3 in the multiplier M_2. As a result, the output from the device is a signal

$$h_- \cdot (h_+ * (u \cdot h_-)) = e^{-2i\omega t} \mathcal{F}_{\mu^{-1}}[\psi] ,$$

which contains the Fourier transform of the function ψ. The practical device contains filters in some of its components to isolate the necessary carrier frequencies as well as the so-called blocking lines, which coordinate the arrival times of the signals at the Rayleigh-wave transducers and the multipliers.

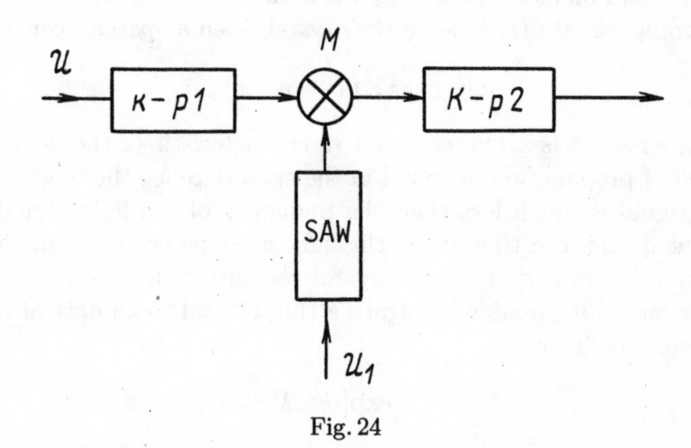

Fig. 24

The design depicted in Fig. 24 works similarly.

4.5. The Acousto-Optical Cell. The propagation of elastic waves in a transparent crystal is accompanied by a change in the refractive index of the crystal. This effect is called the *photoelastic effect.* Neglecting, as usual, the vector character of the processes occurring in the crystal, we can assume that the elastic wave field $w(t,r)$ generates a variable refractive index

$$n(t,r) = n_0 + \gamma w(t,r) ,$$

where n_0 is the equilibrium value of this index and γ is a certain constant—a characteristic of the substance. If the crystal in question is, in addition, piezoelectric, then by creating an electric field in it, we can influence the conditions under which a light beam passes through the crystal. These considerations are the foundation of the so-called *acousto-optical cells.* Acousto-optical cells can be regarded as optical filters controlled by an electronic signal. By inputting an electronic signal into the crystal it is possible to vary the characteristics of such a filter widely and rapidly. For this reason acoustic cells used as components of optical Fourier processors significantly espand their potential and make it possible to form a new class of devices for processing both electronic and optical signals. These devices have received the name of *acousto-optical processors.*

In this section we shall discuss the functioning of acoustic cells. In doing this we shall confine ourselves to general information, leaving aside many real properties of their operation. This, however, does not prevent us from giving some picture of the main uses of such cells to construct the Fourier transform and perform related operations on signals.

An acousto-optical cell should be thought of as a transparent piezocrystal plate to one of whose faces electric contacts are attached suitably and on whose opposite face an acoustic wave absorber is set up (cf. Fig. 25). The direction of the light beam incident on the crystal is shown by the arrow in the figure. If an electronic signal $u(t)$ falls on the crystal, then a *spatial acoustic wave*

$$w(t,r) = Au(t - v^{-1}x) ,$$

arises in it, where A is a real constant (a characteristic of the device), and v is the speed of propagation of sound in the crystal. Since the frequency of the electronic signal is much less than the frequency of the light signal, we can assume that during the time when the light wave passes through the crystal the refractive index $n(t,r)$ in the crystal remains constant (independent of time). This makes it possible to regard a thin crystal as an optical filter with transmittance coefficient

$$s(t,r) = B\exp[ik\gamma\Delta w(t,r)] ,$$

where Δ is the thickness of the crystal. This transmittance coefficient depends on time t, which, however, must be interpreted as an external parameter in the present situation. If, as usual

$$u = \mathrm{Re}\,[e^{-i\omega t}\psi(t)] ,$$

then

$$s(t,r) = B\exp[ik\gamma\Delta A\mathrm{Re}\,[e^{-i\omega(t-v^{-1}x)}\psi(t - v^{-1}x)]] .$$

This means that the acousto-optical cell is a phase grating whose parameters vary relatively weakly along the x-axis.

The properties of phase gratings were discussed in § 3. There we proposed distinguishing between thin and thick crystals. In the present case of thin crystals we shall confine ourselves to the following approximation for the transmittance coefficient

$$s(t,r) \approx B[1 + ik\gamma\Delta \cdot A\mathrm{Re}\,[e^{-i\omega(t-v^{-1}x)}\psi(t - v^{-1}x)]] .$$

4.6. Acousto-Optical Processors. We shall discuss the designs of several devices in which acousto-optical cells are used.

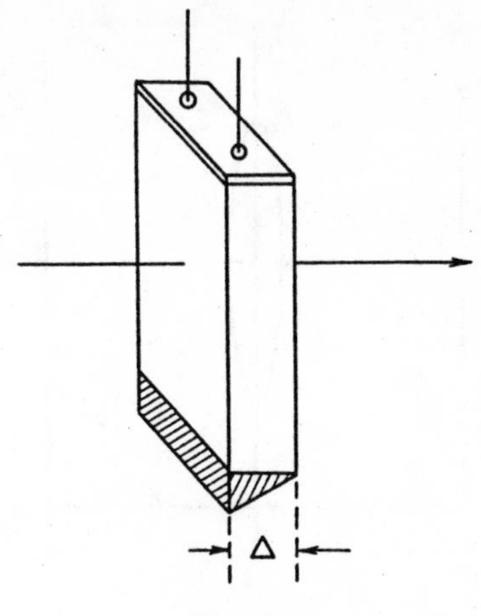

Fig. 25

The Fourier Transform. The Fourier transform of a radio-electronic signal can be obtained using the design depicted in Fig. 26 and containing an acousto-optical cell and a cylindrical lens. In the plane P_3 this system forms spatially separated Fourier images of the functions $\psi(t - v^{-1}x)$ and $\psi^*(t - v^{-1}x)$. The Fourier images depend on the time t and therefore must be measured at fixed instants of time. It is clear also that because of the boundedness of a cross-section of a beam it is actually the Fourier transforms of the functions $\psi(t - v^{-1}x)\chi(x)$ and $\psi^*(t - v^{-1}x)\chi(x)$ that are formed, where χ is the characteristic function of the entrance aperture. Frequently it is just such a local Fourier transform that is of particular interest in analyzing radio-electronic signals.

A Cell as a Lens. Assume that an electronic signal of the form

$$u(t) = \cos\left(\omega t - \frac{1}{2}\mu t^2\right)$$

is input to the cell depicted in Fig. 25, so that

$$\varphi(t) = e^{-\frac{i}{2}\mu t^2} .$$

For two noncentered rays emanating from the cell the transmittance coefficients turn out to be equal (up to constant multiples) to

$$e^{-i\frac{1}{2}\mu(t-v^{-1}x)^2} \text{ and } e^{i\frac{1}{2}\mu(t-v^{-1}x)^2} .$$

This result can be interpreted as follows. On each of these rays the cell acts on the optical signal as a gathering (resp. scattering) *lens* with focal length $F = k\mu^{-1}$, and the axis of the lens moves with velocity v along the x-axis.

Convolvers and Correlators. There exists a large set of different acousto-optical designs by means of which it is possible to perform the convolution and

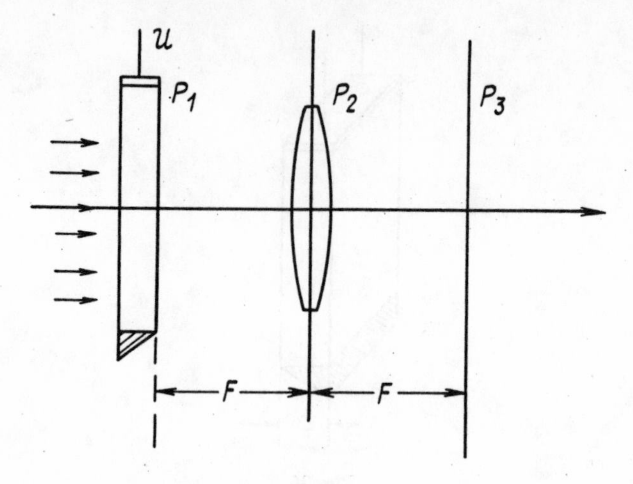

Fig. 26

the cross-correlation directly, without turning these operations into Fourier images. These designs, in contrast, can be used successfully to construct the Fourier transform independently. Each different version of such designs has its own technical or technological advantages or disadvantages, which of course we shall not discuss. We limit ourselves to a description of one typical design. This design is depicted in Fig. 27. In this design a holographic filter with transmittance coefficient

$$s(x) = 1 + \mathrm{Re}\,[r(x)e^{ik_1x}]$$

and an acoustic cell to which a signal

$$u(t) = \mathrm{Re}\,[e^{-i\omega t}\psi(t)]\ ,$$

$\omega = vk_1$, is input, are set up one immediately behind the other. We shall isolate the terms containing the exponential $\exp(ik_1x)$:

$$\frac{1}{2}e^{ik_1x}[r(x) + ik\Delta\gamma A\psi(t - v^{-1}x)\exp(-i\omega t)]\ .$$

In the plane P_3 the lens separates the contributions of the rays corresponding to different powers of the exponent $\exp(ik_1x)$, or more precisely, their Fourier transforms. The aperture shown in the figure admits only the isolated terms. The *photodetector* measures the full intensity of the signal that has passed through the aperture, i.e., it computes its L_2-norm. Since the Fourier transform preserves the L_2-norm, the electronic signal measured at the exit from the photodetector is (up to a constant factor)

$$\int dx|r(x) + ik\Delta\gamma A\psi(t - v^{-1}x)\exp(-i\omega t)|^2\ .$$

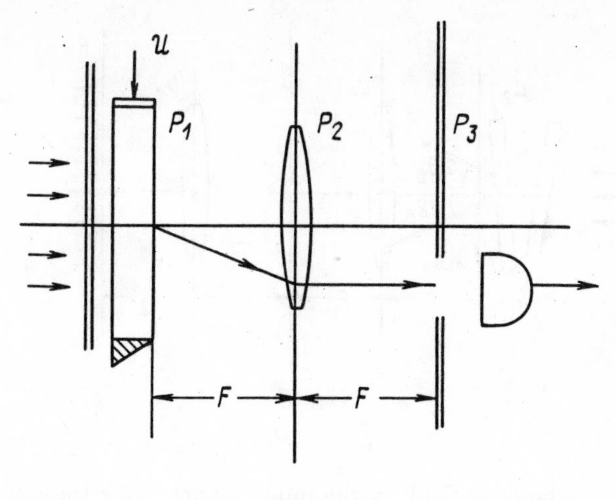

Fig. 27

It consists of two terms with carrier frequencies 0 and ω. These can be separated by means of a standard electronic filter. The second of them has the form

$$2\mathrm{Re}\, ik\gamma\Delta A e^{-i\omega t}\int \psi(t-v^{-1}x)r(x)\,dx$$

and thus gives the convolution $\psi * r$.

Reversing the transparent, we obtain the cross-correlation instead of the convolution.

The design just described makes it possible to measure the convolution of an optical and an electronic signal. Setting up a second cell instead of the holographic filter, one can measure the convolution of two electronic signals.

From the applied point of view finding the cross-correlation of two electronic signals is an extremely important operation, widely applied, as already noted, in distinguishing weak signals against background noise. In conclusion we shall give a very brief description of yet another design that uses thick acousto-optical cells in which the cross-correlation is formed. In contrast to the preceding design, which was based on the procedure of "spatial integration," the new design is based on the procedure of "time integration." Specific estimates show that it makes possible an enlargement of the interval of integration in the expression for the cross-correlation by nearly three orders of magnitude. The design is depicted in Fig. 28.

This design contains two thick crystals set at the Bragg angle θ_B to the axis of the incident wave and located one directly behind the other. Behind the crystals an optical device is set up that carries out spatial filtering. In the final plane of this design there is a chain of photodetectors that measure the full magnitude of the charge accumulated in each photodetector under

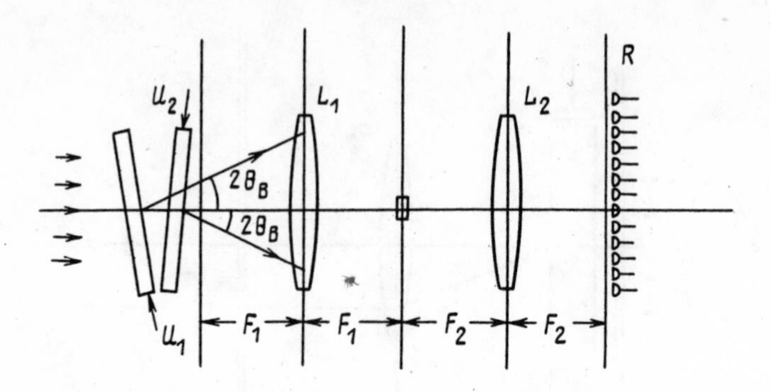

Fig. 28

the influence of the light field; in the intermediate plane there is a mask that blocks the central ray.

Electronic signals u_1 and u_2 are input to the crystals and generate modulated plane waves in these crystals which propagate toward each other (cf. Fig. 28). A plane light wave is also directed into the system. According to (3.7) each of the crystals splits a light ray propagating along the axis of the system into two rays. The electronic signals and the corresponding acoustic fields in the crystals are weak, and therefore the "direct" rays pass through the crystals with almost no transformation, not noticing the acoustic fields. The "reflected" rays carry weak signals proportional to the acoustic fields in the corresponding crystals, cf. (3.7). The ray reflected in the first crystal passes the second crystal with almost no changes, since it meets it at an angle that differs significantly from the Bragg angle. The spatial filtering device outputs an optical field of the form

$$u(t,x) = B\{\mathrm{Re}\,[u_1(t - v^{-1}x)]\exp(-ik_0 x\sin 2\theta_B) + \\ + \mathrm{Re}\,[u_2(t + v^{-1}(x - L))]\exp(ik_0 x\sin 2\theta_B)\}\,,$$

where L is the length of the crystal.

The chain of photodetectors makes it possible to measure the charges on them, i.e., the quantity

$$Q(x) = \int_0^T |u(t,x)|^2\,dt\,,$$

where T is the measuring time. The time T is limited by the properties of the photodetectors. If we isolate the components of the expression $Q(x)$ that contain various rapid spatial oscillations, then the cross-correlation is formed when the oscillation is $\exp(-2ik_1x)$ (with the integration restricted to the interval $[0,T]$). It is not difficult to separate from the general field.

Comments on the References

Because of the popular character of the present article the list of literature below contains only three types of sources: textbooks or popularizations in textbook style, surveys intended for a wide circle of specialists, and a few original works referred to in the text, mostly early ones. Only a few textbooks and surveys are given. The principle of selection was very simple: all were to some degree used in the writing of the article and seemed to the author to be suitable for mathematicians and specialists in analysis and mathematical physics. The author has the impression that the literature represented here gives a true picture of the state of the problems in the early 1980's. As bibliographical sources we can recommend the monograph *Radio Signal Processing by Acousto-Electronic and Acousto-optical Devices*, (Parygin and Balakshii 1987), (Casasent 1978), (Jack et al. 1980), (Lee 1981), *Optical Computing Technics*, and (Steward 1983). For the convenience of the reader we distinguish the following groups in the list of literature: Fourier optics—textbooks and popularizations (Born and Wolf 1964), (Casasent 1978), (Lee 1981), (Steward 1983); Fourier optics—surveys (Vasilenko and Taratorin 1966), (Gurevich et al. 1978), (Parygin and Balakshii 1987), (Soroka 1981), (Tarasov 1985), (Timofeev et al. 1985), (Born and Wolf 1964), (Casasent 1978), (Collier et al. 1971), (Cowley 1975), (Duffieux 1964), (Goodman 1968), (Lee 1981), (Maréchal and Françon 1962), *Optical Computing Techniques*, (Stark 1982), (Steward 1983), (Stroke 1966); Fourier optics—early works (Khalfin et al. 1973), (Abbe 1904), (Elias et al. 1952), (Elias 1953), (Maréchal 1953), (O'Neill 1956), (Porter 1906), (Shannon 1949), (Vander Lugt 1963), (Wiener 1949), (Zernike 1942); acoustic Fourier processors (Krasil'nikov and Krylov 1964), *Processing of Radio signals by Acousto-electronic and Acousto-Optical Devices*, (Dieulesaint and Royer 1974), (Jack et al. 1980), (Matthews 1977), *Surface acoustic waves*; Acousto-Optical Fourier processors (Kulakov 1978), (Magdich and Molchanov 1978), *Processing of Radio Signals by Acousto-Electronic and Acousto-Optical Devices*, (Parygin and Balakshii 1987), (Marom 1977), (Stark 1982).

References*

Abbe, E. (1904): Beiträge zur Theorie des Microskops und der microscopischen Wahrnehmung. Arch. f. Microskop. Anat. *9*, 413–468

Born, M., Wolf, E. (1964): Principles of Optics. Pergamon Press, Oxford

Casasent, D., Ed. (1978): Optical Data Processing. Applications. In: Topics in Applied Physics *23*. Springer-Verlag, Berlin - Heidelberg - New York

Collier, R.J., Burckhardt, C.B., Lin, L.H. (1971): Optical Holography. Academic Press, New York

Dieulesaint, E., Royer, D. (1974): Ondes Elastiques dans les Solides. Masson, Paris

Duffieux, P.M. (1964): L'intégrale de Fourier et ses Applications à l'Optique. Masson, Paris

* For the convenience of the reader, references to reviews in *Zentralblatt für Mathematik* (Zbl.), compiled using the MATH database, have, as far as possible, been included in this bibliography.

Elias, P., Grey, D., Robinson, D. (1952): Fourier treatement of optical processes. J. Opt. Soc. Amer. *42*, 127–134
Elias, P. (1953): Optics and Communication Theory. J. Opt. Soc. Amer. *43*, 229–232
Goodman, J.W. (1968): Introduction to Fourier Optics. McGraw-Hill, New York
Gurevich, S.B., Konstantinov, V.B., Sokolov, V.K., Chernykh, D.F. (1978): Transmission and Processing of Information by Holographic Methods. Soviet Radio, Moscow (Russian)
Jack, M.A., Grant, P.M., Collins, J.H. (1980): The theory, design and applications of surface acoustic wave Fourier-transform processors. Proc. IEEE *68*(4), 450–468
Khalfin, L.A., Pavlichuk, T.A., Shul'man, M.Ya. (1973): Experiments on the Improvement of Image Quality. Optics and Spectroscopy, *35*(4), 766–768 (Russian)
Krasil'nikov, V.A., Krylov, V.V. (1964): Introduction to Physical Acoustics. Nauka, Moscow (Russian)
Kulakov, S.V. (1978): Acousto-Optical Devices of Spectral and Correlation Signals. Nauka, Leningrad (Russian)
Lee, S.H., Ed. (1981): Optical Information Processing. Fundamentals. In: Topics in Applied Physics *48*, Springer-Verlag, Berlin - Heidelberg - New York
Magdich, L.N., Molchanov, V.Ya. (1978): Acousto-optical Devices and their Application. Soviet Radio, Moscow (Russian)
Maréchal, A. (1953): Contrastes des images de quelques objets types dans un instrument stigmatique. Rev. Opt. *32*, 649
Maréchal, A., Françon, M. (1962): Diffraction structure des images. Rev. d'Opt. Théor. et Instr.
Marom, E., Ed. (1977): Optical Communications Using Surface Acoustic Waves. Pergamon Press, Oxford
Matthews, Ed. (1977): Surface Wave Filters. Wiley, New York
McLachlan, N.W. (1951): Theory and Applications of Mathieu Functions. Clarendon Press, Oxford
O'Neill, E.L. (1956): Spatial filtering in optics. IEEE Trans. Inf. Theory, IT-2, 56–65
Optical Computing Technics. Proc. IEEE *65*(1)
Parygin, V.N., Balakshij, V.N. (1987): Optical Information Processing. Moscow University Press (Russian)
Porter, A.B. (1906): On the diffraction theory of microscopic vision. Phil. Mag., Ser. 6, *11*, 154–166
Radio Signal Processing by Acousto-Electronic and Acousto-Optical Devices. Nauka, Leningrad 1983 (Russian)
Shannon, C. (1949): The Mathematical Theory of Communication. University of Illinois Press, Urbana, Zbl. 41,258
Soroka, L.M. (1981): Hilbert Optics. Nauka, Moscow (Russian)
Stark, H., Ed. (1982): Applications of Optical Fourier Transforms. Academic Press, New York
Steward, E.G. (1983): Fourier Optics. An Introduction. Ellis Horwood, Ltd., New York
Stroke, G.W. (1966): An Introduction to Coherent Optics and Holography. Academic Press, New York
Surface Acoustic Waves. Proc. IEEE *64*(5), (1976)
Tarasov, L.V. (1985): Lasers: Reality and Hope. Nauka, Moscow (Russian)
Timofeev, Yu.P., Fridman, S.A., Fok, M.V. (1985): Transformation of Light. Nauka, Moscow (Russian)
Vander Lugt, A.V. (1963): Signal Detection by Complex Spatial Filtering. Repts. Rad. Lab. Univ. Mich., No. 4594-22-C
Vasilenko, G.N., Taratorin, A.M. (1966): Reproduction of Images. Radio and Network, Moscow (Russian)

Wiener, N. (1949): The Extrapolation, Interpolation, and Smoothing of Stationary Time Series. Technology Press and Wiley, New York, Zbl. 36,79

Zernike, F. (1942): Phase contrast: a new method for the microscopic observation of transparent objects. Physica *9*, 688–698, 974–986

III. The Uncertainty Principle in Harmonic Analysis

V. P. Havin and B. Jöricke

Translated from the Russian
by Roger Cooke

Contents

Introduction 180

Chapter 1. The Uncertainty Principle Without Complex Variables ... 181

§1. Functions and Charges with Semibounded Spectrum. The Theorem of F. and M. Riesz 181
1.1. Plus- and Minus-functions. Hardy Classes 181
1.2. The Theorem of F. and M. Riesz 182
1.3. The Sharpness of the Theorem of F. and M. Riesz 183
1.4. Two Quantitative Improvements of (Part I of) the Theorem of F. and M. Riesz 184
1.5. Jensen's Inequality (A Quantitative Improvement of Part II of the F. and M. Riesz Theorem) 187
1.6. $\mathcal{R}$-Sets and $\mathcal{D}$-sets 190
§2. Singular Measures and Fourier Transforms that Tend to Zero 192
2.1. Some Classes of Measures and Two-sided Sequences 192
2.2. The Cantor Set and the Cantor Measure. The Theorem of Bari and Salem 193
2.3. Riesz Products 195
2.4. The Ivashev-Musatov Theorem 198
§3. Hilbert-Space Methods 201
3.1. Definitions and Statement of the Problem 202

3.2. Annihilation of Projections 203
3.3. The Amrein-Berthier Theorem 204
3.4. Strong Annilihation of Subsets of the Circle and Sparse Spectra 206
3.5. Supports that are Strongly Annihilating with any Bounded Spectrum 206

Chapter 2. Complex Methods 207

§1. First Examples of the Application of Complex Methods 208
1.1. Limiting Rate of Decay of the Fourier Transform of a Rapidly Decreasing Function. Djrbashyan's Theorem 208
1.2. High-order Zeros and Sparse Spectrum. Mandelbrojt's Theorem 211
§2. The Uncertainty Principle for Plus-functions from the New Point of View 214
2.1. Plus-charges on the Circle $\mathbb{T}$ 214
2.2. Jensen's Inequality and the Theorem of F. and M. Riesz Theorem (Complex Approach) 217
2.3. Hardy Classes in a Half-plane. Plus-charges on the Line 218
§3. Divergence of the Integral of the Logarithm and Certain Forms of the Uncertainty Principle for Functions and Charges with Rapidly Decreasing Amplitudes 220
3.1. Infinite-order Zeros and Decay of Amplitude 221
3.2. Sketch of the Proof 222
3.3. The Connection with the Moment Problem and Weighted Approximation 223
3.4. One-sided Decay of Amplitudes. Beurling's Theorem 224
3.5. The Levinson-Cartwright Theorem 225
§4. One-sided Decay of Amplitudes and Convergence of the Logarithmic Integral 228
4.1. Vol'berg's First Theorem 228
4.2. Almost-analytic Functions. Vol'berg's Second Theorem 229
4.3. Dyn'kin's Theorem on Almost-analytic Continuation and Reduction of the First Vol'berg Theorem to the Second 229
4.4. On the Proof of Vol'berg's Second Theorem 231
4.5. The Borichev Approach 233
§5. Some Forms of the Uncertainty Principle for Charges with a Spectral Gap 235
5.1. Spectral Gaps and Orthogonality to Certain Entire Functions. The Pollard Function 236
5.2. Applications to the Uncertainty Principle for Charges with a Spectral Gap 237
5.3. Spectral Gaps and Sparseness of Support 238

5.4. Sapogov's Problem (on the Characteristic Function of a Set with a Spectral Gap) ... 240
5.5. De Brange's Theorem on Extreme Points ... 241
5.6. On Fabry's Theorem ... 241
§6. Some Methods of Constructing Small Functions with Small Fourier Transform ... 242
6.1. Outer Functions ... 242
6.2. Determining Majorants and the Logarithmic Integral ... 243
6.3. One-sided Polynomial Decay of Amplitudes and Smallness of the Support. Hruščev's Theorem ... 243
§7. Two Theorems of Beurling and Malliavin ... 245
7.1. Statement of the Problem ... 245
7.2. Admissible Majorants ... 246
7.3. On the Second Beurling-Malliavin Theorem ... 248
Commentary on the References ... 249
References ... 252

Introduction

The Fourier transform assigns to a function x defined on the real line $\mathbb{R}$ its "spectral representation" $\hat{x}$:

$$\hat{x}(\xi) = \frac{1}{2\pi} \int_{\mathbb{R}} x(t) e^{-it\xi}\, dt \quad (\xi \in \mathbb{R}) .$$

In many applications x can be interpreted as a "function of time." The functions x and $\hat{x}$ determine each other completely. Therefore the system of equations

$$\left.\begin{array}{l} x = f\,, \\ \hat{x} = g\,, \end{array}\right\}$$

(with unknown function x) is strongly overdetermined. In other words the temporal and spectral properties of the function x cannot be prescribed arbitrarily. One of the manifestations of their rigid interdependence, which is encountered in solving many mathematical, engineering, and physical problems, is the following: the more concentrated in time the function x is, the more smeared its spectral representation $\hat{x}$, and conversely. Putting it another way,

$$\text{if } x \not\equiv 0, \text{ then the functions } x \text{ and } \hat{x} \text{ cannot both be very small}\,, \qquad (*)$$

the word "small" being understood in a very broad sense. Thus we shall say that if both quantities $|x|$ and $|\hat{x}|$ decay very rapidly at infinity or if both have compact support, then they must both vanish identically. There are many ways of giving a precise meaning to the heuristic assertion (*). The fundamental quantum-mechanical uncertainty principle of Heisenberg (more exactly, its mathematical model) is one of them. In this case x and $\hat{x}$ stand for "coordinate" and "impulse".

In this article the term "uncertainty principle" does not have a very specific meaning, but is used to signify a large collection of mathematical facts that illustrate the assertion (*).

This collection, however, is not complete, since it omits entirely the operator interpretations of the uncertainty principle of Heisenberg connected with commutation relations (Tikhomirov 1986, von Neumann 1932). It also omits other known connections with mathematical physics (cf. in particular (Fefferman 1983)). We confine ourselves to purely analytic variations on the theme (*). All of these lie strictly within classical harmonic (and partly complex) analysis.

Chapter 1
The Uncertainty Principle Without Complex Variables

The most profound and delicate results relating to the uncertainty principle have been obtained using complex analysis. However we shall begin by talking about versions of the uncertainty principle established by purely real-variable methods.

§1. Functions and Charges with Semibounded Spectrum. The Theorem of F. and M. Riesz

We shall agree to use the letter X to denote either the real line $\mathbb{R}$ or the unit circle $\mathbb{T} := \{\zeta \in \mathbb{C} : |\zeta| = 1\}$ and the letter Ξ to denote either $\mathbb{R}$ or $\mathbb{Z}$ respectively. Thus the Fourier transform $\hat{f}$ of a function f defined on X is defined on Ξ:

$$\hat{f}(\xi) = \int_X f e^{-i\xi x}\, dm(x) \quad (\xi \in \Xi)\,.$$

Here m denotes *Lebesgue measure* on $\mathbb{R}$ if $X = \mathbb{R}$ and *normalized Lebesgue measure* on $\mathbb{T}$ if $X = \mathbb{T}$ (so that $m(\mathbb{T}) = 1$).

The symbol $M(X)$ will denote the set of all complex countably additive set functions defined on the Borel σ-algebra of the space X. Elements of the set $M(X)$ will be called *charges* and nonnegative charges will be called *measures.*

The topic of the present section is the uncertainty principle for charges $\nu \in M(X)$ possessing the following property: there exists a number $A \in \mathbb{R}$ such that $\hat{\nu}(\xi) = 0$ for all $\xi \in \Xi \cap (-\infty, A)$ (or for all $\xi \in \Xi \cap (A, +\infty)$). Putting it another way, we shall be interested in charges with semibounded spectrum. We shall usually assume that $A = 0$, since the general case can easily be reduced to this one.

1.1. Plus- and Minus-functions. Hardy Classes. We shall agree to call a distribution $f \in S'(\mathbb{R})$ a *plus-function* (resp. *minus-function*) if $\hat{f}|_{(-\infty,0)} = 0$ (resp. $\hat{f}|_{(0,+\infty)} = 0$). In the case $X = \mathbb{T}$ the concept of plus- and minus-functions is introduced as follows: a distribution $f \in S'(\mathbb{T})$ is called a *plus-function* if $\hat{f}|_{\mathbb{Z}\cap(-\infty,0)} = 0$ and a *minus-function* if $\hat{f}|_{\mathbb{Z}\cap(0,+\infty)} = 0$. The set of all plus-functions belonging to $L^p(X, m)$ $(1 \le p \le +\infty)$ is denoted $H^p(X)$ and is called the *Hardy class* (with index p).

Plus-functions naturally arise in the mathematical description of physical situations connected with the representation of the past and future and with the principle of causality ("the consequences of an action cannot occur before the action itself"). Their very definition contains a certain requirement of spectral smallness: for a plus-function half of the spectrum (the negative half) is missing. It is therefore not surprising that a nonzero plus-function cannot be very small. It is remarkable, however, that one can obtain a very simple and complete characterization of the size of a set on which no nonzero *plus-charge* (function in a Hardy class) can vanish. And such a characterization turns out to be the property of having positive length. Results illustrating the uncertainty principle by no means always have such a definitive form.

1.2. The Theorem of F. and M. Riesz. We shall write

$$\nu_1 \ll \nu_2 ,$$

if $\nu_1, \nu_2 \in M(X)$ and the measure $|\nu_1|$ is absolutely continuous with respect to the measure $|\nu_2|$.

A) Let ν be a nonzero plus-charge. Then

$$\nu \ll m \text{ and } m \ll \nu .$$

This is the proposition known as the *Theorem of F. and M. Riesz.* It follows from it that any plus-charge coincides (as a distribution) with some function in $L^1(X)$ and hence belongs to $H^1(X)$. Therefore the theorem of F. and M. Riesz can also be stated as follows:

I. Every plus-charge is absolutely continuous with respect to Lebesgue measure;
II. A function of class $H^1(X)$ that vanishes on a set of positive measure vanishes identically.

B) We shall prove assertion II in a special case. Suppose that $f \in H^2(\mathbb{T})$ and let $E := \{t \in \mathbb{T} : f(t) = 0\}$. We shall show that if $m(E) > 0$, then $f = 0$. To do this we consider the closure M (in $L^2(\mathbb{T})$) of the linear span of the sequence of functions $(z^j f_j)_{j=0}^{\infty}$, where z denotes the identity mapping of the circle $\mathbb{T}$. We shall verify that

$$zM = M . \tag{1}$$

The inclusion $zM \subset M$ is obvious. The set zM is closed in $L^2(\mathbb{T})$, since the operator $h \mapsto zh$ is unitary. Therefore it suffices to prove that the set zM is dense in M. Assuming the contrary, we find a nonzero function $g \in M$ orthogonal to zM and thus to all functions $z^j g$ $(j = 1, 2, \ldots)$. In such a case $\widehat{|g|^2}(j) = \int_{\mathbb{T}} \bar{z}^j |g|^2 \, dm = 0$ for any integer $j \neq 0$, so that the function $|g|^2$ is constant; if $m(E) > 0$, then g (as an element of the set M) vanishes almost everywhere on E, and so $g = 0$. Thus (1) is proved. Therefore if $h \in M$, then $h = zh_1$, where $h_1 \in M$; but then $\hat{h}(0) = 0$. In this case

$h \in M \Rightarrow \hat{h}(0) = \hat{h}(1) = 0$, etc. In particular $\hat{f}(j) = 0$ ($j \in \mathbb{Z}_+$) and $f = 0$ if $m(E) > 0$.

C) We shall discuss the proof of assertion I (for $X = T$). We define a *plus-polynomial* to be any function of the form $\sum_{N \geq k \geq 0} c_k z^k$. Let ν be a plus-charge. Then the charge $z^{-1}\nu$ is orthogonal to all powers z^j with $j \in \mathbb{Z}_+$ and hence to all plus-polynomials:

$$\int_{\mathbb{T}} pz^{-1}\, d\nu = 0, \quad \text{if } p \text{ is a plus-polynomial}.$$

Assume that the characteristic function χ_A of a set $A \subset \mathbb{T}$ can be approximated by plus-polynomials $(p_j)_{j=1}^{\infty}$ in the following sense:

$$\lim_{j\to\infty} p_j(\zeta) = \chi_A(\zeta) \quad \text{for any } \zeta \in \mathbb{T}, \quad \sup_j \|p_j\|_\infty < +\infty .$$

Then obviously $\int_A z^{-1}\, d\nu = 0$.

It follows from the proposition proved in B) that the construction of a sequence such as (p_j) is not feasible if $m(A) > 0$ and $m(\mathbb{T} \setminus A) > 0$. Indeed, the sequence (p_j) would have to converge to χ_A in $L^2(\mathbb{T})$, and then $\chi_A \in H^2(\mathbb{T})$. However, if the set A is closed and $m(A) = 0$, it is possible to construct the required sequence (p_j) (more or less explicitly). Therefore $z^{-1}\nu \ll m$ and then $\nu \ll m$ also.

We confine ourselves to these remarks. The details of the proof can be found in (Doss 1981) (cf. also (Bari 1961) and (Privalov 1950)). We shall return to the theorem of F. and M. Riesz in Sect. 2 of this chapter, where we shall discuss its "complex" proof, which was historically the first proof of it.

1.3. The Sharpness of the Theorem of F. and M. Riesz. Assertions II of 1.2A) are sharp in the following sense: if a set $E \subset \mathbb{T}$ is closed and $m(E) = 0$, there exists a continuous plus-function (of class $H^1(\mathbb{T})$) equal to zero on all of E. Thus the length of a set turns out to be the correct characterization of its size when speaking about the uncertainty principle for plus-functions. True, it is necessary to stipulate that this assertion holds only as long as one is speaking about functions in the classes $H^p(X)$ and continuous plus-functions. For classes of distributions larger than $H^1(X)$ or smaller than $C(X) \cap H^1(X)$ (for example for $C^1(X) \cap H^1(X)$) one must look for characterizations of the size of sets other than length (cf. for example, Sect. 1.5E) below).

Here we limit ourselves to the class $H^1(\mathbb{T}) \cap C(\mathbb{T})$ (i.e., the class of continuous plus-functions on $\mathbb{T}$); we shall denote it by the symbol $C_+(\mathbb{T})$ $(:= C_+)$.

A) Instead of $f\big|_A$, where $A \subset \mathbb{T}$ and f is a function defined on $\mathbb{T}$ we shall write $R_A f$.

Theorem. (Rudin-Carleson) *If the set A is closed and $m(A) = 0$, then*

$$R_A(C_+(\mathbb{T})) = C(A) . \tag{2}$$

Equality (2) means that any function that is continuous on the set A can be extended to a continuous plus-function. What was said above about the existence of nonzero continuous functions of class $H^1(\mathbb{T})$ vanishing on a given closed set of length zero follows easily from the Rudin-Carleson theorem.

B) It is not difficult to deduce the Rudin-Carleson theorem from the theorem of F. and M. Riesz using certain general considerations. The space C_+ is a closed subspace of the space $C(\mathbb{T})$ $(:= C)$. Therefore the dual space $(C_+)^*$ of the space C_+ is isometrically isomorphic to the quotient space $C^*/C_+^\perp$, where $C_+^\perp := \{l \in C^* : l|_{C_+} = 0\}$. The space C^* is isometrically isomorphic to the space $M(\mathbb{T})$ $(:= M)$ (that is, there is an isomorphism J that assigns to a charge $\mu \in M$ the functional $l_\mu \in C^*$, where $l_\mu(f) := \int_{\mathbb{T}} f\, d\mu$ $(f \in C)$). We set $M_+^0 := J^{-1}(C_+^\perp)$. The isomorphism J makes it possible to identify $(C_+)^*$ with the quotient space M/M_+^0. It is easy to see that M_+^0 consists of plus-charges that, according to the theorem of F. and M. Riesz, are absolutely continuous with respect to m. Therefore for any charge $\mu \in M$ concentrated on a set of length zero we have

$$\|\mu + M_+^0\| \geq \operatorname{var} \mu$$

(here $\mu + M_+^0$ is an element of the quotient space M/M_+^0 and contains μ). By a theorem of Banach (Kantorovich and Akilov 1977, p. 459) Eq. (2) is equivalent to the following assertion:

$$\|R_A^* l\| \geq c\|l\|, \quad \text{for any functional } l \in (C(A))^* , \tag{3}$$

where c is a positive number independent of l (we regard R_A as a transformation from C_+ into $C(A)$). If $l \in C(A)^*$, there exists a charge $\mu \in M$ concentrated on A such that $l(f) = l_\mu(f)$ $(f \in C(A))$, $\|l\| = |\mu|(A) = \operatorname{var} \mu$. Further $(R_A^* l)(f) = l(R_A f) = \int_{\mathbb{T}} f\, d\mu = (\mu + M_+^0)(f)$ for any function $f \in C_+$ (we are identifying the class $\mu + M_+^0$ with the functional $l_\mu \in C_+^*$). Therefore $\|R_A^* l\| \geq \operatorname{var} \mu = \|l\|$, and assertion (3) is proved (moreover $c = 1$); this guarantees the possibility of extending any function $f \in C(A)$ to a function $g \in C_+$ such that $\|g\|_C \leq (1 + \varepsilon)\|f\|_{C(A)}$ for any $\varepsilon > 0$. In reality one can say that $\varepsilon = 0$; moreover there exists a linear operator $S : C(A) \to C_+$ such that $\|S\| = 1$ and $R_A S = \mathrm{Id}_{C(A)}$ (Pelczynski 1962). For the Rudin-Carleson theorem see (Carleson 1957, Gamelin 1969, Hoffman 1962, Rudin 1956).

1.4. Two Quantitative Improvements of (Part I of) the Theorem of F. and M. Riesz. We begin with a far-reaching improvement due to Pigno and Smith

(1982); the statement of this theorem (cf. C)) is prepared for in subsections A) and B) and discussed in D). The second improvement (the de Leeuw-Katznelson Theorem) is discussed in E).

A) With each function $\varphi : \mathbb{Z} \to \mathbb{C}$ we associate a linear functional F_φ defined on the set $\mathcal{F}(\mathbb{Z})$ of functions of an integer argument having finite support:

$$F_\varphi(f) = \sum_{n\in\mathbb{Z}} \varphi(n)f(n) \quad (f \in \mathcal{F}(\mathbb{Z})) .$$

The *spectral mass* (s-mass for short) of the function φ on the set E of integers is defined as the quantity

$$\mathcal{M}^E_{\text{Spec}}(\varphi) := \sup\{|F_\varphi(\hat{p})| : p \in \mathcal{P}(\mathbb{T}),\ \operatorname{Spec} p \subset E,\ \|p\|_\infty \le 1\} \tag{4}$$

(here $\mathcal{P}(\mathbb{T})$ denotes the set of trigonometric polynomials). We shall denote the full s-mass $\mathcal{M}^{\mathbb{Z}}_{\text{Spec}}(\varphi)$ simply by $\mathcal{M}_{\text{Spec}}(\varphi)$. The quantity $\mathcal{M}^E_{\text{Spec}}(\hat{\nu})$, where $\nu \in M$ $(= M(\mathbb{T}))$ will be called the *spectral mass of the charge* ν on the set E and denoted $\mathcal{M}^E_{\text{Spec}}(\nu)$. If we had replaced $\|p\|_\infty$ in (4) (i.e., $\|p\|_{C(\mathbb{T})}$) by $\|p\|_1 := \sum |\hat{p}(n)|$ or by $\|p\|_2 := \left(\sum |\hat{p}(n)|^2\right)^{\frac{1}{2}}$, the corresponding supremum would have coincided with $\sup\limits_E |\varphi|$ or $\left(\sum\limits_{n\in E} |\varphi(n)|^2\right)^{\frac{1}{2}}$. But because of the occurrence of the "hard-to-compute" (using $\hat{p}$) uniform norm $\|p\|_\infty$ no reasonably "explicit" expression for the s-mass $\mathcal{M}^E_{\text{Spec}}(\varphi)$ in terms of the values of the function φ on the set E is possible (although $\mathcal{M}^E_{\text{Spec}}(\varphi)$ depends only on $\varphi|_E$).

B) It is easy to see that $\mathcal{M}^E_{\text{Spec}}(\varphi)$ increases along with E and $\sup |\varphi| \le \mathcal{M}^E_{\text{Spec}}(\varphi) \le \left(\sum\limits_{n\in E} |\varphi(n)|^2\right)^{\frac{1}{2}}$. Moreover $\mathcal{M}_{\text{Spec}}(\nu) = \inf\{\|\nu - \lambda\|_M : \operatorname{Spec}\lambda \subset \mathbb{Z} \setminus (-E)\}$, $\mathcal{M}_{\text{Spec}}(\nu) = \|\nu\|_{M(\mathbb{T})}$. We set $\mathcal{M}^N_{\text{Spec}} := \mathcal{M}^{\mathbb{Z}\setminus[-N,N]}_{\text{Spec}}$ $(N \in Z_+)$. The following assertion can be regarded as an improvement of the theorem of the *Riemann-Lebesgue theorem*, which asserts that the Fourier coefficients of an integrable function tend to zero. For any charge $\nu \in M$

$$\lim_{N\to\infty} \mathcal{M}^N_{\text{Spec}}(\nu) = \|\nu_s\|_M = \mathcal{M}_{\text{Spec}}(\nu_s)$$

(here ν_s is the m-singular part of the charge ν). If $\nu \in M_s := \{\mu \in M : \mu = \mu_s\}$, then $\mathcal{M}^N_{\text{Spec}}(\nu) = \mathcal{M}_{\text{Spec}}(\nu)$ for any $N \in \mathbb{Z}_+$.

C) We agree to call the set $\{n \in \mathbb{Z} : a \le n \le b\}$, where $a, b \in \mathbb{Z}$, a *segment* and denote it by $[a, b]$. If the segments S_1 and S_2 are situated so that $\max S_1 < \min S_2$, we shall write $S_1 < S_2$. The set $B := [-b, -a] \cup [a, b]$ $(a, b \in \mathbb{N},\ a \le b)$ will be called a *block*; B^+ will denote the segment $[a, b] = B \cap \mathbb{N}$.

The Pigno-Smith theorem on spectral masses of a plus-charge (Pigno and Smith 1982).

Theorem. *Let* $(B_k)_{k=1}^{\infty}$ *be a sequence of blocks such that* $B_1^+ < B_2^+ < \cdots$, *and*

$$\min B_{k+1}^+ \geq 2\max B_k^+ \quad (k \in \mathbb{N}) . \tag{5}$$

Then

$$\nu \in M_+ \Rightarrow \sum_{k=1}^{\infty} \frac{\mathcal{M}_{\mathrm{Spec}}^{B_k}(\nu)}{k} < 12\|\nu\|_M . \tag{6}$$

D) We shall show how assertion I of 1.2A) follows from this theorem. We use the equality $\mathcal{M}_{\mathrm{Spec}}^{N}(\nu) = \mathrm{dist}\,(\nu, \mathcal{P}_N)$ $(N \in \mathbb{N},\, \nu \in M)$, where $\mathcal{P}_N := \{p \in \mathcal{P}(\mathbb{T}) : \mathrm{Spec}\, p \subset [-N, N]\}$ (cf. B)). It follows from this equality that for any $\nu \in M$ and $N \in \mathbb{N}$ there exists a polynomial $P_N \in \mathcal{P}$ such that

$$\mathrm{Spec}\, p_N \cap [-N, N] = \varnothing, \qquad \left|\int p_N\, d\nu\right| \geq \frac{1}{2}\mathrm{dist}\,(\nu, \mathcal{P}_N), \qquad \|p_N\|_C \leq 1 .$$

Let $B(N)$ be a block such that $\min(B(N))^+ = N$, $\mathrm{Spec}\, p_N \subset B(N)$. Obviously $(1/2)\mathrm{dist}\,(\nu, \mathcal{P}_N) \leq \mathcal{M}_{\mathrm{Spec}}^{B(N)}(\nu)$ $(N \in \mathbb{N})$. We define the subscripts N_k by induction, setting $N_1 := \max B(1)$, $N_{k+1} := 2\max B(N_k)$, $k = 1, 2, \ldots$. The blocks $B_k := B(N_k)$ satisfy condition (5), and it follows from (6) that $\lim\limits_{k\to+\infty} \mathrm{dist}(\nu, \mathcal{P}_{N_k}) = 0$ if $\nu \in M_+$. Thus any plus-charge is m-absolutely continuous, being the limit (in variation) of a sequence of charges of the form pm, where $p \in \mathcal{P}$. Estimate (6) means that the spectral masses of a plus-charge concentrated on blocks that move rapidly away from the origin must be small enough so that, for example, the series on the left-hand side of the estimate (6) always converges. What is important here is only the rate at which the blocks move away from the origin, not their length. It is useful to compare what has just been said with the last assertion of B). In connection with the Pigno-Smith theorem we note the earlier work (Stein 1966).

E) It follows from assertion I of the F. and M. Riesz theorem that the coefficients of any plus-charge (and any minus-charge) tend to zero. This property has a certain stability, as can be seen from the following proposition: We set

$$r := \{\varphi \in M : \lim_{|n|\to+\infty} \hat{\varphi}(n) = 0\} ,$$

$$r_+ := \{\varphi \in M : \lim_{n\to+\infty} \hat{\varphi}(n) = 0\} \quad (M := M(\mathbb{T})) .$$

It can be shown that $r = r_+$. This assertion in turn can be significantly strengthened. We set

$$l_+(\varphi) := \overline{\lim_{n\to+\infty}}\, |\hat{\varphi}(n)| , \quad l_-(\varphi) := \overline{\lim_{n\to-\infty}}\, |\hat{\varphi}(n)| \quad (\varphi \in M) ,$$

$$\mathcal{DLH}(\varepsilon) := \sup\{l_+(\varphi) : \|\varphi\|_M \leq 1,\ l_-(\varphi) \leq \varepsilon\} \quad (\varepsilon > 0) .$$

Theorem. (de Leeuw-Katznelson) $\lim_{\varepsilon\downarrow 0} \mathcal{DLH}(\varepsilon) = 0.$

The proof can be found in (de Leeuw, Katznelson 1970).

Thus the functions $\hat{\varphi}|_{\mathbb{Z}_+}$ and $\hat{\varphi}|_{\mathbb{Z}_-}$, where $\varphi \in M$, are in a peculiar equilibrium. However, in general $l_+(\varphi) \neq l_-(\varphi)$. (cf. (Graham, McGehee 1979)).

1.5. Jensen's Inequality (A Quantitative Improvement of Part II of the F. and M. Riesz Theorem). The uncertainty principle is connected with a large class of theorems whose hypotheses mention convergence or divergence of the integral of the logarithm of the absolute value of some function. In order to appreciate such theorems properly, we remark that if a nonnegative function G is integrable with respect to a measure μ but $\ln G$ is not integrable with respect to that measure, then $\int \ln G\, d\mu = -\infty$, since $\ln^+ G \leq G$ and $\int \ln^+ G\, d\mu < +\infty$. Roughly speaking, the function $\ln G$, where $G \in L^1(\mu)$ can fail to be integrable only because the function G is small on a set of rather large measure.

A) Here is the simplest assertion from the class mentioned above, in whose statement a certain "integral of the logarithm" occurs.

Theorem. *If* $f \in H^1(\mathbb{T})$ *and* $\int_{\mathbb{T}} \ln|f|\, dm = -\infty$, *then* $f = 0$.

This theorem is a significant strengthening of assertion II of the F. and M. Riesz theorem: a nonzero function of class $H^1(\mathbb{T})$ not only cannot vanish on a set of positive length, it cannot even be too small on such a set (for example, it cannot tend to zero too rapidly when approaching any point of the circle $\mathbb{T}$).

B) The theorem stated in subsection A) is a consequence of the following remarkable inequality:

$$f \in H^1(\mathbb{T}) \Rightarrow \ln|\hat{f}(0)| \leq \int_{\mathbb{T}} \ln|f|\, dm\,. \qquad (J_{\mathbb{T}})$$

Estimate $(J_{\mathbb{T}})$ is called *Jensen's inequality.*

According to the classical inequality between the arithmetic and geometric means (which is also called Jensen's inequality),

$$f \in L^1(\mathbb{T}) \Rightarrow \exp \int \ln|f|\, dm \leq \int |f|\, dm\,.$$

Thus $(J_{\mathbb{T}})$ is a significant strengthening of the trivial estimate $|\hat{f}(0)|\left(= \left|\int f\, dm\right|\right) \leq \int |f|\, dm$, applied to integrable plus-functions.

The theorem of A) follows easily from $(J_{\mathbb{T}})$: if $f \in H^1 := H^1(\mathbb{T})$ and $f \neq 0$, then $\hat{f}(n) \neq 0$ for some $n \in \mathbb{Z}_+$ and $\hat{f}(j) = 0$ for $0 \leq j \leq n-1$; since $g := \bar{z}^n f \in H^1$, we have by $(J_{\mathbb{T}})$

$$\ln |\hat{f}(n)| = \ln |\hat{g}(0)| \leq \int_{\mathbb{T}} \ln |g| \, dm = \int_{\mathbb{T}} \ln |f| \, dm \,,$$

and $\hat{f}(n) = 0$ if $\int_{\mathbb{T}} \ln |f| \, dm = -\infty$.

C) Let us sketch the proof of inequality $(J_{\mathbb{T}})$. It suffices to verify that

$$f \in C_+ \Rightarrow \ln |\hat{f}(0)| \leq \int_{\mathbb{T}} \ln |f| \, dm \tag{7}$$

(the set C_+ is dense in H^1 with respect to the L^1-norm). The set C_+ forms a closed subalgebra of the algebra C ($= C(\mathbb{T})$), and the functional $f \mapsto \hat{f}(0)$ is multiplicative on C_+. It is easy to deduce from these assertions that $(\widehat{\exp f})(0) = \exp \hat{f}(0)$ if $f \in C_+$. We remark also that for every real-valued polynomial $t \in \mathcal{P}$ one can choose a real polynomial $\tilde{t} \in \mathcal{P}$ such that $t + i\tilde{t} \in \mathcal{P}_+$ (it suffices to set $\tilde{t} := -i \sum \operatorname{sgn} k \cdot \hat{t}(k) z^k$). Let us prove (7). Given a number $\varepsilon > 0$, we find a real polynomial $t \in \mathcal{P}$ such that $t - \varepsilon < \ln(|f| + \varepsilon) < t + \varepsilon$. We set $s := t + i\tilde{t}$ (so that $s \in C_+$). We then have: $|fe^{-s}| = |f|e^{-t} < e^{\ln(|f|+\varepsilon)-t} < e^{\varepsilon}$ and therefore $|\widehat{fe^{-s}}(0)| = \left| \int_{\mathbb{T}} fe^{-s} \, dm \right| \leq e^{\varepsilon}$. But $|(\widehat{fe^{-s}})(0)| = |\hat{f}(0)| \, |e^{-\hat{s}(0)}| = |\hat{f}(0)| e^{-\hat{t}(0)}$, so that

$$|\hat{f}(0)| \leq e^{\varepsilon + \hat{t}(0)} = e^{\varepsilon + \int t \, dm} \leq e^{2\varepsilon + \int \ln(|f|+\varepsilon) \, dm} \,.$$

In this "real-variable" proof of the inequality $(J_{\mathbb{T}})$ the complex origins of the inequality are carefully hidden (cf. Sect. 2 of Chapt. 2 below). In this form it admits various generalizations (cf. (Barbey, König 1977), (Gamelin 1969), (Hoffman 1962)).

D) Using the change of variable $x \mapsto \dfrac{x-i}{x+i}$, which maps the line $\mathbb{R}$ onto the circle $\mathbb{T}$, one can obtain an analogue of Jensen's inequality for plus-functions on the line. To write this analogue we introduce the Poisson measure Π on $\mathbb{R}$:

$$\Pi := \pi(1+x^2)^{-1} m \,,$$

where m is Lebesgue measure on $\mathbb{R}$. If f is a plus-function of class $L^1(\Pi)$, then

$$\ln \left| \int_{\mathbb{R}} f \, d\Pi \right| \leq \int_{\mathbb{R}} \ln |f| \, d\Pi \tag{$J_{\mathbb{R}}$}$$

The function f here may fail to belong to $L^1(m)$ (but of course it belongs to $L^1_{\mathrm{loc}}(m)$). It can be considered a slowly growing distribution with spectrum in $[0, +\infty)$ (i.e., $f[h] = 0$ for any plus-function $h \in S(\mathbb{R})$).

Inequality ($J_{\mathbb{R}}$) implies the following versions of the uncertainty principle: (a) Let $p \in [1, +\infty]$. If $f \in H^p(\mathbb{R})$ and $\int\limits_{\mathbb{R}} \ln |f|\, d\Pi = -\infty$, then $f = 0$. (b) If $\mu \in M(\mathbb{R})$, and $\operatorname{supp} \mu$ is bounded from above or below, while $\int\limits_{\mathbb{R}} \ln |\hat{\mu}|\, d\Pi = -\infty$, then $\hat{\mu} = 0$.

E) It is easy to deduce certain forms of the uncertainty principle for smooth plus-functions from the inequality ($J_{\mathbb{T}}$).

Let Y be a subset of the space C ($:= C(\mathbb{T})$). A closed set $K \subset \mathbb{T}$ is said to be Y-*determining* (or a set of class $\mathcal{U}(Y)$) if

$$f \in Y, \quad f\big|_K = 0 \Rightarrow f = 0\,.$$

It follows from the theorem in subsection A) of Sect. 1.1.3 and the theorem in subsection A) of Sect. 1.1.2 that $\mathcal{U}(C_+)$ coincides with the set of all closed subsets $K \subset \mathbb{T}$ of nonzero length.

The symbol C^α, where $\alpha \in \mathbb{N}$, will be used to denote the set of functions that are α times continuously differentiable on the circle $\mathbb{T}$. If $\alpha \in (0, +\infty) \setminus \mathbb{N}$, then $C^\alpha := \{f \in C^\alpha : D^{[\alpha]} f$ satisfies a Lipschitz condition of order $\alpha - [\alpha]\}$. Here $(Df)(e^{i\theta}) = \frac{d}{d\theta}\big(f(e^{i\theta})\big)$. We now consider sets of the class $\mathcal{U}(C^\alpha_+)$, where $C^\alpha_+ := C^\alpha \cap C_+$. Obviously $\mathcal{U}(C_+) \subset \bigcap\limits_{\alpha > 0} \mathcal{U}(C^\alpha_+)$. However, this inclusion is strict: there exist countable sets belonging to all the classes $\mathcal{U}(C^\alpha_+)$. The class $\mathcal{U}(C^\alpha_+)$ is independent of α and coincides with $\mathcal{U}(C^\infty_+)$. It can be completely described using purely metric terms. To do this we set $\rho_K(\zeta) := \frac{1}{2}\operatorname{dist}(\zeta, K)$ ($K \subset \mathbb{T}$, $\zeta \in \mathbb{T}$). We shall say that a closed set $K \subset \mathbb{T}$ satisfies a *Carleson condition* (and write $K \in \mathrm{Carl}$) if

$$J(K) := \int\limits_{\mathbb{T}} \ln(1/\rho_K)\, dm = +\infty\,. \tag{8}$$

Condition (8) means that K is quite massive. It is satisfied, for example, by all closed sets K of positive length, but by other sets as well; this will become clear after the reformulations of condition (8) that follow.

Consider the "h-inflation" $K_h := \{\zeta \in \mathbb{T} : \operatorname{dist}(\zeta, K) < h\}$ and the set $\mathcal{L}(K)$ of arcs of the circle $\mathbb{T}$ that are complementary to K. The following assertions are equivalent: (a) $K \in \mathrm{Carl}$; (b) $\int_0^1 m(K_h) h^{-1}\, dh = +\infty$; (c) $m(K) > 0$ or

$$\sum_{l \in \mathcal{L}(K)} m(l) \ln\left(\frac{1}{m(l)}\right) = +\infty\,.$$

The following remark is due to Beurling: for any $\alpha > 0$, Carl $\subset \mathcal{U}(C_+^\alpha)$,

If $K \subset$ Carl, $\alpha \in (0,1)$, $f \in C_+^\alpha$, and $f|_K = 0$, then $|f(\zeta)| \leq A(\operatorname{dist}(\zeta, K))^\alpha = A_1(\rho_K(\zeta))^\alpha$ ($\zeta \in \mathbb{T}$). Therefore

$$\ln|f| \leq \ln A' + \alpha \ln \rho_K, \quad \int_{\mathbb{T}} \ln|f|\, dm \leq \ln A' - \alpha J(K) = -\infty ,$$

so that $f = 0$ (cf. A)).

Actually Carl $= \mathcal{U}(\bigcup_{\alpha>0} C_+^\alpha) = \mathcal{U}(C_+^\infty)$ (Carleson 1952).

1.6. $\mathcal{R}$-sets and $\mathcal{D}$-sets. We denote the d-dimensional torus by $\mathbb{T}^d$: $\mathbb{T}^d = \underbrace{\mathbb{T} \times \cdots \times \mathbb{T}}_{d}$, and the lattice of points in $\mathbb{R}^d$ with integer coordinates by $\mathbb{Z}^d$. We denote the set of charges on the torus $\mathbb{T}^d$ by $M(\mathbb{T}^d)$.

A) A set $F \subset \mathbb{Z}^d$ is called an $\mathcal{R}$-set (more precisely an $\mathcal{R}^d$-set) if

$$\mu \in M(\mathbb{T}^d), \ \operatorname{Spec} \mu \subset F \Rightarrow \mu \ll m^d \ (:= \underbrace{m \times \cdots \times m}_{d}) ;$$

it is called a $\mathcal{D}$-set if

$$\left.\begin{array}{l} \mu \in M(\mathbb{T}^d)\ , \ \operatorname{Spec} \mu \subset F\ , \\ |\mu|(E) = 0 \text{ for some } E \subset \mathbb{T}^d \text{ with } m^d(E) > 0 \end{array}\right\} \Rightarrow \mu = 0\ .$$

Using this terminology we can restate the theorem of F. and M. Riesz as follows: $\mathbb{Z}_+$ is an $\mathcal{R}$-set and a $\mathcal{D}$-set.

B) An interesting approach to the concept of an $\mathcal{R}$-set was proposed by J. Shapiro (1978) and A.B. Aleksandrov (1981). This approach uses certain estimates of the Fourier coefficients as functionals in L^p for $p \in (0,1)$.

We shall agree to call a charge $\mu \in M(\mathbb{T}^d)$ an F-*charge* if $\operatorname{Spec} \mu \subset F$. The term "$F$-function" will have an analogous meaning. We define an F-polynomial as a linear combination of functions z^k where $k \in F$. The symbols M^F, $(L^2)^F$, and $\mathcal{P}^F$ will denote respectively the set of all F-charges, F-functions of class L^1, and F-polynomials.

Let p be a positive number. The set $F \subset \mathbb{Z}^d$ will be called a p-*set* if any functional $f \mapsto \hat{f}(k)$ ($k \in \mathbb{Z}^d$) is continuous on the set $\mathcal{P}^F$ with respect to the topology induced from L^p ($= L^p(\mathbb{T}^d, m^d)$). (If $p \geq 1$ this topology is defined using the usual L^p-norm; for $p \in (0,1)$ it is defined by the metric $\rho_p : \rho_p(f,g) := \int_{\mathbb{T}} |f - g|^p$). The metric vector space $L^p(\mathbb{T}^d)$ is not locally convex if $p \in (0,1)$. Moreover no continuous linear functional (except the zero functional) can be defined on it. This, however, does not exclude the possibility of linear functionals that are continuous on certain subspaces of the space L^p. Thus, for example, for $d = 1$ and $p \in (0,1)$ the functional $f \mapsto \hat{f}(0)$ is L^p-continuous on the subspace $H^1(\mathbb{T})$ ($\subset H^p(\mathbb{T})$):

$$|\hat{f}(0)|^p \leq \exp\left(p\int \ln|f|\,dm\right) \leq \int |f|^p\,dm \quad (f \in H^1(\mathbb{T})$$

(here the first inequality follows from ($J_{\mathbb{T}}$) and the second means that the geometric mean does not exceed the arithmetic mean). It can be shown that any set of integers that is bounded above or below is a p-set (for any $p \in (0,1)$).

We agree to call a set $F \subset \mathbb{Z}^d$ a *stable p-set* if $F \cup \{k\}$ is a p-set for any $k \in \mathbb{Z}^d$.

Theorem. (Aleksandrov 1981, J. Shapiro 1978) *Any stable p-set is an $\mathcal{R}$-set.*

Part I of the F. and M. Riesz theorem (cf. Sect. 1.1.1) is a corollary of this assertion.

C) The Aleksandrov-Shapiro approach makes it possible to construct certain multidimensional $\mathcal{R}$-sets. With a set $F \subset \mathbb{Z}^d$ and a number $l \in \mathbb{Z}$ we associate the section $F_l := \{m \in \mathbb{Z}^{d-1} : (l,m) \in F\}$. The set $\{l \in \mathbb{Z} : F_l \neq \varnothing\}$ will be called the *projection* of the set F. Let $p \in (0,1)$. If all sections of the set $F \in \mathbb{Z}^d$ are $\mathcal{R}_{d-1}$-sets, and its projection is a stable p-set, then F is an $\mathcal{R}_d$-set (J. Shapiro 1978). It follows from this theorem that a set $F \subset \mathbb{Z}^2$ that fits into a sector of angle less than π is an $\mathcal{R}_2$-set. Moreover if $F \subset \mathbb{Z}^2$, the projection of F consists of nonnegative integers, and any section is bounded above or below, then $F \in \mathcal{R}_2$.

A *strong $\mathcal{R}$-set* (a set of class $s\mathcal{R}$) is a set $F \subset \mathbb{Z}$ such that

$$F_0 \in \mathcal{R} \Rightarrow F \cup F_0 \in \mathcal{R}\,.$$

Very little is known about the class $s\mathcal{R}$. It contains the set of all prime numbers and the set of all squares of integers. Any geometric progression whose ratio is the reciprocal of a positive integer forms a strong $\mathcal{R}$-set. Thus $\mathbb{Z}_+ \cup A \in \mathcal{R}$ if $A = \{-m^k : k \in \mathbb{Z}_+\}$ and m is a natural number. This assertion can be strengthened by replacing A by any set B of negative numbers that is *Hadamard lacunary*:

$$\exists q > 1 : n, n' \in B\,,\ |n'| > |n| \Rightarrow \frac{|n'|}{|n|} > q\,.$$

(This fact is a corollary of a theorem of Rudin (1960).)

D) Up to now the discussion in this section has involved spectral properties relating to assertion I of the F. and M. Riesz theorem. Assertion II can be written as follows:

$$\mathbb{Z}_+ \in \mathcal{D}_1$$

(the classes $\mathcal{D}_d$ were defined in subsection A) above).

Combining the F. and M. Riesz theorem with Fubini's theorem, one can show easily that a set $F \subset \mathbb{Z}^2$ belongs to the class $\mathcal{D}_2$ if its projection on the x-axis and all sections of it are semibounded. The class $\mathcal{D}_1$ contains the set $\mathbb{Z}_+ \cup A$ studied in subsection C). Any finite set F of integers is a strong $\mathcal{D}_1$-set: $F \cup F_0 \in \mathcal{D}_1$ if $F_0 \in \mathcal{D}_1$.

§2. Singular Measures and Fourier Transforms that Tend to Zero

The properties mentioned in the title of this section are opposed to each other: if a nonzero charge $\mu \in M(\mathbb{T})$ is singular with respect to the measure m (more briefly, it belongs to the class M_s), then its support is small. This, in accordance with the uncertainty principle, prevents it from having spectral smallness, which we shall understand here as follows:

$$\lim_{|n|\to+\infty} \hat{\mu}(n) = 0 . \tag{1}$$

Nevertheless these properties are sometimes compatible. We shall describe below certain ways of constructing nonzero charges of class M_s that are r-charges (i.e., subject to condition (1)). The term "r-charge" will be used in honor of Rajchman, who was the first to draw attention to this topic.

2.1. Some Classes of Measures and Two-sided Sequences. We begin with the class $\widehat{M}$ of functions of the form $n \mapsto \hat{\mu}(n)$ $(n \in \mathbb{Z})$, where $\mu \in M$ $(= M(\mathbb{T}))$. There is no intuitive description of the class $\hat{M}$ comparable in simplicity and clarity with, say, the known descriptions of the classes $\hat{L}^2$ or $\widehat{C}^\infty$. We shall give the details of the relationship of the class $\widehat{M}$ with the classes l^p $(= l^p(\mathbb{Z}))$ Obviously $\widehat{M} \subset l^\infty$ and $l^\infty \setminus \widehat{M} \neq \varnothing$ (e.g. the function $n \mapsto \operatorname{sgn} n$ does not belong to $\widehat{M}$). Moreover $l^p \setminus \widehat{M} \neq \varnothing$ for any $p > 2$ (while obviously $l^2 \subset \widehat{M}$). We now take up the class c_0 of all double sequences that tend to zero:

$$c \in c_0 \Leftrightarrow \lim_{|n|\to+\infty} c(n) = 0 .$$

By the Riemann-Lebesgue theorem $M_a \subset r := \{\mu \in M : \hat{\mu} \in c_0\}$ (here M_a denotes the set of m-absolutely continuous charges $\mu \in M$). On the other hand a nonzero discrete charge cannot be an r-charge. A charge $\mu \in M$ will be called *discrete* (or of class M_d) if it can be represented as $\sum_{t\in\mathbb{T}} c_t\delta_t$, where $(c_t)_{t\in\mathbb{T}}$ is a summable family of complex numbers $(\sum |c_t| < +\infty)$ and δ_t is a unit mass concentrated at the point t. The fact that $M_d \cap r = \{0\}$ follows from *Wiener's formula*, which makes it possible to detect the presence or absence of point masses by purely spectral means:

$$\sum_{a\in\mathbb{T}} |\nu(\{a\})|^2 = \lim_{n\to+\infty} \frac{1}{2n+1} \sum_{|k|\le n} |\hat{\nu}|(k) \quad (\nu \in M) \tag{2}$$

It follows from (2) that $r \subset M_c$, where M_c denotes the set of *continuous charges* (i.e., charges containing no point masses):

$$\mu \in M_c \Leftrightarrow \mu(\{t\}) = 0 \quad \forall t \in \mathbb{T} .$$

The equality $M_d \cap r = \{0\}$ also follows from the fact that $\widehat{M_d}$ consists of functions that are almost periodic on the group $\mathbb{Z}$, and therefore

$$\mu \in M_d \Rightarrow \|\hat{\mu}\|_\infty = \overline{\lim_{n\to+\infty}} |\hat{\mu}(n)| = \overline{\lim_{n\to-\infty}} |\hat{\mu}(n)| .$$

Here we meet yet another manifestation of the interaction of the positive and negative parts of the spectrum (cf. subsection 1.1.4E).

The problem of greatest interest is that of the relation of the class r to the class M_{sc} of singular continuous measures: $M_{sc} := M_s \cap M_c$. We shall see that $M_{sc} \setminus r \neq \{0\}$ and $r \cap M_{sc} \neq \{0\}$.

We conclude this section by stating a criterion for continuity of a measure (Zygmund 1959). To do this we associate with a double sequence $\varphi : \mathbb{Z} \to \mathbb{C}$ the sums $s_n^\varphi = \sum_{|k|\leq n} \varphi(k) z^k$. Assume that $s_{n_k}^\varphi(\zeta) \geq 0$ $(k \in \mathbb{N},\ \zeta \in \mathbb{T})$ for some strictly increasing sequence of subscripts n_k. Then $\varphi = \hat{\mu}$, where μ is a measure of class M_c (it coincides with the weak-star limit of the measures $s_{n_k}^\varphi m$).

2.2. The Cantor Set and the Cantor Measure. The Theorem of Bari and Salem.

A) We recall the construction of a Cantor set. Fix a number $\xi \in (0, \frac{1}{2})$ and consider the operation k_ξ that transforms any closed interval I into the union of two closed intervals I_1 and I_2 whose lengths are $\xi|I|$ and which are situated as in Fig. 1.

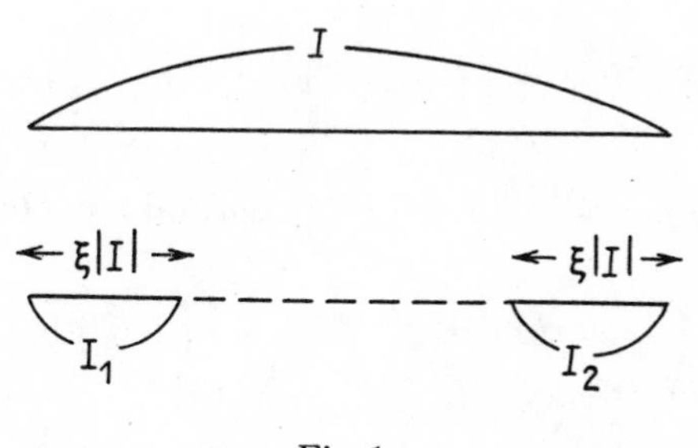

Fig. 1

Let $I = [0, 1]$. Applying the operation k_ξ to each of the closed intervals I_k $(:= I_k^{(1)})$, we obtain four closed intervals $I_k^{(2)}$, $k = 1, 2, 3, 4$, whose lengths are ξ^2. Repeating this construction, at the jth step we obtain closed intervals $I_k^{(j)}$, $k = 1, 2, \ldots, 2^j$, $|I_k^{(j)}| = \xi^j$. We set

$$K_j = \bigcup_{k=1}^{2^j} I_k^{(j)}, \quad K_\xi := \bigcap_{j=1}^{\infty} K_j .$$

The set K_ξ is called a *Cantor set* (with ratio ξ). It is perfect, totally disconnected, and of length zero.

B) With the set K_ξ there is associated the so-called *Cantor measure* $\tilde{\varkappa}_\xi$. To define it we consider a homogeneous distribution $\varkappa_{\xi,j}$ of unit mass on the set K_j (i.e., a distribution having constant density): for any Borel set $E \subset I = [0, 1]$

$$\varkappa_{\xi,j}(E) = \frac{|E \cap K_j|}{|K_j|} .$$

The sequence of measures $(\varkappa_{\xi,j})_{j=1}^{\infty}$ converges in the weak-star topology to a probability measure that we shall denote by the symbol $\tilde{\varkappa}_\xi$. It is concentrated on K_ξ. Its value on the closed interval $[a, b] \subset I$ is $g_\xi(b) - g_\xi(a)$, where g_ξ is a nondecreasing continuous function ($g_\xi(t) := \tilde{\varkappa}_\xi([0, t])$). Its derivative equals zero almost everywhere and its graph (the so-called "Cantor staircase") is a classical object of the theory of functions of a real variable. We emphasize that the measure $\tilde{\varkappa}_\xi$ is continuous and singular.

We now move it from the closed interval I to the circle $\mathbb{T}$, by setting $\varkappa_\xi(A) := \tilde{\varkappa}_\xi(\mathcal{E}^{-1}(A))$ for any Borel set $A \subset \mathbb{T}$, where $\mathcal{E}(t) := \exp(2\pi i t)$. Obviously

$$\varkappa_\xi \in M_{sc} , \quad \varkappa_\xi \geq 0, \quad \varkappa_\xi(\mathbb{T}) = 1 .$$

We continue to use the name Cantor measure with ratio ξ for the measure $\varkappa_\xi$. Its spectral properties depend largely on the arithmetical nature of the ratio ξ.

C) A simple computation shows that

$$\hat{\varkappa}_\xi(m) = (-1)^m \prod_{j=1}^{\infty} \cos\left(\pi(1 - \xi)\xi^j m\right) \quad (m \in \mathbb{Z}) .$$

Assume that $\xi = 1/q$, $q > 2$. Then for any natural number N

$$|\hat{\varkappa}_\xi(q^N)| = \prod_{j=N}^{\infty} |\cos \pi(q - 1)q^{N-j-1}| = \prod_{j=1}^{\infty} |\cos \pi(q - 1)q^{-j}| > 0 ,$$

so that the factors of this last infinite product are positive and tend rapidly to 1. Therefore

$$\varkappa_{\frac{1}{q}} \in M_{sc} \setminus r \quad (q = 3, 4, \ldots) .$$

If $\xi = \frac{p}{q}$ where p and q are relatively prime natural numbers and $p \geq 2$, then $\varkappa_\xi \in M_{sc} \cap r$ (*Bari's theorem*). The question whether a Cantor measure belongs to the class r is thereby completely solved for rational ratios ξ. The definitive solution is given by the following theorem.

Theorem. (Salem 1963, Katznelson 1976). *The inclusion*

$$\varkappa_\xi \in M_{sc} \setminus r$$

is equivalent to the following assertion: the number $\theta := \frac{1}{\xi}$ *is a zero of an irreducible polynomial* $p = x^n + a_{n-1}x^{n-1} + \cdots + a_0$ *with integer coefficients* a_j *all of whose other zeros lie inside the unit disk.*

Numbers ξ having this property are called *Pisot-Vijayaraghavan numbers* (Cassells 1957).

2.3. Riesz Products. We shall now study yet another method of constructing measures belonging to $M_{sc} \setminus r$ or $r \cap M_{sc}$. The measure ρ_a to be discussed (the so-called *Riesz product*) does not have so explicit a construction as the Cantor measure. Its Fourier transform $\hat{\rho}_a$ appears first, and only then will ρ_a be described as a set function.

A) The letter a will denote a family of points of the closed unit disk:

$$a : k \mapsto a(k) \ \ (k \in E) \ , \ \ \sup |a| \leq 1 \tag{4}$$

(here the index k ranges over a certain set E of natural numbers).

The trigonometric polynomial

$$f_k := 1 + \mathrm{Re}\,(a(k)z^k) = \frac{\overline{a(k)}z^{-k}}{2} + 1 + \frac{a(k)z^k}{2}$$

is nonnegative. We set $p_n := f_1 \cdot \ldots \cdot f_n$, and

$$\rho_{a,n} := \rho_n m \ , \tag{5}$$

where E_n is the set of the first n elements of the set $E : E_n = \{k_1, k_2, \ldots, k_n\}$, $k_j < k_{j+1}$. If E is sufficiently sparse, as we shall see at once, the probability measures $\rho_{a,n}$ are weak-star convergent to a certain measure $\rho_a \in M_c$. This measure is called the Riesz product (associated with the family a).

B) We assume that

$$\inf \left\{ \frac{n}{m} : \, m, n \in E, \, m < n \right\} \geq 3 \, . \tag{6}$$

We recall that the symbol $\mathcal{F}(\mathbb{Z})$ denotes the set of all functions of an integer argument with finite support. We set $\mathcal{E} = \{-1, 0, 1\}$ and let $\mathcal{F}_E = \{\varepsilon \in \mathcal{F}(\mathbb{Z}) : \varepsilon(\mathbb{Z}) \subset \mathcal{E}, \ \ \mathrm{supp}\,\varepsilon \subset E\}$. Finally we set

$$\sigma(\varepsilon) := \sum_{n \in \mathbb{Z}} \varepsilon(n)n = \sum_{n \in K} \pm n, \quad (\varepsilon \in \mathcal{F}_E) \, , \tag{7}$$

where $K \subset E$ is a certain finite set of integers. The set of all sums of the form $\sigma(\varepsilon)$, where $\operatorname{supp} \varepsilon \subset K$ will be denoted $[K]$. The following fact will play an important role below: under condition (6) the mapping $\varepsilon \mapsto \sigma(\varepsilon)$ $(\varepsilon \in \mathcal{F}_E)$ is one-to-one and onto. Moreover

$$[K] \subset \left(-\frac{3}{2} \max K, \frac{3}{2} \max K\right),$$

for any finite set $K \subset E$.

C) The sums $\sigma(\varepsilon)$ arise naturally in computing the spectrum of the measure ρ_n (cf. (5)). Indeed

$$p_n = \prod_{k \in E_n} \sum_{l \in \mathcal{E}} \frac{a_l'(k)}{2} z^{lk},$$

where $\mathcal{E} = \{-1, 0, 1\}$, $a_{-1}'(k) := \overline{a(k)}$, $a_1'(k) := a(k)$, $a_0'(k) := 2$. Multiplying out the sums $\sum\limits_{l \in \mathcal{E}}$, we obtain:

$$p_n = \sum_{\varepsilon \in \mathcal{F}_{E_n}} A(\varepsilon) z^{\sigma(\varepsilon)} = \sum_{k \in [E_n]} \hat{p}_n(k) z^k$$

(the index ε can be replaced by the index k because of the property of the sums σ mentioned above). Thus

$$\operatorname{Spec} \rho_{a,n} \subset [E_n] \subset \left(-\frac{3}{2} k_n, \frac{3}{2} k_n\right).$$

If $k \in [E_n]$, then $k = \sigma(\varepsilon)$ $(\varepsilon \in \mathcal{F}_{E_n})$ and

$$\hat{p}_n(k) = A(\varepsilon) = \prod_{k \in E_n} \frac{a_\varepsilon'(k)}{2}. \tag{8}$$

In particular if $k \in E_n$, the representation $k = \sigma(\varepsilon)$ has the form $k = 1 \cdot k$ and $\hat{p}_n(k) = a(k)/2$. Thus

$$\hat{p}_n\big|_{E_n} = \frac{a}{2}\Big|_{E_n}.$$

D) Let us compare p_{n+1} with p_n: $p_{n+1} = p_n f_{n+1} = p_n + q_n$, where

$$q_n := p_n(f_n - 1) = p_n\left(\frac{\overline{a(k_{n+1})} z^{-k_{n+1}}}{2} + \frac{a(k_{n+1}) z^{k_{n+1}}}{2}\right).$$

Therefore

$$\min\left|\operatorname{Spec} q_n\right| > k_{n+1} - \frac{3}{2} k_n \geq 3k_n - \frac{3}{2} k_n = \frac{3}{2} k_n \geq \max\left|\operatorname{Spec} \rho_{a,n}\right|,$$

so that the transition from p_n to p_{n+1} reduces to writing in new terms of the form cz^m, where $|m| > \max|\operatorname{Spec} \rho_{a,n}|$. Thus there exists a sequence

$\varphi : \mathbb{Z} \to \mathbb{C}$ such that $\hat{p}_n(k) = \varphi(k)$, if $|k| < \frac{3}{2}k_n$ and $\sum_{|k|<\frac{3}{2}k_n} \varphi(k)z^k = p_n \geq 0$ $(n = 1, 2, \ldots)$. Therefore (cf. the end of Sect. 1.2.1) $\varphi = \hat{\rho}_a$, where $\rho := \rho_a$ is a certain continuous probability measure ($\rho(\mathbb{T}) = 1$, $\rho \in \mathrm{Meas} \cap M_c$) and the sequence $(\rho_{a,n})$ is weak-star convergent to ρ. This is what is called a *Riesz product*. The construction described above gives a precise meaning to the formal equality

$$\rho_a = \left(\prod_{k\in E} f_k\right) \cdot m \,.$$

We remark that $m = \rho_0$.

E) We now establish the following criterion for a Riesz product to be singular:

$$\sum_{k\in E} |a(k)|^2 = \infty \Rightarrow \rho_a \in M_{sc} \,. \tag{9}$$

The following more precise assertion (Peyrière 1975) also holds: *if*

$$\sum_{k\in E} |a(k) - b(k)|^2 = \infty \,, \tag{10}$$

then the measures ρ_a and ρ_b are mutually singular, i.e., there exist sets $S_a, S_b \subset \mathbb{T}$ such that $\rho_a(S_a) = \rho_b(S_b) = 1$, $S_a \cap S_b = \varnothing$. Setting $b = 0$, we obtain (9).

The system of functions $(e_{a,k})_{k\in E}$, where $e_{a,k} := z^k - \frac{\overline{a(k)}}{2}$, is orthogonal in $L^2(\rho_a)$ (this follows from the equalities $4\hat{\rho}_a(l-k) = a(l)\overline{a(k)}$ $(l, k \in E)$ and $2\hat{\rho}_a\big|_E = a$). Moreover $\int |e_{a,k}|^2 \, d\rho_a = 1 - |a(k)|^2/4 \leq 1$. If the sequence $c : E \to \mathbb{C}$ belongs to $l^2(E)$ (i.e., $\sum_{k\in E} |c(k)|^2 < \infty$), then the series $\sum_{k\in E} c(k)e_{a,k}$ and $\sum_{k\in E} c(k)e_{b,k}$ converge in $L^2(\rho_a)$ and $L^2(\rho_b)$ (respectively). Therefore there are sets S_a and S_b and a sequence of indices (N_j) increasing to infinity such that the limits $\lim_{j\to\infty} \sum_{|k|\leq N_j} c(k)e_{a,k}(\zeta)$ and $\lim_{j\to\infty} \sum_{|k|\leq N_j} c(k)e_{b,k}(\zeta)$ exist and are finite, the first for all $\zeta \in S_a$ and the second for all $\zeta \in S_b$, where $\rho(S_a) = \rho(S_b) = 1$. If S_a and S_b have the common point ζ, then the following limit exists and is finite:

$$\lim_{j\to\infty} \sum_{|k|\leq N_j} c(k)(e_{a,k}(\zeta) - e_{b,k}(\zeta)) = \frac{1}{2} \lim_{j\to\infty} \sum_{|k|\leq N_j} c(k)\overline{(a(k) - b(k))} \,.$$

But under condition (10) the sequence $c \in l^2(E)$ can be chosen so that $c \cdot (\overline{a} - \overline{b}) = |c|\,|a-b|$, and $\sum_{k\in E} |c(k)|\,|a(k) - b(k)| = \infty$.

F) If the sequence a is bounded away from zero, then $\rho_a \in M_{sc}$ and $\hat{\rho}_a\big|_E = a/2$, so that $\rho_a \notin r$. But if $\lim\limits_{k\to+\infty} |a(k)| = 0$ and $a \notin l^2(E)$, then for large k arbitrarily small factors will occur among the factors in (8), while all the factors are at most one in absolute value. In this case $\rho_a \in r \cap M_{sc}$.

2.4. The Ivashev-Musatov Theorem. How fast can the Fourier coefficients of a nonzero charge in class $M_s \cap r$ tend to zero? The inclusions

$$\text{(a)}\ \mu \in M_s\,, \quad \text{(b)}\ \hat{\mu} \in l^2(\mathbb{Z}) \tag{11}$$

are obviously compatible only when $\mu = 0$. It turns out that even trivial weakenings of condition (11b) suffice to remove this barrier. Thus, for example, there exist nonzero measures possessing property (11a) and such that $\hat{\mu} \in l^{2+0}(\mathbb{Z})$ $(:= \bigcap\limits_{\varepsilon>0} l^{2+\varepsilon}(\mathbb{Z}))$. This fact follows (as a very special corollary) from the basic result of the present section.

A) A sequence $\Phi : \mathbb{N} \to [0, +\infty)$ is called an *s-majorant* if there exists a nonzero charge $\mu \in M_s$ for which

$$|\hat{\mu}(n)| \le \Phi(|n|) \quad (n \in \mathbb{Z} \setminus \{0\})\,. \tag{12}$$

If condition (12) holds, we shall say that the charge μ is *subordinate* to the sequence Φ. The Ivashev-Musatov theorem asserts that any sufficiently regular positive sequence not belonging to $l^2(\mathbb{N})$ is an s-majorant. The regularity conditions are essential (cf. Subsect. K below).

B) A bounded function $\Phi : [1, +\infty) \to [0, +\infty)$ is called *fluent* if there is a number $K > 0$ such that

$$x \le y \le 2x \Rightarrow K^{-1}\Phi(x) \le \Phi(y) \le K\Phi(x)\,.$$

C) The Ivashev-Musatov Theorem. *Let Φ be a fluent sequence. Assume that*

$$\sum_{n\in\mathbb{N}} \big(\Phi(n)\big)^2 = +\infty\,. \tag{13}$$

If Φ is nonincreasing or such that

$$\lim_{n\to+\infty} n\Phi(n) = +\infty\,, \tag{14}$$

then Φ is an s-majorant. Moreover, there exists a nonzero measure of class M_s concentrated on a compact set of length zero and subordinate to the sequence $\big(\Phi(n)\big)_{n\in\mathbb{N}}$.

The simplest examples of s-majorants, according to this theorem, are the sequences

$$(n^{-\frac{1}{2}}), \quad \big((n\ln(n+1))\big)^{-\frac{1}{2}}\,, \quad \big((n\ln(n+1)\ln\ln(n+2))^{-\frac{1}{2}}\big)\,,$$

etc.

We shall now attempt to give some idea of the proof (cf. D)–J)). Throughout the following we shall assume that the function Φ is fluent and satisfies conditions (13) and (14).

D) One can see that it suffices to learn how to concentrate a probability measure subordinate to the sequence $(\Phi(n))$ on a compact set of arbitrarily small length. The question thus reduces to proving the following assertion. For any $\varepsilon > 0$ there exists a function ρ $(= \rho_{\Phi,\varepsilon})$ in $C^\infty(\mathbb{T})$ such that

$$\left.\begin{array}{c}\text{(a) } \rho \geq 0; \text{ (b) } m(\operatorname{supp}\rho) \leq \varepsilon\,; \text{ (c) } \int_{\mathbb{T}} \rho\, dm = 1\,; \\ \text{(d) } |\hat{\rho}(n)| \leq \Phi(|n|) \quad (n \in \mathbb{Z} \setminus \{0\})\end{array}\right\}. \tag{15}$$

E) The proposition of D) cannot be proved using only masses concentrated on a short arc. If conditions (a)–(d) could be satisfied for any $\varepsilon > 0$ using a function ρ concentrated on an arc of length ε, using the weak-star compactness of the unit sphere in M, we would obtain a point mass subordinate to the sequence Φ, which is impossible if $\inf \Phi = 0$. Shortening the arc on which the function ρ is concentrated, we would inevitably "spread the hump" of the function $\hat{\rho}$ (i.e., we would increase the number of n for which the amplitude $|\hat{\rho}(n)|$ is large), in complete accordance with the uncertainty principle. But condition (b) does not require that ρ be concentrated on an *arc* of small length; all that is necessary is that ρ be concentrated on a *set* of small length.

F) The following device is known for dispersing the spectrum and the support simultaneously: taking a smooth function g concentrated on a small arc l with center at 1, one must form the composition $\rho := g \circ z^N$, where N is a large natural number. The support of the function ρ is contained in the union of N arcs, each of length $m(l)/N$, with centers at the points $\exp(2\pi i k/N)$ $(k = 0, 1, \ldots, N-1)$. Thus $m(\operatorname{supp}\rho) \leq m(l)$. At the same time $\rho = \sum \hat{g}(n) z^{Nn}$, so that $\operatorname{Spec}\rho \subset N\mathbb{Z}$. Moreover $\hat{\rho}(0) = (\hat{g} \circ z^N)(0) = \hat{g}(0)$, i.e., $\int \rho\, dm = \int g\, dm$. Thus if g satisfies conditions (a)–(c), then $\rho := g \circ z^N$ also satisfies these conditions. The function ρ has many zero amplitudes. To be sure, property (d) is not yet guaranteed, since at the initial points of the progression $N\mathbb{Z}$ which enlarges further as N increases, the amplitudes are still quite large (for large $|n|$ the inequality $|\hat{\rho}(n)| \leq \Phi(|n|)$ is true, since $\rho \in C^\infty$ and $\Phi(|n|) \geq c|n|^{-m}$; this last estimate follows from the fluency of the function Φ).

G) The procedure just described for "simultaneous dispersal" can be improved. We define a *regime of motion* (over the circle $\mathbb{T}$) to be a function h subject to the following conditions

$$\left.\begin{array}{c} h : [0, 2\pi] \to [0, +\infty) \quad h \text{ strictly increasing}\,, \\ h(0) = 0\,,\ h(2\pi) = 2\pi N \ (N \in \mathbb{N})\,,\ h \in C^\infty([0, 2\pi]) \end{array}\right\}. \tag{16}$$

We set

$$\rho_h(e^{it}) := g(E_h(t)), \quad E_h(t) := e^{ih(t)} \ (t \in [0, 2\pi]) .$$

As time flows from 0 to 2π, the point $E_h(t)$ traverses the circle $\mathbb{T}$ counterclockwise exactly N times. The linear regime $h : t \mapsto Nt$ defines a uniform motion over $\mathbb{T}$ (a motion at constant velocity). We have seen in Subsect. F) that in this case the length of the supports and the integrals over $\mathbb{T}$ coincide for the functions ρ_h and g, while the spectrum of ρ_h is significantly sparser than Spec g.

The main idea of the proof of the lemma of D) is that if a regime h defines an accelerated motion with very large initial velocity, then:

1) the lengths of the supports and the integrals of the functions ρ ($= \rho_h$) and g nearly coincide;
2) all the amplitudes $|\hat{\rho}(n)|$ ($n \neq 0$) happen to be very small (i.e., the measure ρm is subordinate to the majorant Φ).
3) A function ρ ($= \rho_h$) possessing the properties (15) can be defined by the equality

$$\rho(e^{it}) = \varkappa(g \circ E_h)(t) ,$$

where h is some convex regime, and the number $\varkappa$ is close to 1. The function $g \in C^\infty$ can be chosen so that $g \geq 0$, $\int_{\mathbb{T}} g = 1$, $\operatorname{supp} g \subset l$, where l is a compact arc, $m(l) < \frac{\varepsilon}{2}$, and $1 \notin l$. The principal difficulty lies in choosing a regime h under which condition (d) holds.

It is easy to see that

$$\hat{\rho}_h(n) = \varkappa \sum_{p \in \mathbb{Z}} \hat{g}(p) I(ph, n) \ \ (n \in \mathbb{Z}) , \tag{17}$$

where $I(H,k) := \dfrac{1}{2\pi} \displaystyle\int_0^{2\pi} e^{i(H(t)-kt)} \, dt$.

H) Simple estimates of the sum (17) make it possible to reduce the lemma of D) to proving the following assertion: *If ψ is a fluent sequence and*

$$\text{(a)} \ \lim_{n \to \infty} \psi(n) n = +\infty , \quad \text{(b)} \ \sum \big(\psi(n)\big)^2 = +\infty , \tag{18}$$

then for any number $A > 0$ there exists a convex regime h such that $h'(0) > A$ and

$$\left.\begin{aligned} p, n \in \mathbb{Z} \setminus \{0\} &\Rightarrow |I(ph, n)| \leq \tfrac{a}{|n|} + bK^2 |p|^m \psi(|n|) , \\ |n| < \tfrac{A}{2}, \ p \in \mathbb{Z} \setminus \{0\} &\Rightarrow |I(ph, n)| \leq c\big(h'(0)\big)^{-1} \end{aligned}\right\} . \tag{19}$$

Here a, b, and c are absolute constants and K and m depend only on ψ. To prove the lemma of D) it is necessary to apply the estimates (19) to the sequence $\psi := C\Phi$ with a suitable constant C.

J) Estimating the integral $I(ph, n)$ by the stationary phase method ("Van der Corput's Lemma," cf. (Zygmund 1959)), we find a fixed point $t_{p,n}$ of the function $t \mapsto ph(t) - kt$ and arrive at the conclusion that we can get by with a regime h for which $h''(t) \approx \text{const}\,(\psi(n/p))^{-2}$ near $t_{p,n}$ $(h'(t_{p,n}) = n/p)$. This relation must hold for any rational fraction n/p, forcing us to seek h as a solution of the following Cauchy problem:

$$h(0) = 0,\ \ h'(0) = A\ ,\ \ h''(t) = \alpha \cdot \bigl(\psi(h'(t))\bigr)^{-2}\ \ (t \in [0, 2\pi])\ . \tag{20}$$

The following danger arises in doing this: if the quantity $\psi(u)$ is very small for large u, the velocity h' will increase rapidly and the point h will not reach 2π; for h' will become infinite somewhere in $(0, 2\pi)$. This does not happen, however, by virtue of condition (18b)! Indeed, $\displaystyle\int_\lambda^{+\infty} \bigl(\psi(u)\bigr)^2\, du = +\infty$, so that the function $\phi : v \mapsto \displaystyle\int_A^v \psi^2$ maps $[A, +\infty)$ one-to-one onto $[0, +\infty)$. We denote by θ the function inverse to φ and we set $h(s) := \displaystyle\int_0^s \theta\Bigl(\frac{v}{\alpha}\Bigr)\, dv$ $(s \in [0, 2\pi])$. The function h is a solution of the problem (20). It is easily seen that for some $\alpha \in (0, 1]$ the number $\displaystyle\int_0^{2\pi} \theta\Bigl(\frac{v}{\alpha}\Bigr)\, dv$ is a multiple of 2π, so that h is a convex regime. Its initial velocity (equal to A) is under our control, and for large A, in the final analysis, it guarantees that the measure of the support of the function ρ is small. But condition (15d) can be deduced from (19) using Van der Corput's lemma, as already mentioned.

K) Let us discuss the role of the regularity conditions on the majorant in the Ivashev-Musatov theorem. We mean the condition that the function $\varPhi$ be fluent and condition (14) (or the condition that the function $\varPhi$ be nonincreasing). It can be shown (Katznelson 1976) that if $\mu \in M$, $\operatorname{Spec} \mu \subset E$, and E satisfies condition (6), then $\mu = fm$, where $f \in L^2$. Therefore a sequence $\varPhi : \mathbb{N} \to \mathbb{R}_+$ equal to zero everywhere outside a such a set E cannot be an s-majorant (even when condition (13) holds).

The role of the condition of fluency is illustrated by the following assertion: there exists a convex nonincreasing sequence $\bigl(\varPhi(n)\bigr)$ such that (13) holds, but the closed support of any nonzero charge subordinate to it coincides with $\mathbb{T}$ (Körner 1977).

§3. Hilbert-Space Methods

Here we shall describe an approach to the uncertainty principle based on the L^2 theory of the Fourier transform. A geometric (indeed an elementary geometric) point of view will predominate: the main conclusions will be based on the concept of the angle between two subspaces of Hilbert space. The investigation will be conducted in abstract language almost to the very end. Only at the last stage will the results be translated (with a minimum of analytic techniques) into the language of Fourier theory.

3.1. Definitions and Statement of the Problem. The letters X, Ξ, and m will have the same meaning as in Sect. 1.1 above; $L^2(X)$ will have the same meaning as $L^2(X,m)$, and the symbol $L^2(\Xi)$ will denote $L^2(\mathbb{R},m)$ or $l^2(\mathbb{Z})$, depending on the meaning of the letter X. Sometimes X will be denoted $\mathbb{R}^d$; then the role of the measure m will be played by the measure $m^d := \underbrace{m \times \cdots \times m}_{d}$; in this case $L^2(\Xi) := L^2(\mathbb{R}^d, m^d)$.

We shall be interested in the interrelationship of the sets $\{f \neq 0\}$ and $\{\hat{f} \neq 0\}$, where $f \in L^2(X)$; if $X = \mathbb{R}^d$, then $\hat{f}$ is interpreted in accordance with the Plancherel theorem. We now define the concepts of support and spectrum corresponding to this new context.

A) Let (Y,μ) be a measure space, f a function defined on Y, and $E \subset Y$. Assume that $\mu(\{y \in E : f(y) = 0\}) = 0$ and $\mu(\{y \in Y \setminus E : f(y) \neq 0\}) = 0$. Then E is called the *essential support* of the function f and is denoted $\operatorname{ess\,supp} f$. It is uniquely determined up to a set of measure zero. The essential support of an element f of the space $L^p(Y,\mu)$ is defined to be the essential support of any function representing f.

Corresponding to the element f of the space $L^2(X)$, in addition to its essential support $\operatorname{ess\,supp} f$ is its *essential spectrum* $\operatorname{ess\,Spec} f := \operatorname{ess\,supp} \hat{f}$ (in this situation $\hat{f}$ is interpreted as an element of the space $L^2(\mathbb{R}^d, m^d)$ if $X = \mathbb{R}^d$; when $X = \mathbb{T}$, the set $\operatorname{ess\,Spec} f$ coincides with the usual spectrum $\operatorname{Spec} f$ $(\subset \mathbb{Z})$.

B) Consider a pair of sets (S, Σ), where $S \subset X$. We shall say that the pair (S,Σ) is *annihilating* (or *mutually annihilating*, or that it is an *α-pair*) if

$$f \in L^2(X),\ \operatorname{ess\,supp} f \subset S\,,\ \operatorname{ess\,Spec} f \subset \Sigma \Rightarrow f = 0\,. \tag{1}$$

We shall be interested in yet stronger properties of the pair (S,Σ) ("annihilation with an estimate"):

$$\exists c > 0 : f \in L^2(X) \Rightarrow \|f\|^2_{L^2(X)} \leq c\Big(\int_{X\setminus S} |f|^2\,dm + \int_{\Xi\setminus\Sigma} |\hat{f}|^2\,dm\Big)\,. \tag{2}$$

In this case we shall say that the pair (S,Σ) is *strongly annihilating* (or that it is a *strong α-pair*).

Estimate (2) expresses a certain stability of the uncertainty principle. It follows from this estimate that the simultaneous vanishing of the function f outside S and the function $\hat{f}$ outside Σ entails the total vanishing of the function f. But in addition (2) means that the simultaneous smallness of the functions f (outside S) and $\hat{f}$ (outside Σ) entails the "global" smallness of

f. The concepts of annihilation and strong annihilation are distinct (the sets $[0, +\infty)$ and $[0, 1]$ form an α-pair but not a strong α-pair).

The description of the annihilating and strongly annihilating pairs is the main topic of the present section. In the following subsection we shall take up certain questions involving the location of subspaces of a Hilbert space and then study the subspaces $\mathcal{E}(S)$, $\widehat{\mathcal{E}}(\Sigma)$ ($\subset L^2(X)$), where

$$\begin{aligned} \mathcal{E}(S) &:= \{f \in L^2(X) : \operatorname{ess\,supp} f \subset X\}, \\ \widehat{\mathcal{E}}(S) &:= \{f \in L^2(X) : \operatorname{ess\,Spec} f \subset X\}. \end{aligned} \tag{3}$$

3.2. Annihilation of Projections. We shall use the letter H to denote a certain Hilbert space. The symbol P_N will denote the orthogonal projection of the space H onto the subspace N. The word "subspace" will always mean a closed subspace. The identity operator on H will be denoted by the letter I, and the inner product of vectors v and w by the symbol $\langle v, w\rangle$.

A) If the subspaces $\mathcal{L}$ and $\mathcal{M}$ are such that $\mathcal{L} \cap \mathcal{M} = \{0\}$, we shall say that they (or the projections $P_{\mathcal{L}}$ and $P_{\mathcal{M}}$) are *annihilating* (or that they form an *α-pair.*) The pair $(\mathcal{L}, \mathcal{M})$ (or $(P_{\mathcal{L}}, P_{\mathcal{M}})$) is called a *strong* α-pair if

$$\|P_{\mathcal{L}} P_{\mathcal{M}}\| < 1. \tag{4}$$

Every strong α-pair is an α-pair.

B) We now list several propositions that are equivalent to inequality (4), setting $P := P_{\mathcal{L}}$, $Q := P_{\mathcal{M}}$, $R := I - PQ$:

(a) $\sup\{\langle v, w\rangle : v \in \mathcal{L},\ w \in \mathcal{M},\ \|v\| = \|w\| = 1\} < 1$ (the angle between the subspaces $\mathcal{L}$ and $\mathcal{M}$ is positive);
(b) the operator R is invertible;
(c) $\mathcal{L} \cap \mathcal{M} = \{0\}$ and the linear manifold $\mathcal{L} + \mathcal{M}$ is closed;
(d) $\exists c > 0 : \|(I - P)v\| \geq c\|v\|$ for any $v \in \mathcal{M}$;
(e) $\exists c > 0 : c\|v\| \leq \|(I - P)v\| + \|(I - Q)v\|$ for all $v \in H$.

C) The pair of projections (P, Q) is called *compact* if the operator PQ is compact. Any compact α-pair is a strong α-pair.

If the operator $S := PQ$ is compact, it follows that there exists a vector v such that $\|v\| = 1$, $\|Sv\| = \|S\|$; if $\|S\| = 1$, then $1 = \|Sv\| \leq \|Qv\| \leq \|v\| = 1$ and $\|Qv\| = \|v\|$, so that $Qv = v$, $Sv = Pv$, $\|Sv\| = \|Pv\| = \|v\|$, $Pv = v$, and $v \in \mathcal{L} \cap \mathcal{M}$, which is impossible, since (P, Q) is an α-pair. Hence $\|S\| < 1$.

D) If (P, Q) is a strong α-pair, then the operator $P \times Q : H \to \mathcal{L} \times \mathcal{M}$, which assigns to the vector $v \in H$ the pair (Pv, Qv) is surjective. Thus the system

$$Pr = p, \quad Qr = q \tag{5}$$

is solvable for any $p \in \mathcal{L}$, $q \in \mathcal{M}$. One solution is the vector

$$r := (I - Q)R^{-1}p + (I - P)(R^*)^{-1}q. \tag{6}$$

The solution (6) has minimal norm among the solutions of the system (5).

E) To conclude this "abstract" section we consider the following problem: What values can the quantity $\|Qh\|$ assume under the hypothesis that $\|h\| = 1$ and $\|Ph\| = a$, where a is a prescribed number in the interval $[0,1]$? A fairly complete answer to this question can be given when the α-pair (P,Q) is compact and controllable. The latter means that one can map any vector h into a vector almost orthogonal to $\mathcal{M}$ by rotating the space H while leaving $\mathcal{L}$ and $H \ominus \mathcal{M}$ invariant. More precisely, there exists a family $\{U(\tau)\}_{\tau\in\mathbb{T}}$ of unitary operators commuting with P such that for any $h \in H$ the vector $U(\tau)h$ depends continuously on the parameter τ ranging over the connected topological space $\mathbb{T}$, the identity operator is among the operators $U(\tau)$, and for any $h \in H$ the quantity $\|QU(\tau)h\|$ assumes arbitrarily small values. Moreover we assume that $PQ \neq 0$, $P \neq I$, $Q \neq I$, $\dim \mathcal{L} = \dim \mathcal{M} = \infty$. The question just posed can then be stated as follows: What is the image W of the unit sphere of the space H under the mapping $h \mapsto (\|Ph\|, \|Qh\|)$ (from H into $\mathbb{R}^2$)?

We set $c := \sqrt{\|QPQ\|}$ and consider the curve

$$\gamma := \{(x,y) \in J \times J : \arccos x + \arccos y = \arccos c\} .$$

Using the letter A to denote the union of all rectangles of the form $(0,x] \times (0,y)$, where $(x,y) \in \gamma$, we can write the following answer to the question:

$$(\mathring{J} := (0,1)) \ :$$

$$A \cup (\{0\} \times \mathring{J}) \cup (\mathring{J} \times \{0\}) \subset W \subset A \cup (\{0\} \times J) \cup (J \times \{0\}) . \tag{7}$$

The inclusion (7) gives a complete picture of the interior of the set W. This result can be interpreted as an abstract uncertainty principle: under the condition $\|h\| = 1$ the two numbers $\|Ph\|$ and $\|Qh\|$ cannot both be close to 1. Assertion (7) can be proved using simple (essentially two-dimensional) considerations in the plane $\{\lambda \cdot Ph + \mu \cdot Qh : \lambda, \mu \in \mathbb{C}\}$ (cf. (Khurgin and Yakovlev 1971)).

In solving this problem a system (ψ_j) of normalized eigenvectors of the compact self-adjoint (positive) operator PQP arises. The system of vectors $(P\psi_j)$ is obviously orthogonal ($\|P\psi_j\|^2 = \lambda_j$, where λ_j is the jth eigenvalue of the operator PQP). Such "doubly orthogonal systems" have been studied by Krasichkov-Ternovskij (1968) and by Khurgin and Yakovlev (1971).

3.3. The Amrein-Berthier Theorem. We turn again to harmonic analysis. We shall apply the theory discussed in Sect. 1.3.2 to the projections P_S and $\widehat{P}_\Sigma$ on the subspaces $\mathcal{E}(S)$ and $\widehat{\mathcal{E}}(\Sigma)$ (cf. 1.3.1) where S and Σ are Lebesgue-measurable parts of the space $\mathbb{R}^d$. Thus we now have $Pf = \chi_S f$, $Qf = \mathcal{F}^{-1} P_\Sigma \mathcal{F} f$, where χ_S is the characteristic function of the set S, $\mathcal{F}f := \hat{f}$, and

$$\hat{f}(\xi) = (2\pi)^{-d/2} \int_{\mathbb{R}^d} f(u) e^{-i\langle u,\xi \rangle} \, du$$

(the integral being interpreted in accordance with Plancherel's theorem). Let us agree to write $|A|$ instead of $m^d(A)$.

A) The Amrein-Berthier Theorem (Amrein, Berthier 1977). *If*

$$|S| + |\Sigma| < +\infty \,, \tag{8}$$

then the sets S and Σ are strongly annihilating.

Let us sketch the proof. The operator $K := PQ$ is an integral operator (with a Hilbert-Schmidt kernel). Therefore it suffices to verify that (P,Q) is an α-pair (cf. Subsect. 1.3.2C)). The subspace $\mathcal{L} \cap \mathcal{M}$ is contained in the eigenspace N of the operator K corresponding to the eigenvalue 1. It is finite-dimensional; a direct estimate shows that $\dim N \le c\sqrt{|S|\,|\Sigma|}$. Let $f \ne 0$, and $f \in \mathcal{L} \cap \mathcal{M}$. Any translate of the function f remains in $\mathcal{M}$ $(= \widehat{\mathcal{E}}_\Sigma)$. On the other hand, by carefully moving ess sup f out of S, one can construct arbitrarily many linearly independent translates of the function f whose essential supports remain in some fixed set $S^* \supset S$, and $|S^*| < |S| + 1$. The dimension of the subspace N^* corresponding to the pair (S^*, Σ) can thus be assumed arbitrarily large, contradicting the estimate given above.

B) We can now apply all the "abstract" results of Sect. 1.3.2 to the pair $(\mathcal{E}_S, \widehat{\mathcal{E}}_\Sigma)$. Thus under condition (8) for any pair of functions $p, q \in L^2$ $(= L^2(\mathbb{R}^d))$ there exists a function $r \in L^2$ such that

$$r\big|_S = p\big|_S \,, \quad \hat{r}\big|_\Sigma = \hat{q}\big|_\Sigma \,. \tag{9}$$

We can now free ourselves completely from the constricting action of the uncertainty principle: under hypothesis (8) the functions $r\big|_S$ and $\hat{r}\big|_\Sigma$ can be chosen completely arbitrarily. The solution of the problem (9) reduces to inverting the operator $I - PQ$ (cf. (6)).

C) Under condition (8) the projections P and Q form a controllable pair (we assume that $|S| > 0$ and $|\Sigma| > 0$). Assertion 2E) then becomes the *Slepyan-Pollak theorem.*

The quantity $\|Ph\|/\|h\|$ will be called the portion of the energy of the function h reaching the set S, and the quantity $\|Qh\|/\|h\|$ the portion of its spectral energy reaching the set Σ. The Slepyan-Pollak theorem answers the following question: What portion of the spectral energy can reach the set Σ under the condition that the portion of energy reaching S equals a?

D) Assume that condition (8) holds. Then what is the value of the constant c occurring in inequality (2)? Can it be expressed completely in terms of the Lebesgue measure of the sets S and Σ? The following theorem gives a kind of answer to this question.

Nazarov's Theorem. *Let $d = 1$. Under condition* (8) *one can set $c = A\exp(A|S||\Sigma|)$ in* (2), *where A is a positive constant.*

Let g be a nonnegative Lebesgue-measurable function defined on $\mathbb{R}$. A nonincreasing function $\Phi : [0, +\infty) \to [0, +\infty]$ will be called a *majorant* of the distribution of the function g if $\big|\{t \in \mathbb{R} : g(t) > y\}\big| \le \big|\{t \ge 0 : \Phi(t) > y\}\big|$ for all $y > 0$. The following assertion is a corollary of Nazarov's theorem: *if $f \in L^2(\mathbb{R})$ and Φ and Ψ are majorants of the distributions of the functions $|f|$ and $|\hat{f}|$ (respectively), then*

$$\theta(a,b) := \int_a^{+\infty} \Phi^2 + \int_b^{+\infty} \Psi^2 \ge \frac{1}{A} \exp(-Aab)\|f\|_2^2, \quad (a, b > 0)\,. \tag{10}$$

In particular $\varliminf_{a,b\to+\infty} [\theta(a,b) \exp Aab] > 0$ if $f \ne 0$. Therefore

$$p, q > 0\,, \quad p^{-1} + q^{-1} < 1, \quad \Phi(y) = O(e^{-y^p})\,, \quad \Psi(y) = O(e^{-y^q}) \Rightarrow f = 0\,.$$

In connection with this fact see Morgan's theorem below (Sect. 2.1.1).

Nazarov showed that inequality (10) remains true if Ψ is replaced by Ψ_-, where Ψ_- is a majorant of the distribution of the function $|\hat{f}|\big|_{(-\infty,0)}$, (i.e., $\big|\{t < 0 : |\hat{f}(t)| > y\}\big| \le \big|\{t \ge 0 : \Psi(t) > y\}\big|$ for all $y > 0$.

3.4. Strong Annihilation of Subsets of the Circle and Sparse Spectra. An interesting class of strong α-pairs is provided by Zygmund's theorem on lacunary spectra. It refers to the case $X = \mathbb{T}$, $\Xi = \mathbb{Z}$ and to the sets $S \subset \mathbb{T}$ and $\Sigma \subset \mathbb{Z}$. The projections P and Q are as follows: $P_S f = \chi_S f$, $Qf = \chi_\Sigma \hat{f}$.

A set A of integers will be called *Q-sparse* if it is contained entirely in $\mathbb{N}$ or $-\mathbb{N}$ and $\inf\{|n|/|m| : n, m \in A, |n| > |m|\} \ge Q$. A set $\Lambda \subset \mathbb{Z}$ is called *lacunary* if there exist numbers $N \in \mathbb{N}$ and $Q > 1$ such that both of the sets $\Lambda \cap [N, +\infty)$ and $\Lambda \cap (-\infty, -N]$ are Q-sparse.

Zygmund's Theorem (1959). *If $mS < 1$ and the set Σ is lacunary, then (S, Σ) is a strong α-pair.*

Under the hypotheses of Zygmund's theorem the phenomenon of "compulsory smoothness" is observed: functions with a sparse (lacunary) spectrum are either nowhere smooth or everywhere smooth. More precisely, if $f \in L^2(\mathbb{T})$ and the set $\operatorname{Spec} f$ is lacunary, and if f coincides with some function of class $C^{n+1}(\mathbb{T})$ on a set S of positive length, then $f \in C^n(\mathbb{T})$. And if f coincides on S with a function that is analytic on an arc, containing S, then f is analytic on the entire circle $\mathbb{T}$. Analogous assertions hold under the hypotheses of the Amrein-Berthier theorem.

3.5. Supports that are Strongly Annihilating with any Bounded Spectrum. None of the descriptions of strong α-pairs given in Sects. 1.3.3 and 1.3.4 was a complete characterization. Thus for example having established that any support of finite volume is strongly annihilating with any spectrum of finite volume (Sect. 1.3.3), we were not in a position to characterize the set of supports that form a strong α-pair with any spectrum of finite volume (it can be shown that the volume of such a support may be infinite).

In this section we shall exhibit a complete description of the supports that are strongly annihilating with any bounded spectrum.

Let $\mathcal{E}$ be a subspace in L^2 ($= L^2(\mathbb{R}^d)$). A set $S \subset \mathbb{R}^d$ is called $\mathcal{E}$-essential if

$$\exists c > 0 \ : \ f \in \mathcal{E} \Rightarrow c\|f\|_2^2 \leq \int\limits_S |f|^2 \, dm^d \, .$$

We shall call a set S *essential* if it is $\widehat{\mathcal{E}}(\Sigma)$-essential (cf. (3)) for any bounded set $\Sigma \subset \mathbb{R}^d$ (this means that $S' := \mathbb{R}^d \setminus S$ is strongly annihilating with any ball).

The set $S \subset \mathbb{R}^d$ is called *relatively dense* if it occupies a fixed (positive) portion of any cube that is congruent to some fixed cube:

$$\exists \gamma > 0 \ : \ \text{cube}\, K \subset \mathbb{R}^d \, , \ x \in \mathbb{R}^d \Rightarrow |(K + x) \cap S| \geq \gamma \, .$$

The Logvinenko-Sereda Theorem (1974). *The following assertions are equivalent:*

(a) *the set S is essential;*
(b) *the set S is relatively dense.*

The proof that (a) $\Rightarrow$ (b) is quite simple. Indeed, if $\mathcal{E}$ is a nonzero subspace of L^2 that is translation-invariant, then any $\mathcal{E}$-essential set is relatively dense, so that it remains only to set $\mathcal{E} = \widehat{\mathcal{E}}(\Sigma)$. The proof of the fundamental part ((b) $\Rightarrow$ (a)) can be found in the paper of Logvinenko-Sereda (1974) (and also in (Gorin 1985) and (Jöricke, Havin 1985b)).

Chapter 2
Complex Methods

The complex point of view opens wide prospects for our theme: in its different manifestations the uncertainty principle arises as a corollary of the various uniqueness theorems in which complex analysis is so rich. Here is a crude example: the fact that any charge $\mu \in M(\mathbb{R})$ with bounded support and bounded spectrum vanishes identically follows immediately from the fact that $\hat{\mu}$ coincides on $\mathbb{R}$ with some entire function if $\operatorname{diam} \operatorname{supp} \mu < +\infty$. It is clear that this primitive consideration can be developed into much more precise results.

In this part we gather results that illustrate the possibilities of the "complex ideology" in application to different forms of the uncertainty principle.

§1. First Examples of the Application of Complex Methods

1.1. Limiting Rate of Decay of the Fourier Transform of a Rapidly Decreasing Function. Dzhrbashyan's Theorem. Consider functions $a : X \to [0, +\infty]$ and $\alpha : \Xi \to [0, +\infty]$, where X and Ξ have the same meaning as in §1 of Chapter 1. We shall call the functions a and α *majorants.* We call a pair of majorants *sufficient* if

$$f \in L^1(X)\,, \quad |f(x)| \leq a(x)\,(x \in X)\,, \quad |\hat{f}(\xi)| \leq \alpha(\xi)\,(\xi \in \Xi) \Rightarrow f = 0\,.$$

(The requirement that $f \in L^1$ is not essential here; it can be replaced by any other condition under which $\hat{f}$ is a function that is defined everywhere or almost everywhere in Ξ.)

A majorant pair (a, α) is obviously sufficient if $X = \mathbb{R}$ and both functions a and α vanish outside a bounded set, or if

$$a(x) = \begin{cases} 0, & x < 0, \\ +\infty, & x \geq 0, \end{cases} \quad \int\limits_{\mathbb{R}} \log \alpha(t) \frac{dt}{1+t^2} = -\infty$$

(cf. Sect. 1.5D) of Chapter 1). Here we consider in detail the case when $X = \mathbb{R}$ and the majorants a and α are strictly positive and finite and tend to zero at infinity. In this situation the source of many specific examples of sufficient pairs is *Dzhrbashyan's Theorem,* whose proof makes it possible to give an effective illustration of the complex approach to the uncertainty principle using minimal tools.

A) Let M and N be continuous real-valued functions defined on the half-line $[0, +\infty) =: \mathbb{R}_+$. If $ab \leq M(a) + N(b)$, $(a, b \geq 0)$, these functions are said to be *mutually complementary in the sense of Young.* If $M \in C(\mathbb{R}_+)$, $M(0) = 0$, $M \geq 0$, and $\lim\limits_{x \to +\infty} \dfrac{M(x)}{x} = +\infty$, there exists a nonnegative increasing and convex function N complementary to M, and $\lim\limits_{x \to +\infty} \dfrac{N(x)}{x} = +\infty$ (Natanson 1974). An example of mutually complementary functions of great importance for us is provided by the pair $M = Cx^p$, $N = (C_p)^{-1/(p-1)} q^{-1} x^q$, $(C > 0, p > 1, q = p/(p-1))$.

B) Let M and N be mutually complementary nonnegative functions. Assume that

$$\text{(a)}\ f \in L^1(\mathbb{R})\,, \quad \text{(b)}\ fe^M \in L^1(\mathbb{R}_+)\,. \tag{1}$$

The last inclusion means that f is "small" on the half-line $\mathbb{R}_+$. This provides the possibility of extending the function f continuously from the real axis $\mathbb{R}$ to a function F defined in the closed upper half-plane and holomorphic in the open half-plane $\mathbb{C}_+ := \{\zeta \in \mathbb{C} : \operatorname{Im} \zeta > 0\}$. Indeed, $|f(x)e^{-ix\zeta}| = |f(x)|e^{x\eta}$ $(x \in \mathbb{R}, \zeta = \xi + i\eta \in \mathbb{C})$. Therefore, setting $c_- := \|f\|_{L^1(\mathbb{R})}$, $c_+ := \|fe^M\|_{L^1(\mathbb{R}_+)}$, we obtain for $\eta > 0$

$$\int_{-\infty}^{+\infty} |f(x)e^{-ix\zeta}|\,dx \le c_- + \int_0^{+\infty} |f(x)|e^{x\eta}\,dx \le c_- + c_+ e^{N(\eta)}\,. \qquad (2)$$

We set $F(\zeta) := (2\pi)^{-1} \int\limits_{\mathbb{R}} f(x)e^{-ix\zeta}\,dx$ ($\zeta \in \mathbb{C}_+$). It can be seen from (2) that the integral that defines F converges uniformly on any bounded portion of the half-plane $\mathbb{C}_+ \cup \mathbb{R}$ so that the function F is analytic in $\mathbb{C}_+$, continuous on $\mathbb{C}_+ \cup \mathbb{R}$, and $F|\mathbb{R} = \hat{f}$. In the course of this reasoning we have established that $|F(\zeta)|$ does not grow too quickly as $|\zeta|$ increases:

$$|F(\zeta)| \le e^{N(\eta)}(c_- + c_+)\,. \qquad (3)$$

But in that case, in accordance with well-known principles of complex analysis, $|F|$ cannot decay too quickly (for example, along $\mathbb{R}_+$). To realize this principle we use *Carleman's formula.*

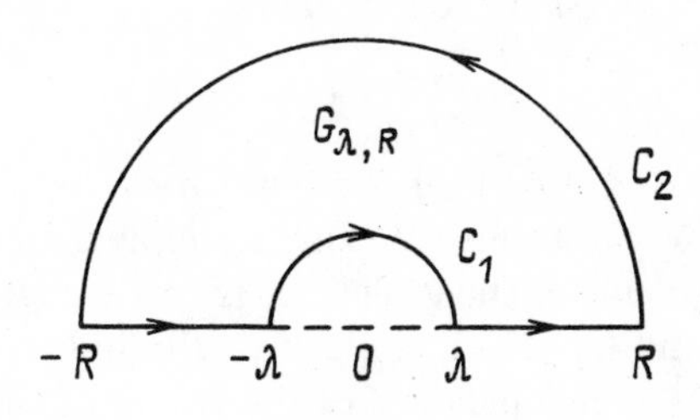

Fig. 2

C) Carleman's formula involves positive numbers λ and R ($\lambda < R$), the domain $G_{\lambda,R}$ shown in Fig. 2, and a function F continuous in the closure of $G_{\lambda,R}$ and holomorphic in $G_{\lambda,R}$. Assume that F does not vanish on the boundary C of the domain $G_{\lambda,R}$, and let $Z(F) := \{a \in G_{\lambda,R} : F(a) = 0\}$.

Fix a branch of the function $\log F$ along a simple closed path whose trajectory coincides with C and whose beginning and end are at the point λ. We consider the contour integral

$$I := \frac{1}{2\pi i} \oint\limits_C \left(\frac{1}{R^2} - \frac{1}{z^2}\right) \log F(z)\,dz$$

and write it in two different ways. First we represent it as the sum of integrals over the arcs C_1 and C_2 and the union of the intervals $[-R, -\lambda]$ and $[\lambda, R]$. Second, integrating by parts, we obtain: $I = \alpha + j$, where $\alpha \in \mathbb{R}$ and $j := \frac{1}{2\pi i} \oint\limits_C \frac{F'(z)}{F(z)}\left(\frac{z}{R^2} + \frac{1}{z}\right) dz$. Computing j by the residue theorem and equating the imaginary parts of the two expressions obtained by the methods above, we obtain

$$A_1 + A_2 + A_3 = A_4 \,, \tag{4}$$

where

$$A_1 := -\operatorname{Im} \frac{1}{2\pi} \int_0^\pi \log F(\lambda e^{i\theta}) \Big(\frac{\lambda e^{i\theta}}{R^2} - \frac{e^{-i\theta}}{\lambda} \Big) \, d\theta \,,$$

$$A_2 := \frac{1}{2\pi} \int_\lambda^R \Big(\frac{1}{x^2} - \frac{1}{R^2} \Big) \log |F(x)F(-x)| \, dx \,,$$

$$A_3 := \frac{1}{\pi R} \int_0^\pi \log |F(Re^{i\theta})| \sin\theta \, d\theta \,,$$

and A_4 is a nonnegative quantity (expressed in terms of the roots of the function F, though its precise form is of no importance to us).

D) Assume that the nonzero function F is defined and continuous in $\mathbb{C}_+ \cup \mathbb{R}$ and holomorphic in $\mathbb{C}_+$. Then the inequality

$$A_1 + A_2 + A_3 \geq 0 \tag{5}$$

holds for any $R > 1$ and $\lambda = 1$ (if there are roots of the function F on the circle $|z| = R$, then (5) is established using a simple passage to the limit). The quantity A_3 takes account of the growth of $|F(z)|$ for large $|z|$. Knowing how to find an upper bound for $|F(z)|$, we can easily find an upper bound for A_3 also. The quantity A_1 is independent of the values of the function F outside the arc $|z| = 1$ and remains bounded as $R \to +\infty$. Therefore, knowing an upper bound for $|F(z)|$ for large $|z|$ we can deduce a lower bound for A_2 from inequality (8) when R is large.

But such an estimate means precisely that the function $|F|$ cannot be too small on large subsets of the real axis.

E) Applying inequality (5) and estimate (3) to the function F, which we defined in B), it is not difficult to arrive at the following conclusion.

Theorem (Dzhrbashyan). *If $f \neq 0$ and conditions* (1) *hold, then*

$$\varliminf_{R \to +\infty} \{ S_+^N(R) + S_-^{l(\hat{f})}(R) \} > -\infty \,, \tag{6}$$

where

$$S_+^N(R) := R^{-1} \int_0^{\pi/2} N(R\sin\theta) \sin\theta \, d\theta, \quad S_-^l(R) := \frac{1}{2R^2} \int_l^R x l(x) \, dx \,,$$

$$l(x) \, (= l(g)(x)) := \int_1^x \frac{\log |g(\xi) g(-\xi)|}{\xi^2} \, d\xi \ \ (x \geq 1) \,.$$

The faster M increases, i.e., the smaller the function f subject to condition (1b) is on $\mathbb{R}_+$, the slower the function N grows (and with it S_+^N), and hence the more rigid the restriction imposed by inequality (6) on the rate of decay of $|\hat{f}(\xi)|$ as $|\xi|$ increases.

F) By making a specific choice of the function M, one can deduce from (6) a variety of very intuitive forms of the uncertainty principle (cf. Dzhrbashyan 1958). We shall limit ourselves here to just one example. If $p > 1$, $r > 1$, and $p^{-1} + r^{-1} < 1$, then the pair of majorants $(e^{-A|x|^p}, e^{-B|\xi|^r})$ is sufficient, for any positive numbers A and B. Indeed, taking account of what was said in Sect. A), one can easily show that

$$|f(x)| = O(e^{-A|x|^p}) \Rightarrow S_+^N(R) = O(R^{q-1}) ,$$
$$|\hat{f}(\xi)| = O(e^{-B|\xi|^r}) \Rightarrow S_-^{l(\hat{f})}(R) < -\text{const}\, R^{r-1} ,$$

and (6) is violated, so that $f = 0$.

It is interesting to compare this result with the "real-variable" theorem of Nazarov (cf. Sect. 3.3 of Chapt. 1).

The condition $1/p+1/r < 1$ is essential here, as is shown by the example of the function $\exp(-x^2/2)$ $(p = r = 2)$. If $1/p + 1/r = 1$, the sufficiency of the corresponding pair of majorants now depends on the relation of the constants A and B. Precise results in this case are due to Morgan (1934). His approach is based on the Phragmén-Lindelöf theorem, and the proof of the sharpness of the result uses the saddle point method.

1.2. High-order Zeros and Sparse Spectrum. Mandelbrojt's Theorem. In this section we shall consider the following version of the uncertainty principle: if a nonconstant function tends to zero very rapidly as a certain point is approached, its spectrum cannot be too sparse. Here is an example of a precise assertion of this type:

$$\left.\begin{array}{l} f \in C(\mathbb{T}),\ \varepsilon > 0,\ f(t) = O(\exp(-|t-1|^{-(1+\varepsilon)})\ (t \to 1) \\ \displaystyle\sum_{n \in \operatorname{Spec} f} |n|^{-\varepsilon-1/2} < +\infty \end{array}\right\} \Rightarrow f = 0 .$$

This is a very special case of Mandelbrojt's theorem, mentioned in the title of this section.

A) With any set Λ of integers we associate the quantities

$$S(p, \Lambda) := \sum_{\lambda \in \Lambda \setminus \{0\}} |\lambda|^{-p} \ \ (p \in (0,1]),\ \sigma(\Lambda) := \inf\{p :\ S(p, \Lambda) < +\infty\} .$$

The number $\sigma(\Lambda)$ is called the *exponent of convergence* of the set Λ. The smaller $\sigma(\Lambda)$ the sparser the set Λ (in some sense).

Theorem (Mandelbrojt (Levin 1956; Levin, Livshits 1941; Mandelbrojt 1937)). 1) *If the function f of class $L^1(\mathbb{T})$ is such that $\sigma := \sigma(\operatorname{Spec} f) < 1$ and*

$$\int_0^\varepsilon |f(e^{it})|\,dt = O(\exp(-\varepsilon^{-\rho})) \quad (\varepsilon \downarrow 0)\,, \tag{8}$$

where $\rho > \sigma(1-\sigma)^{-1}$, *then* $f = 0$; 2) *for any* $\sigma \in (0,1)$ *there exists a nonzero function* f *of class* $C(\mathrm{T})$ *such that* $\sigma(\operatorname{Spec} f) = \sigma$ *and*

$$f(e^{it}) = O(\exp(-t^{-\rho})) \quad (|t| \downarrow 0)\,, \tag{9}$$

where $\rho := \sigma(1-\sigma)^{-1}$.

The proof, whose major points we shall describe (cf. Sects. B)–D) below), is borrowed from Levin (1956) (where certain generalizations are also discussed, along with the history of the problem). An interesting "purely real" approach to Mandelbrojt's theorem (and to other forms of the uncertainty principle of the same type) was proposed by A.S. Belov (1976).

B) With the function f we associate a 2π-periodic function $\varphi : t \mapsto f(e^{it})$. The "complexification" of the problem is carried out by using the Laplace transform Φ of the function φ:

$$\Phi(\zeta) := \int_0^{+\infty} \varphi(t) e^{-t\zeta}\,dt = \sum_{n\in\mathbb{Z}} \frac{\hat{f}(n)}{\zeta - in} =: F(\zeta)\,(\operatorname{Re}\zeta > 0)\,.$$

We remark that F is defined everywhere in $\mathbb{C}$ as a meromorphic function with poles at the points in, where $n \in \operatorname{Spec} f$. A simple estimate shows that

$$F(\zeta) = O(|\zeta| \log|\zeta|) \quad \left(|\zeta| \to +\infty,\ \operatorname{dist}(\zeta, i\mathbb{Z}) > \frac{1}{4}\right). \tag{10}$$

The sparseness of the spectrum of the function f means, roughly speaking, that F has "few poles." How does a "high-order zero" of the function f (cf. (8)) affect F?

If the function φ were identically equal to zero in an interval of the form $(0,a)$, where $a > 0$, then a rough estimate of the integral that defines Φ would yield

$$|\Phi(\xi + i\eta)| \le C(\varepsilon) e^{\varepsilon a} e^{-\xi a} \quad (\xi > 0,\ \eta \in \mathbb{R}, \varepsilon > 0)\,.$$

Thus the gap $(0,a)$ in $\operatorname{supp}\varphi$ manifests itself in the fact that $|\Phi(\zeta)|$ tends to zero rather rapidly as $\operatorname{Re}\zeta \uparrow +\infty$. One can also use information about the "high-order zero" of the function f at the point 1 (i.e., condition (8)) to estimate the decay of the function Φ:

$$|\Phi(\xi + i\eta)| \le Ce^{-c\xi^\lambda} + O(e^{-2\pi\xi})\,, \quad (\xi \uparrow +\infty)\,, \quad \lambda := \rho/(\rho+1)\,. \tag{11}$$

Assume that $f \ne 0$. Then (by known properties of the Laplace transform) $F \ne 0$ also. By (10) the meromorphic function F does not grow too rapidly. If f has a "high-order zero" (condition (8)), then the quantity $|F(\xi+i\eta)| = |\Phi(\xi+i\eta)|$ is very small for large $\xi > 0$. At this point the principle enunciated in Sect. 1.1 comes into force: if a nonzero function is not too large, it cannot tend to zero very rapidly. A rapid decay of the function F (as the independent variable moves off to the right) is possible only when the function has a sufficient number of poles, and that is prevented by the fact that the number $\sigma(\operatorname{Spec} f)$ is small.

C) The realization of these considerations is based on the *Poisson-Jensen formula*:

$$\int_0^R \frac{N(t)}{t}\,dt = \frac{1}{2\pi}\int_0^{2\pi} \log|F(Re^{i\theta})|\,d\theta - \log|F(0)| \ (R>0)\,. \tag{12}$$

Here the role of F can be played by any meromorphic function that is analytic and nonzero at the origin; $N(t) := n_F(t) - p_F(t)$, where $n_F(t)$ is the number of zeros of the function F in the disk $t\mathbb{D}$ and $p_F(t)$ the number of poles. It follows from the identity (12) that for any $R>0$

$$\int_0^R \frac{p_F(t)}{t}\,dt \geq -\frac{1}{2\pi}(J_-^F(R) + J_+^F(R)) + \log|F(0)|\,, \tag{13}$$

where

$$J_-^F(R) := \int\limits_{|\theta|\leq\frac{\pi}{2}} \log|F(Re^{i\theta})|\,d\theta\,, \quad J_+^F := \int_{\pi/2}^{3\pi/2} \log|F(Re^{i\theta})|\,d\theta\,.$$

We can now sharpen the assertion about the number of poles made in B): if for large R the integral $J_+^F(R)$ is not too large (cf. (10)), and the function $|F|$ is very small on a significant portion of the arc $\{z = Re^{i\theta} : |\theta| < \pi/2\}$ (cf. (11)), then $J_-^F(R)$ is a negative number of very large absolute value (much larger than $|J_+^F(R)|$), and as a result of (13) the average $p_F(t)/t$ cannot be too small on the whole interval $(0,R)$.

D) Such are the general features of the line of reasoning in the proof of assertion 1) of Mandelbrojt's theorem. The reasoning itself suggests a way of seeking the function f needed in the hypotheses of assertion 2). It is first necessary to construct an "image" of it – a meromorphic function F with simple poles at $i\Lambda$, where $\Lambda \subset \mathbb{Z}$ is the spectrum of the unknown function f (with a prescribed exponent of summability σ). We set $F := 1/B$, where $B := \prod\limits_{\lambda\in\Lambda_\sigma}\left(1+\frac{z^2}{\lambda^2}\right)$ and Λ_σ is the set of numbers of the form $[n^\sigma]$ $(n\in\mathbb{N})$. We then define a 2π-periodic function $t\mapsto f(e^{it})$ by the Riemann-Mellin formula (i.e., we find the "Laplacian primitive" of the function F). The "high-order zero" needed (i.e., estimate (9)) follows from the rapid decay of the quantity $|F(\zeta)|$ as $|\zeta|\to+\infty$ with $|\operatorname{Re}\zeta| > c$ (namely $|F(\zeta)| = O(\exp(-A|\zeta|^\sigma))$).

§2. The Uncertainty Principle for Plus-functions from the New Point of View

We now apply the complex approach to questions already touched upon in §1 of Chapt. 1. Plus-functions arise here as the boundary values of functions that are analytic in the disk or the half-plane. This point of view historically preceded the approach discussed in the first chapter. It is based on the following observation: if $f \in L^1(\mathbb{T})$ and $\operatorname{Spec} f \subset \mathbb{Z}_+$, then the series $\sum_{n\in\mathbb{Z}} \hat{f}(n)\zeta^n = \sum_{n\geq 0} \hat{f}(n)\zeta^n$ converges in the open unit disk $\mathbb{D} := \{\zeta \in \mathbb{C} : |\zeta| < 1\}$, and its sum f_+ is analytic in that disk. In exactly the same way a function of a real variable $f \in L^1(\mathbb{R})$ whose spectrum lies on the half-line $[0, +\infty)$ is connected with a function of a complex variable $f_+ : \zeta \mapsto \int_0^{+\infty} \hat{f}(\xi)e^{i\xi\zeta}\, d\xi$ that is analytic in the open upper half-plane $\mathbb{C}_+ := \{\zeta \in \mathbb{C} : \operatorname{Im}\zeta > 0\}$. In both cases $f(\zeta)$ is the limit (in a certain sense) $\lim_{\zeta'\to\zeta} f_+(\zeta')$ for $\zeta \in X$. The requirement that the function f be integrable is not really essential (for example, one could consider a finite charge instead of f).

2.1. Plus-charges on the Circle $\mathbb{T}$. We begin with the definition of some important classes of functions that are analytic in the disk $\mathbb{D}$.

A) With a function f defined in $\mathbb{D}$ and a number $r \in (0, 1)$ we associate the function $f_r : \zeta \mapsto f(r\zeta)$ $(|\zeta| < 1/r)$. The symbol $\|\cdot\|_p$ will denote the L^p-norm (with respect to normalized Lebesgue measure m on the circle $\mathbb{T}$). The symbol $\mathcal{O}(G)$ will denote the set of functions that are analytic in the domain G.

B) The *Nevanlinna class* N, by definition, consists of all functions $f \in \mathcal{O}(\mathbb{D})$ for which

$$\sup_{0<r<1} \int_{\mathbb{T}} \log^+ |f(rz)|\, dm(z) \ \left(= \sup_{0<r<1} \|\log^+ |f_r|\,\|_1\right) < +\infty$$

($\log^+ t := \log t$ if $t > 1$ and $\log^+ t = 0$ if $0 \leq t \leq 1$).

C) To each number $p > 0$ there corresponds a Hardy class $H^p(\mathbb{D}) := \{f \in \mathcal{O}(\mathbb{D}) : \sup_{0<r<1} \|f_r\|_p < +\infty\}$. The symbol $H^\infty(\mathbb{D})$ is normally used to denote the set of functions that are analytic and bounded in $\mathbb{D}$. It is easy to see that

$$0 < p_1 < p_2 \Rightarrow H^\infty \subset H^{p_2} \subset H^{p_1} \subset N\,.$$

We shall now list the basic propositions of a beautiful and important theory that makes it possible in the end to identify the class of analytic functions $H^p(\mathbb{D})$, $p \in [1, +\infty]$, with the corresponding class $H^p(\mathbb{T})$ of plus-functions (in the sense of Sect. 1.1) belonging to $L^p(\mathbb{T})$. We begin by describing the subsets of the disk $\mathbb{D}$ on which a nonzero function of class $H^p(\mathbb{D})$ or N can vanish.

D) We define a *divisor* to be an integer-valued function of a complex variable. The *divisor of the function* $f \in \mathcal{O}(G)$ $(f \not\equiv 0)$ is defined as the divisor that vanishes throughout the domain G except at the zeros of the function f and equals the multiplicity of the zero a of the function f if $f(a) = 0$; this divisor is denoted by the symbol $\operatorname{dvs} f$. The symbol $\operatorname{Dvs} X$, where $X \subset \mathcal{O}(G)$ denotes the set of divisors of the nonzero functions belonging to the class X. Finally we shall say that a divisor Δ defined on $\mathbb{D}$ satisfies the *Blaschke condition* (and write $\Delta \in \operatorname{Bl}(\mathbb{D})$) if

$$\sum_{a\in\mathbb{D}} (1-|a|)\Delta(a) < +\infty .$$

Given the Poisson-Jensen formula (cf. §1) it is not difficult to deduce the following result: for any $p \in (0, +\infty)$

$$\operatorname{Dvs} H^\infty = \operatorname{Dvs} H^p = \operatorname{Dvs} N = \operatorname{Bl}(\mathbb{D}) . \tag{1}$$

E) The proof of the inclusion $\operatorname{Bl}(\mathbb{D}) \subset \operatorname{Dvs} H^\infty$ uses the following important construction. Set

$$b_a := \frac{|a|}{a}\frac{a-z}{1-z\bar{a}} \quad (a \in \mathbb{D}\setminus\{0\})$$

(so that $\sup\limits_{\mathbb{D}} |b_a| \le 1$ and $|b_a| \equiv 1$ on $\mathbb{T}$). If $m \in \operatorname{Bl}(\mathbb{D})$, then the product $B_m := z^{m(0)} \prod\limits_{a\in\mathbb{D}\setminus\{0\}} (b_a)^{m(a)}$ converges uniformly inside $\mathbb{D}$, $B_m \in H^\infty$, and $\operatorname{dvs} B_m = m$. The function B_m is called the *Blaschke product* corresponding to the divisor m.

F) Blaschke products make it possible to "display the zeros" of functions of classes N and H^p. If $f \in H^p(\mathbb{D})$ or $f \in N$ and $f \not\equiv 0$, then $f = Bg$, where $B = B_{\operatorname{dvs} f}$ is a Blaschke product, while $g \in H^p(\mathbb{D})$ (resp. $g \in N$) and g has no zeros.

The following proposition is a key result.

G) It follows from the Cauchy-Bunyakovskij inequality that $gh \in H^1(\mathbb{D})$ if $g, h \in H^2(\mathbb{D})$, so that $H^2(\mathbb{D}) \cdot H^2(\mathbb{D}) \subset H^1(\mathbb{D})$. It is less obvious that the inclusion sign here can be replaced by equality: for any function $f \in H^1(\mathbb{D})$ there exist functions $g, h \in H^2(\mathbb{D})$ such that $f = gh$.

Proof. Let $G := f/B$, where $B := B_{\operatorname{dvs} f}$. The function G does not vanish, and hence it coincides with g^2, where $g \in \mathcal{O}(\mathbb{D})$. It is obvious that $g \in H^1(\mathbb{D})$, since $G \in H^2(\mathbb{D})$; finally $f = gh$, where $h := Bg$.

H) To each charge $\mu \in M(\mathbb{T})$ there corresponds an *integral of Cauchy type* $\mathcal{C}(\mu)$ and a *Poisson integral* $\mathcal{P}(\mu)$:

$$\mathcal{C}(\mu)(\zeta) := \int_{\mathbb{T}} \frac{z\,d\mu(z)}{z-\zeta}, \quad \mathcal{P}(\mu)(\zeta) := \int_{\mathbb{T}} \frac{1-|\zeta|^2}{|z-\zeta|^2}\,d\mu(z) \quad (\zeta \in \mathbb{C}\setminus\mathbb{T}). \tag{2}$$

The function $\mathcal{C}(\mu)$ is analytic, and the function $\mathcal{P}(\mu)$ is harmonic in $\mathbb{C}\setminus\mathbb{T}$. We shall write $\mathcal{C}(\varphi)$ and $\mathcal{P}(\varphi)$ respectively instead of $\mathcal{C}(\varphi m)$ and $\mathcal{P}(\varphi m)$. We shall need the following properties of the Poisson integral:

(a) $\|\mathcal{P}(\mu)_r\|_1 \le \operatorname{var}\mu$, $\|\mathcal{P}(\varphi)_r\|_p \le \|\varphi\|_p$, $(0<r<1, \mu \in M(\mathbb{T}), 1 \le p \le +\infty$, $\varphi \in L^p(\mathbb{T}))$;
(b) the limit $\lim\limits_{r\uparrow 1}(\mathcal{P}(\mu))_r$ exists m-almost everywhere on the circle $\mathbb{T}$ and coincides with $d\mu/dm$; if $\mu = \varphi m$, where $\varphi \in L^1(\mathbb{T})$, then $\lim\limits_{r\uparrow 1}(\mathcal{P}(\varphi))_r = \varphi$ m-almost everywhere;
(c) if $p \in [1,+\infty)$ and $\varphi \in L^p(m)$ or if $p=+\infty$, and $\varphi \in C(\mathbb{T})$, then $\lim\limits_{r\uparrow 1}\|(\mathcal{P}(\varphi))_r - \varphi\|_p = 0$.

Assertion (a) follows immediately from the identity $\mathcal{P}(1)\big|\mathbb{D} = 1$ and Hölder's inequality; assertion (c) is a classical result in the case $p=+\infty$ and $\varphi \in C$; the full assertion (c) follows from this classical result. For the proof of (b) see (Privalov 1950).

I) Let $t \in \mathbb{T}$. Expanding the Cauchy kernel $(t-\zeta)^{-1}$ in powers of ζ in $\mathbb{D}$ and in $\mathbb{D}_- := \mathbb{C}\setminus(\mathbb{T}\cup\mathbb{D})$, we conclude that the Taylor series of the function $\mathcal{C}(\mu)$ coincides in $\mathbb{D}$ with the series $\sum\limits_{n\ge 0}\hat{\mu}(n)z^n$, and its Laurent series in $\mathbb{D}_-$ coincides with the series $\sum\limits_{n<0}(-\hat{\mu}(n)z^n)$. Using this observation and the identity

$$\mathcal{C}(\mu)(\zeta) - \mathcal{C}(\mu)(\tilde{\zeta}) = \mathcal{P}(\mu)(\zeta) \quad (\zeta \in \mathbb{D}, \quad \tilde{\zeta} := \zeta|\zeta|^{-2}),$$

it is easy to prove the equivalence of the following assertions:

(a) μ is a plus-charge (i.e., $\hat{\mu}\big|(\mathbb{Z}\setminus\mathbb{Z}_+) = 0$);
(b) $\mathcal{C}(\mu)\big|\mathbb{D}_- = 0$;
(c) $\mathcal{C}(\mu)\big|\mathbb{D} = \mathcal{P}(\mu)\big|\mathbb{D}$.

We remark further that $\mathcal{P}(\mu)\big|\mathbb{D} \in H^1(\mathbb{D})$ if μ is a plus-charge.

J) We can now take a new look at plus-functions of class $L^1(\mathbb{T})$ (i.e., at functions of class $H^1(\mathbb{T})$), which we studied in §1 of Chapt. 1.

Let $f \in \mathcal{O}(\mathbb{D})$ and

$$f(\zeta) = \sum_{n\ge 0} a_n\zeta^n \quad (\zeta \in \mathbb{D}), \quad a := (a_n)_{n\ge 0} = \left(\frac{f^{(n)}(0)}{n!}\right)_{n\ge 0}. \tag{3}$$

If $f \in H^1(\mathbb{D})$, then

(a) the series $\sum\limits_{n\geq 0} a_n z^n$ coincides on the circle $\mathbb{T}$ with the Fourier series of some plus-function $f^* \in L^1(\mathbb{T})$:

$$\exists\, f^* \in L^1(\mathbb{T}) : \hat{f}\big|_{\mathbb{Z}_+} = a, \quad \hat{f}\big|_{(\mathbb{Z}\setminus\mathbb{Z}_+)} = 0\,;$$

(b) $\lim\limits_{r\uparrow 1} \|f_r - f^*\|_1 = 0$;
(c) $f(\zeta) = \mathcal{C}(f^*)(\zeta) = \mathcal{P}(f^*)(\zeta) \quad (\zeta \in \mathbb{D})$;
(d) $\lim\limits_{r\uparrow 1} f(r\zeta) = f^*(\zeta)$ for m-almost every $\zeta \in \mathbb{T}$;
(e) the operator $f \mapsto f^*$ is a one-to-one mapping of $H^1(\mathbb{D})$ onto $H^1(\mathbb{T})$.

Thus the functions of the class $H^1(\mathbb{T})$ are the boundary values of the functions of class $H^1(\mathbb{D})$.

We sketch the proof of assertions (a)–(e):

Proof. It follows from Parseval's equality that $\|f_r\|_2^2 = \sum\limits_{n\geq 0} |a_n|^2 r^{2n}$ $(0 < r < 1)$. Therefore the inclusion $f \in H^2(\mathbb{D})$ is equivalent to the finiteness of the sum $\sum\limits_{n\geq 0} |a_n|^2$. Hence by the Riesz-Fischer theorem to any function $f \in H^2(\mathbb{D})$ there corresponds a function $f^* \in L^2(\mathbb{T})$ defined by equalities (a); it is obvious that $\lim\limits_{r\uparrow 1} \|f_r - f^*\|_2 = 0$ (hence (b) follows). In the general case (i.e., for $f \in H^1(\mathbb{D})$) we use the assertion of G): $f = gh$, where $g, h \in H^2(\mathbb{D})$; we set $f^* = g^* h^*$. After this is done assertions (a) and (b) can be established immediately, and (c) follows from the obvious identities $f_r\big|\mathbb{D} = \mathcal{C}(f_r)\big|\mathbb{D} = \mathcal{P}(f_r)\big|\mathbb{D}$ $(0 < r < 1)$. Assertion (d) follows from (c) (cf. assertion (b) in Sect. H). We now prove (e). Let $\varphi \in H^1(\mathbb{T})$; we set $f := \mathcal{C}(\varphi)\big|\mathbb{D}$. It is clear that $f \in \mathcal{O}(\mathbb{D})$, $f = \mathcal{P}(\varphi)\big|\mathbb{D}$ (cf. I)), and by H) we have $\|f_r\|_1 \leq \|\varphi\|_1$ $(0 < r < 1)$. Hence $f \in H^1(\mathbb{D})$. Further $\lim\limits_{r\uparrow 1} \|f_r - \varphi\|_1 = \lim\limits_{r\uparrow 1} \|(\mathcal{P}(\varphi))_r - \varphi\|_1 = 0$. Therefore $\varphi = f^*$. Thus the operator $f \mapsto f^*$ maps $H^1(\mathbb{D})$ onto $H^1(\mathbb{T})$.

2.2. Jensen's Inequality and the Theorem of F. and M. Riesz (Complex Approach)

A) Let $f \in \mathcal{O}(\mathbb{D})$, $f(0) \neq 0$, $0 < r < 1$. It follows from identity (12) of §1 that

$$\log|f(0)| \leq \int\limits_{\mathbb{T}} \log|f(rz)|\, dm(z) \ (0 < r < 1)\,. \tag{4}$$

Furthermore $\log a = \log^+ a - \log^+(1/a)$ $(a > 0)$. Therefore the following inequality is a consequence of (4):

$$j(r) := \log|f(0)| + \int\limits_{\mathbb{T}} \log^+ \frac{1}{|f(rz)|}\, dm(z) \leq \int\limits_{\mathbb{T}} \log^+|f(rz)|\, dm(z) \ (0 < r < 1)\,.$$

We assume that $f \in H^1(\mathbb{D})$ and apply Fatou's lemma; we thereby obtain:

$$\log|f(0)| + \int_{\mathbb{T}} \log^+ \frac{1}{|f^*|}\, dm \le \underline{\lim}_{r\uparrow 1} j(r) \le \lim_{r\uparrow 1} \int_{\mathbb{T}} \log^+ |f_r|\, dm = $$

$$= \int_{\mathbb{T}} \log^+ |f^*|\, dm \le \int_{\mathbb{T}} |f^*|\, dm < +\infty\,. \qquad (5)$$

The equality in (5) follows from assertion (b) in Sect. J of §2.1 and the inequality $|\log^+ a - \log^+ b| \le |a - b|$ $(a, b \ge 0)$. We have thus proved that if $f \in H^1(\mathbb{D})$ and $f(0) \ne 0$, then $\log|f^*| \in L^1(\mathbb{T})$ and

$$\log|f(0)| \le \int_{\mathbb{T}} \log|f^*|\, dm\,. \qquad (6)$$

Therefore

$$f \in H^1(\mathbb{D}),\ \int_{\mathbb{T}} \log|f^*|\, dm = -\infty \Rightarrow f = 0\,. \qquad (7)$$

Proof. Let $f \in H^1(\mathbb{D})$, $f \ne 0$, and $f_1 := z^{-n} f$, where n is the index of the first nonzero Taylor coefficient of the function f. Then $f_1 \in H^1(\mathbb{D})$, $f_1(0) \ne 0$, $|f_1^*| = |f^*|$, and therefore $\log|f^*| \in L^1(\mathbb{T})$.

B) It follows immediately from (7) (by virtue of Sect. 2.1) that $\varphi \in H^1(\mathbb{T})$ and $\int_{\mathbb{T}} \log|\varphi|\, dm = -\infty \Rightarrow \varphi = 0$. Thus we have again obtained assertion II of the theorem of F. and M. Riesz (cf. §1 of Chapt. 1). We now give a complex proof of assertion I: any plus-charge $\mu \in M_+(\mathbb{T})$ is m-absolutely continuous.

Proof. If $\mu \in M_+(\mathbb{T})$, then $f := \mathcal{C}(\mu)|\mathbb{D} \in H^1(\mathbb{D})$ (cf. Sect. 2.1I)). Therefore $f = \mathcal{C}(f^*)|\mathbb{D}$, where $f^* \in H^1(\mathbb{T})$. From the equality of the series $\sum_{n\ge 0} \hat{\mu}(n) z^n$ and $\sum_{n\ge 0} \hat{f}^*(n) z^n$ (cf. Sect. 2.1I)) in $\mathbb{D}$ it follows that $\hat{\mu} = \hat{f}^*$, i.e., that $\mu = f^* m$.

One must not be misled by the brevity of these proofs. The path that leads up to them is quite a long and winding one. To traverse it one must leave the circle $\mathbb{T}$ and establish, using the Poisson-Jensen formula (cf. (14) of §1) and the machinery of Blaschke products, that the set $H^1(\mathbb{T})$ is isomorphic to the set of analytic functions $H^1(\mathbb{D})$. In the first chapter we obtained the theorem of F. and M. Riesz without leaving the circle $\mathbb{T}$ and ignoring the complex plane in which it is embedded.

2.3. Hardy Classes in a Half-plane. Plus-charges on the Line. We shall use the symbol f_η, where f is a function defined in the upper half-plane

$\mathbb{C}_+ := \{\zeta \in \mathbb{C} : \operatorname{Im}\zeta > 0\}$ and $\eta > 0$, to denote the function $\zeta \mapsto f(\zeta + i\eta)$ $(\operatorname{Im}\zeta > -\eta)$. Let $p \in [1, +\infty)$. We set

$$H^p(\mathbb{C}_+) := \{f \in \mathcal{O}(\mathbb{C}_+) : \sup_{\eta>0} \|f_\eta\|_p < +\infty\}$$

(where $\| \ \|_p$ denotes the norm in $L^p(\mathbb{R}, m)$ and m is Lebesgue measure on $\mathbb{R}$). The set of such functions that are analytic and bounded in $\mathbb{C}_+$ is denoted $H^\infty(\mathbb{C}_+)$.

A) Any function $f \in \mathcal{O}(\mathbb{C}_+)$ can be transferred to the disk $\mathbb{D}$ using the substitution $Z : w \mapsto i\dfrac{1-w}{1+w}$. The transferred function will be denoted F:

$$F(w) := f(Z(w)) = f\Big(i\frac{1-w}{1+w}\Big) \quad (w \in \mathbb{D}) . \tag{8}$$

We set $\Phi_p := f(Z) \cdot (Z')^{1/p}$. It can be shown that

$$f \in H^p(\mathbb{C}_+) \Leftrightarrow \Phi_p \in H^p(\mathbb{D}) . \tag{9}$$

This connection makes it possible to construct a theory of the classes $H^p(\mathbb{C}_+)$ parallel to the theory of $H^p(\mathbb{D})$ classes (cf. Sect. 2.1). In particular it follows easily from this theory that

$$H^1(\mathbb{C}_+) = H^2(\mathbb{C}_+) \cdot H^2(\mathbb{C}_+)$$

(compare with the assertion of Sect. 2.1G)).

B) To each charge $\mu \in M(\mathbb{R})$ there corresponds an integral of Cauchy type $\mathcal{C}(\mu)$ and an integral of Poisson type $\mathcal{P}(\mu)$:

$$\text{a)} \quad \mathcal{C}(\mu)(\zeta) := \frac{1}{2\pi i} \int_{\mathbb{R}} \frac{d\mu(t)}{t-\zeta} \qquad \text{b)} \quad \mathcal{P}(\mu)(\zeta) = \frac{1}{\pi} \int_{\mathbb{R}} \frac{|\operatorname{Im}\zeta|\, d\mu(t)}{|\zeta - t|^2}$$

$(\zeta \in \mathbb{C} \setminus \mathbb{R})$. Definition a) makes sense also when $\mu = \varphi m$, $\varphi \in L^p(\mathbb{R})$, where $p \in [1, +\infty)$; and definition b) makes sense when $p = +\infty$ (under these conditions μ does not necessarily belong to $M(\mathbb{R})$). The function $\mathcal{C}(\mu)$ is analytic and $\mathcal{P}(\mu)$ is harmonic in $\mathbb{C} \setminus \mathbb{R}$. It is easy to see that

$$\begin{aligned} \mathcal{C}(\mu)(\zeta) &= \int_0^{+\infty} \hat{\mu}(u) e^{iu\zeta}\, du \quad (\zeta \in \mathbb{C}_+) , \\ \mathcal{C}(\mu)(\zeta) &= -\int_{-\infty}^{0} \hat{\mu}(u) e^{iu\zeta}\, du \quad (\zeta \in \mathbb{C}_-) . \end{aligned} \tag{10}$$

Here μ is any charge of class $M(\mathbb{R})$ and $\hat{\mu}$ is its Fourier transform: $\hat{\mu}(u) = (2\pi)^{-1} \displaystyle\int_{-\infty}^{+\infty} e^{-itu}\, d\mu(t)$ $(u \in \mathbb{R})$. Let $\mu \in M(\mathbb{R})$. The following assertions are equivalent:

(a) μ is a plus-charge (i.e., $\operatorname{Spec}\mu \subset [0, +\infty)$);
(b) $\mathcal{C}(\mu)|\mathbb{C}_- = 0$;
(c) $\mathcal{C}(\mu)|\mathbb{C}_+ = \mathcal{P}(\mu)|\mathbb{C}_+$.

C) We now state the analogue of the theorem of Sect. 2.1J). If $f \in H^p(\mathbb{C}_+)$ and $p \in [1, +\infty]$, then

(a) $\lim_{\eta \downarrow 0} f(\xi + i\eta) =: f^*(\xi)$ exists m-almost everywhere;
(b) $f^* \in L^p(\mathbb{R}, m)$;
(c) $f = \mathcal{P}(f^*)|\mathbb{C}_+$.
If in addition $p < +\infty$, then
(d) $\lim_{\eta \downarrow 0} \|f_\eta - f^*\|_p = 0$ and
(e) $f = \mathcal{C}(f^*)|\mathbb{C}_+$.

D) The class $H^p(\mathbb{C}_+)$ is in natural one-to-one correspondence with the class $H^p(\mathbb{R})$ of plus-functions of class $L^p(\mathbb{R})$. This assertion has the following meaning. Let $p \in [1, +\infty]$ and $f \in H^p(\mathbb{C}_+)$.

(a) The boundary function f^* belongs to $H^p(\mathbb{R})$;
(b) the operator $f \mapsto f^*$ is a one-to-one mapping of $H^p(\mathbb{C}_+)$ onto $H^p(\mathbb{R})$.

Reasoning approximately as in Sect. 2.2, one can now easily show that any plus-charge of class $M(\mathbb{R})$ is m-absolutely continuous. Moreover

$$f \in H^p(\mathbb{C}_+) \Rightarrow \log|f(i)| \leq \frac{1}{\pi} \int_{-\infty}^{+\infty} \frac{\log|f^*(t)|}{1+t^2}\, dt \quad (1 \leq p \leq +\infty)\,.$$

It follows from this inequality that

$$\varphi \in H^p(\mathbb{R}),\ \varphi \neq 0 \Rightarrow \int_{-\infty}^{+\infty} \frac{\log|\varphi(t)|}{1+t^2}\, dt > -\infty\,.$$

§3. Divergence of the Integral of the Logarithm and Certain Forms of the Uncertainty Principle for Functions and Charges with Rapidly Decreasing Amplitudes

The integral of the logarithm of the absolute value is encountered so frequently in what follows that it makes sense to introduce a special notation for it:

$$\mathcal{L}(f) := \int_{\mathbb{T}} \log|f|\, dm \quad (f \in L^1(\mathbb{T}, m))$$

or

$$\mathcal{L}(f) := \int_{\mathbb{R}} \log|f|\, d\Pi \quad (f \in L^1(\mathbb{R}, \Pi))\,,$$

where Π denotes the *Poisson measure* in $\mathbb{R} : d\Pi = \pi^{-1}(1+x^2)^{-1}\,dx$. In the preceding section we established that $\mathcal{L}(f) = -\infty \Rightarrow f = 0$ if f is a function with semi-bounded spectrum. Echoes of this fundamental theorem can be distinctly heard in the statements of many other versions of the uncertainty principle, including those where "spectral smallness" is expressed not in the complete vanishing of half of the spectrum of the function f, but only in a rapid decay in the amplitudes of its harmonics. This means that the quantity $|\hat{f}(\xi)|$ becomes rather small when ξ goes to infinity (or just to plus infinity) along $\mathbb{Z}$ or $\mathbb{R}$.

3.1. Infinite-order Zeros and Decay of Amplitude. Throughout the following the letter H will denote a positive function defined and continuous on the ray $[0, +\infty)$.

A) Assume that the function $f \in L^1(\mathbb{R})$ satisfies the following condition of "small amplitudes":

$$\exists c > 0 : |\xi| \geq c \Rightarrow |\hat{f}(\xi)| \leq H(|\xi|)\,. \tag{1}$$

Can a nonzero function f have a zero of infinite order under this condition? More precisely: does there exist a nonzero function f subject to condition (1) and such that

$$f(t) = O(t^n) \ \ (t \to 0) \tag{2}$$

for any $n \in \mathbb{N}$? The answer depends on the smallness of the majorant H at infinity. If $\lim\limits_{\xi \to +\infty} H(\xi)\xi^n = 0$ for any $n > 0$, then (1) implies that $f \in C^\infty$, and (2) assumes the following form:

$$f^{(n)}(0) = 0 \ \ (n \in \mathbb{Z}_+)\,. \tag{3}$$

We shall agree to call a function H a *determining majorant* if (1) and (2) imply that $f = 0$. In the problem of describing the determining majorants the decisive role is played by the integral $\mathcal{L}(H)$ (we assume that the function H is extended to the whole real axis $\mathbb{R}$ as an even function). The equality $\mathcal{L}(H) = -\infty$ is necessary in order for H to be determining (cf. below Sect. 6.2). We shall now show that under a certain additional hypothesis on H it suffices that the integral $\mathcal{L}(H)$ diverge. Here we encounter the frequent situation that occurs when along with the purely quantitative condition that expresses the "smallness" of the majorant H it is necessary to impose on it in addition a condition that is qualitative in nature and expresses a certain regularity of decay. This additional condition is inherent in the situation and cannot be eliminated.

B) In the present case the regularity condition on the majorant is stated as follows: there exist numbers A and $c > 0$ and an increasing positive function $\omega : [c, +\infty) \to \mathbb{R}$ such that

$$\lim_{\xi\to+\infty} \omega(\xi) = +\infty \, , \;\; H(\xi) = \exp(-p(\xi)) \;\; (\xi \geq c) \, , \;\; p(\xi) := A + \int_c^\xi \frac{\omega(u)}{u} \, du \, . \tag{4}$$

This means that $-\log H(\xi)$ is a convex function of $\log \xi$ on the ray $[c, +\infty)$ and the "slope" $d \log H(\xi)/d \log \xi$ increases without bound. The divergence of the integral $\mathcal{L}(H)$ under condition (4) means that

$$\int^{+\infty} \frac{\omega(u)}{u^2} \, du = \infty \, . \tag{5}$$

Under condition (4) the equality

$$\mathcal{L}(H) = -\infty \tag{6}$$

implies that the majorant H is determining.

C) Setting $\omega(u) = ku$ in (4) ($k > 0$), we obtain the simplest example of a determining majorant. In this case the estimate (1) can be written as follows: $\hat{f}(\xi) = O(e^{-k|\xi|})$ ($|\xi| \to +\infty$). This example is uninteresting, since it follows from the last estimate that the function f is analytic on the real axis $\mathbb{R}$; it is obvious that under condition (3) the function f vanishes completely. Substantive examples of determining majorants correspond to the functions

$$\omega(u) := u(\log u \cdot \log\log u \cdot \ldots \cdot \log\log \cdots \log u)^{-1} \, . \tag{7}$$

Equality (5) holds with any set of iterations of the logarithm in (7). Thus the majorant H is determining if

$$H(\xi) \asymp \exp(-\xi(\log \xi \cdot \log\log \xi \cdot \ldots \cdot \log\log \cdots \log \xi)^{-1} \;\; (\xi \uparrow +\infty) \, . \tag{8}$$

For such H it is not difficult to show that certain functions f satisfying condition (1) are not analytic at even one point of the line $\mathbb{R}$.

3.2. Sketch of the Proof

A) Let $n \in \mathbb{Z}_+$ and let the charge $\mu \in M(\mathbb{R})$ be such that $\int_{\mathbb{R}} |x|^n \, d|\mu|(x) < +\infty$. Then we shall say that there exists an nth *moment* $m_n(\mu)$ of the charge μ and write

$$m_n(\mu) = \int_{\mathbb{R}} x^n \, d\mu(x) \, .$$

If $d\mu(x) = g(x) \, dx$, we shall write $m_n(g)$ instead of $m_n(\mu)$.

Consider the Cauchy transform $\mathcal{C}(\mu)$ of the charge μ:

$$\mathcal{C}(\mu)(\zeta) = (2\pi i)^{-1} \int_{-\infty}^{\infty} (x - \zeta)^{-1} \, d\mu(x) \;\; (\zeta \in \mathbb{C} \setminus \mathbb{R}) \, .$$

If the moments $m_j(\mu)$ exist ($j = 0, \ldots, n+1$), then obviously

$$\mathcal{C}(\mu)(\zeta) = -(2\pi i)^{-1}\sum_{j=0}^{n} m_j(\mu)\zeta^{-(j+1)} + \zeta^{-(n+1)}\mathcal{C}(x^{n+1}\mu)(\zeta) \ (\zeta \in \mathbb{C}\backslash\mathbb{R})\,. \quad (9)$$

Consequently

$$z^n\mathcal{C}(\mu) = \mathcal{C}(z^n\mu) \ (n \in \mathbb{Z}_+)\,, \quad (10)$$

if all the moments $m_n(\mu)$ $(n \in \mathbb{Z}_+)$ exist and are equal to zero. Under condition (4) the function H decays faster than any power, and it follows from (1) that $f \in C^\infty$, all the moments $m_n(\hat{f})$ exist, and (3) means that they are all equal to zero. According to (10)

$$\mathcal{C}(\hat{f}) = z^{-n}\mathcal{C}(z^n\hat{f}) \ (n \in \mathbb{Z}_+)\,. \quad (11)$$

By controlling the "free parameter" n, one can show that

$$\int_{\mathbb{R}} \log|\mathcal{C}(\hat{f})(\xi \pm i)|\, d\Pi(\xi) = -\infty\,. \quad (12)$$

The function $\mathcal{C}(\hat{f})$ is holomorphic and bounded in each of the half-planes $\mathbb{C}_+ + i$ and $\mathbb{C}_- - i$. It follows from (12) that $\mathcal{C}(\hat{f}) \equiv 0$ in each of these half-planes (cf. Sect. 2.3D)) and therefore also everywhere in $\mathbb{C}\backslash\mathbb{R}$. But then $f = 0$ (for example, because $\mathcal{P}(f)(\zeta) = \mathcal{C}(f)(\zeta) - \mathcal{C}(f)(\bar{\zeta})$) when $\operatorname{Im}\zeta > 0$ (cf. Sect. 2.3B)).

B) It remains to deduce (12) from (11). Direct estimation shows that

$$2\pi|\mathcal{C}(\hat{f})(\xi \pm i)| \le \tilde{m}_n(H)|\xi|^{-n} \ (\xi \in \mathbb{R},\, n \in \mathbb{Z}_+)\,, \quad (13)$$

where $\tilde{m}_n(H) := 2\int_0^{+\infty} H(u)u^n\,du$. Minimizing the right-hand side of inequality (13) over n, one can show that

$$\mathcal{C}(\hat{f}(\xi \pm i)) = O(\xi^3 H(\xi)) \ (\xi \to +\infty)\,, \quad (14)$$

and (12) follows from (6). We shall not take the time to prove estimate (14); for majorants of the form (8) the minimum over n in (13) is easily estimated directly.

3.3. The Connection with the Moment Problem and Weighted Approximation. With each function h that is positive and continuous on the line $\mathbb{R}$ we associate the set of charges

$$M^h := \left\{\mu \in M_{\text{loc}}(\mathbb{R}) : \int_{\mathbb{R}} \frac{d|\mu|}{dt} < +\infty\right\}.$$

The reasoning of Sect. 3.2 leads to the following uniqueness theorem for the solution of the moment problem. Assume that the function h is even and let $H := h|[0, +\infty)$. If conditions (4) and (5) hold, then

$$\mu, \nu \in M^h\,,\ \ m_n(\nu) = m_n(\mu)\ \ (n \in \mathbb{Z}_+) \Rightarrow \nu = \mu\,. \tag{15}$$

We remark that under condition (4) any charge $\mu \in M^h$ has moments of all orders; the theorem means that it is uniquely determined by its moments.

The dual restatement of this assertion gives the solution of an important problem – the so-called *Bernstein weight problem.* We denote by C_h^0 the set of functions f that are continuous on the whole line $\mathbb{R}$ and such that $h(t)f(t) = o(1)$ as $|t| \to +\infty$. We set $\|f\|_h := \max\limits_{\mathbb{R}} h|f|$. If $\lim\limits_{|t|\to+\infty} h(t)|t|^n = 0$ $(n > 0)$, then the set $\mathcal{P}$ of polynomials is contained in C_h^0. Bernstein's problem is to describe the h for which

$$\operatorname{Clos}_{C_h^0}\mathcal{P} = C_h^0\,, \tag{16}$$

Equality (16) is obviously equivalent to the assertion (15). Therefore (16) follows from (4) and (5).

3.4. One-sided Decay of Amplitudes. Beurling's Theorem

A) Let us return to the problem stated in Sect. 3.1A) and replace the "two-sided" condition (1) (i.e., the condition for ξ of large absolute value) by the condition for one-sided decay of amplitudes:

$$\exists\, c > 0 : \xi \geq c \Rightarrow |\hat{f}(\xi)| \leq H(\xi)\,. \tag{17}$$

Under condition (17) it does not follow from (1) and (2) that $f = 0$, even when $H \equiv 0$. However under condition (6) the estimate (17) nevertheless imposes essential restrictions on the zeros of a function $f \not\equiv 0$: they cannot fill up a set of positive length.

It will be convenient for us to interchange the roles of f and $\hat{f}$ here, subjecting f to a condition of one-sided decay (of type (17)) rather than $\hat{f}$, and studying the set of zeros of the function $\hat{f}$. We shall generalize the statment of the question slightly and instead of a function $f \in L^1(\mathbb{R})$ we shall consider a charge $\mu \in M(\mathbb{R})$. If the charge μ vanishes on a half-line, then $\hat{\mu}$ admits a continuous and bounded extension into the closed (upper or lower) half-plane that is analytic in the interior of the half-plane. Therefore (cf. Sect. 2.3D)) the Fourier transform of a nonzero charge that vanishes on a half-line cannot vanish on a set of positive length. Beurling's theorem stated below shows that this same regularity is also characteristic of charges that decay rapidly along a half-line. The rate of decay of the charge μ along the half-line $\mathbb{R}_+$ will be measured using the function $\rho_\mu (= \rho) : A \mapsto |\mu|[A, +\infty)$. We shall denote the Lebesgue measure of a set $E \subset \mathbb{R}$ by the symbol $|E|$ or $m(E)$; $\hat{\mu}(\sigma)$ is, as always, $(2\pi)^{-1} \int\limits_{\mathbb{R}} e^{-i\sigma u}\, d\mu(u)$.

Theorem (Beurling). *If* $\mu \in M(\mathbb{R})$,

$$\int^{+\infty} \frac{\log \rho_\mu(A)}{A^2}\, dA = -\infty\,, \tag{18}$$

and

$$m\big(\{t\in\mathbb{R}:\ \hat{\mu}(t)=0\}\big)>0\,, \tag{19}$$

then $\mu=0$.

It is remarkable that the "regularity condition" (cf. 3.1A)) is absent here, and the uniqueness of the charge is guaranteed by the divergence of the "integral of the logarithm" (18) alone. Without giving the proof we shall merely outline the strategy for it: Equation (12) is deduced from conditions (18) and (19), in which $\hat{f}$ must be replaced by μ. It will then follow from the boundedness of the functions $C(\mu)$ in the half-planes $\mathbb{C}_+ + i$ and $\mathbb{C}_- - i$ that $\hat{f}$ is identically zero in $\mathbb{C}\setminus\mathbb{R}$ (cf. Sect. 2.3D)), and hence also $\mu=0$ (by formulas (10) of the preceding section).

B) We now give the details of certain discrete analogues of Beurling's theorem. We assume that $f\in C(\mathbb{T})$ and $\sum_{n\in\mathbb{Z}}|\hat{f}(n)|<+\infty$. If

$$\sum_{n=1}^{\infty}\frac{\log\rho_n^1(f)}{n^2}=-\infty\,, \tag{20}$$

where $\rho_n^1(f):=\sum_{k\le -n}|\hat{f}(k)|$, then

$$f\neq 0\Rightarrow m\big(\{\zeta\in\mathbb{T}:\ f(\zeta)=0\}\big)=0\,. \tag{21}$$

For the proof it is necessary to apply Beurling's theorem to the charge $\sum_{n\in\mathbb{Z}}\hat{f}(n)\delta_n$, where δ_n denotes a unit load concentrated at the point n. The L^2-analogue of this theorem has the following appearance: the implication (21) continues to hold if $f\in L^2(\mathbb{T})$ and Eq. (20) holds, in which $\rho_n^1(f)$ must be replaced by $\rho_n^2(f):=\sum_{k\le -n}|\hat{f}(k)|^2$. This fact does not follow directly from A) and requires a separate proof.

3.5. The Levinson-Cartwright Theorem

A) Let us compare the assertions of Sect. 3.4B) and another remarkable result of the same kind. A function H that is analytic in $\mathbb{C}\setminus\mathbb{T}$ plays a role in its statement:

$$H(t)=\sum_{n\ge 0}h_nt^n\ \ (t\in\mathbb{D}_+:=\mathbb{D})\,,\ \ H(t)=\sum_{n<0}h_nt^n\ \ (t\in\mathbb{D}_-:=\mathbb{C}\setminus(\mathbb{D}\cup\mathbb{T}))\,,$$

along with two majorants: an even function $M:\mathbb{R}\to(0,+\infty)$ and a function $\theta:(0,+\infty)\to(0,+\infty)$, where $M\big|(0,+\infty)$ is decreasing and θ is increasing. We shall say that the function H can be extended along an (open) arc $\gamma\subset\mathbb{T}$

if there exists a function $\overline{H}$ that is analytic in the domain $\mathbb{D}_+ \cup \mathbb{D}_- \cup \gamma$ and coincides with H in $\mathbb{R} \setminus \mathbb{T}$.

Theorem (Levinson-Cartwright). *Assume that*

$$\text{(a)} \int\limits_0 \log\log M(u)\,du < +\infty\,, \quad \text{(b)} \int^{+\infty} \frac{\theta(u)}{u^2}\,du = +\infty\,, \tag{22}$$

$$H(t) \le M(|t|-1) \ \left(\frac{1}{2} < |t| < \frac{3}{2}\right), \tag{23}$$

$$|h_{-n,}| \le \exp[-\theta(n)] \ (n \in \mathbb{N})\,. \tag{24}$$

If the function H can be extended along some nondegenerate arc $\gamma \subset \mathbb{T}$, then

$$h_n = 0 \ \text{ for any } n \in \mathbb{Z}\,. \tag{25}$$

Under conditions (22) and (23) assertion (25) remains in effect even when (24) holds not for all natural numbers n but only for $n = n_k$, $k = 1, 2, \ldots$, where $\lim\limits_{k\to+\infty} (k/n_k) = D$ and $m(\gamma) > 1 - D$.

B) This profound result provides yet another form of the uncertainty principle and fits together with Beurling's theorem, or more precisely with the discrete analogues of it (cf. Sect. 3.4B)). In some respects it is significantly stronger than these theorems, and in other respects noticeably weaker. To verify this we begin with the following remark.

If the coefficients h_n admitted a polynomial estimate ($|h_n| = O(|n|^C)$ for some $C > 0$), then the function H could be associated with a distribution Φ_H on the circle $\mathbb{T}$:

$$\Phi_H[\varphi] := \sum_{n\ge 0} h_n \hat{\varphi}(-n) - \sum_{n>0} h_n \hat{\varphi}(-n) \ \ (\varphi \in C^\infty(\mathbb{T}))\,. \tag{26}$$

The function H can easily be recovered, knowing Φ_H:

$$H(t) = \Phi_H[\varphi_t] \ \ (t \in \mathbb{C} \setminus \mathbb{T}), \ \text{ where } \varphi_t(\zeta) := \zeta(\zeta - t)^{-1} \ \ (\zeta \in \mathbb{T})\,. \tag{27}$$

The numbers h_n can be very simply expressed in terms of the Fourier coefficients $\widehat{\Phi}_H(n) : h_n = \operatorname{sgn} n \cdot \widehat{\Phi}_H(n)$ $(n \in \mathbb{Z})$. Thus condition (24) expresses "smallness of the amplitudes" $|\widehat{\Phi}_H(n)|$ as $n \to -\infty$. On the other hand, the requirement that the function H be continuable along the arc γ can now be read as follows:

$$\operatorname{supp} \Phi_H \subset \mathbb{T} \setminus \gamma\,, \tag{28}$$

i.e., as the condition that the distribution Φ_H have "small support." (The equivalence of the inclusion (28) and continuability of the function H follows from (27) and the equality $\Phi_H = \lim\limits_{r\uparrow 1} \Phi_h^r$, where $\Phi_H^r (\in C^\infty(\mathbb{T}))$ denotes the function $\zeta \mapsto H(r\zeta) - H(r^{-1}\zeta)$ $(\zeta \in \mathbb{T})$).

C) The polynomial growth of the quantities $|h_n|$ is equivalent to condition (23) with a power function M ($M(t) \equiv |t|^{-C'}$, $C' > 0$). However the interpretation of the Levinson-Cartwright theorem given in Sect. B (The L-C Theorem for short) can be preserved in the general case also, restricting the domain of definition of the functional Φ_H. The right-hand side of Eq. (26) has meaning if $\varphi \in \mathcal{O}(\mathbb{T})$, where $\mathcal{O}(\mathbb{T})$ denotes the set of functions analytic in $\mathbb{T}$. (In fact $\overline{\lim}_{|n|\to+\infty} |h_n|^{1/|n|} \leq 1$). A continuous linear functional Φ defined on the space $\mathcal{O}(\mathbb{T})$ is called a *hyperfunction*. Continuity of this functional means by definition that $\lim_{j\to\infty} \Phi[\varphi_j] = 0$ for any sequence (φ_j) of elements of the space $\mathcal{O}(\mathbb{T})$ that are analytically continuable from $\mathbb{T}$ to a fixed annulus (independent of j) $K := \{\rho < |t| < R\}$ ($\rho < 1 < R$) and such that $\lim_{j\to\infty} \sup_K |\varphi_j| = 0$. It is easy to see that Φ_H is a hyperfunction and Eq. (27) is preserved. It remains to adopt the following definition: a hyperfunction Φ_H vanishes on an arc γ (i.e., the inclusion (28) holds), if the function H can be continued along γ. After that the L-C theorem becomes a particular form of the uncertainty principle for hyperfunctions (and, in particular, for distributions and charges).

D) Beurling's theorem (in the "discrete" version of Sect. 3.4B)) and the L-C theorem are equivalent if they are stated in terms of charges $\mu \in M(\mathbb{T})$ that vanish on a nondegenerate arc: such a charge is identically zero if (a) $|\hat{\mu}(n)| = O(\exp(-\theta(|n|)))$ ($n \to -\infty$) or (b) $\sum_{n=1}^{\infty} n^{-2}\Big(\log \sum_{k\leq -n} |\hat{\mu}(k)|\Big)$. Each of these theorems implies this assertion. We remark that the convolution $\mu * \alpha$, where the function α belongs to $C^\infty(\mathbb{T})$ and is concentrated in a small neighborhood of unity, vanishes on some nondegenerate arc if μ has this property: moreover under condition (a) $\sum_{k\leq -n} |(\widehat{\mu * \alpha})(k)| \leq C_\alpha \exp(-\theta(n))$. On the other hand, under condition (b) $|\hat{\mu}(n)| \leq \exp(-\theta(n))$, where $\theta(n) := -\log \sum_{k\leq -n} |\hat{\mu}(k)|$.

Beurling's theorem, however, gives more than the L-C theorem, since the former is stated for functions that vanish on an arbitrary subset of positive length, while the latter applies only to functions concentrated outside a nonempty open arc. At the same time the L-C theorem applies to any hyperfunctions Φ_H provided H does not grow too quickly in approaching $\mathbb{T}$ (cf. (23) and (22a))), while Beurling's theorem is directly applicable only to ordinary functions (cf. B)). We remark finally that the known proofs of the theorems under discussion are completely different (the proof of the L-C theorem uses the delicate theory of entire functions of Cartwright class).

§4. One-sided Decay of Amplitudes and Convergence of the Logarithmic Integral

Assume that the quantities $|\hat{f}(n)|$, where $f \in L^1(\mathbb{T})$, $f \neq 0$, tend to zero very rapidly as $n \to -\infty$. Can the logarithmic integral $\mathcal{L}(f) := \int_{\mathbb{T}} \log|f|\, dm$ then diverge? We have two arguments in favor of a negative answer. First, the inequality $\mathcal{L}(f) > -\infty$ under the condition that $f \neq 0$ and $\operatorname{Spec} f \subset \mathbb{Z}_+$ (cf. Sect. 2.2A)). Second, Beurling's theorem, which forbids a nonzero function f to vanish on a set of positive length if condition (20) of §3 holds.

In this section we formulate and discuss Vol'berg's theorem on the logarithmic integral, which with certain stipulations can be considered as a generalized Beurling theorem (in the version of Sect. 3.4B)). At the end of the section we shall discuss yet another approach to this circle of questions proposed by A.A. Borichev.

4.1. Vol'berg's First Theorem

A) The "regularity conditions" about which we have spoken in Sect. 3.1B) will play an important role in this section. We begin with one of them. We shall say that a positive function M defined on an interval of the form $[A, +\infty)$, where $A > 0$ satisfies condition R_1 if the function $t \mapsto t^{-1/2}M(t)$ is increasing on some interval of the form $[A_1, +\infty)$, $A_1 \geq A$.

B)

Theorem. *Assume that the function M is concave and satisfies condition R_1 and that*

$$\int^{+\infty} \frac{M(t)}{t^2}\, dt = +\infty\,. \tag{1}$$

If a nonzero function $f \in L^1(\mathbb{T})$ satisfies the condition

$$\hat{f}(n) = O(\exp(-M(|n|))), \quad (n \to -\infty)\,, \tag{2}$$

then

$$\mathcal{L}(f) := \int_{\mathbb{T}} \log|f|\, dm > -\infty\,. \tag{3}$$

For functions f subject to condition (2), this theorem gives much more than the theorem of Sect. 3.4B). However Beurling's theorem (in the form of 3.4B)) does not assume any "regularity conditions" on the decay in amplitudes $|\hat{f}(n)|$, and therefore the complete theorem still does not follow from Vol'berg's theorem.

4.2. Almost-analytic functions. Vol'berg's Second Theorem

A) The theorem of Sect. 4.1 follows from a certain uniqueness theorem for almost-analytic functions. A function F that is continuously differentiable in the open unit disc $\mathbb{D}$ is said to be *almost-analytic* if there exists a nondecreasing continuous function $h : (0, c) \to (0, 1)$ such that

$$\text{(a)} \quad \int_0 \log\log \frac{1}{h(r)}\, dr = +\infty \,;$$

$$\text{(b)} \quad |\overline{\partial} F(\zeta)| \le h(1 - |\zeta|) \quad (\zeta \in \mathbb{D},\ 1 - |\zeta| < c)\,. \tag{4}$$

Here $\overline{\partial} := \frac{1}{2}\Big(\frac{\partial}{\partial x} + i\frac{\partial}{\partial x}\Big)$ is the *Cauchy-Riemann operator.* Condition (4a) means that the quantity $h(1 - |\zeta|)$ tends to zero rather rapidly, when ζ approaches $\mathbb{T}$. Therefore (cf. 4b)) the almost analytic function F becomes "more analytic" as we examine it closer to $\mathbb{T}$.

It is not difficult to show that if an almost analytic function F "almost belongs to the class N" (cf. Sect. 2.1B)), i.e., if

$$\sup\Big\{\int_{\mathbb{T}} \log^+ |F(r\zeta)|\, dm(\zeta) : 0 < r < 1\Big\} < +\infty\,, \tag{5}$$

then for m-almost all $\zeta \in \mathbb{T}$ the nontangential limit of $F^*(\zeta)$ exists.

B) We now subject the function h that occurs in (4) to the following additional regularity condition:

$$\left.\begin{array}{c}\text{the function } t \mapsto t\log(1/h(t)) \text{ decays on some}\\ \text{interval of the form } (0, c_1)\ (c_1 \le c).\end{array}\right\} \tag{R_2}$$

It follows from (R_2) and (4a) that $\lim\limits_{t\to 0} t\log(1/h(t)) = +\infty$.

Theorem. *Assume that the function h satisfies conditions R_2 and* (5a) *and that the almost-analytic function F satisfies* (6) *and* (5b). *If*

$$m(\{\zeta \in \mathbb{T} : F^*(\zeta) \ne 0\}) > 0\,, \tag{6}$$

then $\mathcal{L}(F^*) > -\infty$.

We shall call this assertion "Vol'berg's second theorem." It implies his first theorem, as we shall verify in the next section.

4.3. Dyn'kin's Theorem on Almost-analytic Continuation and Reduction of the First Vol'berg Theorem to the Second

A) Let us consider a function $\mathcal{M} : [1, +\infty) \to (0, +\infty)$. Assume that

$$\text{(a)} \ \ \mathcal{M}''(t) \le 0 \ \ \text{and} \ \ \text{(b)} \ \ \mathcal{M}(t) = o(t) \ \ (t \to +\infty)\,. \tag{7}$$

The function $m : \xi \mapsto \sup\{M(t) - \xi t : t \geq 1\}$ ($\xi > 0$) is called the *Legendre transform* of the function $\mathcal{M}$ (for this concept cf. (Arnol'd 1974) and (Tikhomirov 1986)). We set $h := \exp(-m)$. It can be shown that conditions (4a) and (1) are equivalent. Moreover conditions R_1 and R_2 are equivalent.

We set $\eta_n(r) := r^n h(1-r)$ $(0 < r < 1)$, $\mu_n^\infty(h) := \sup\limits_{(0,1)} \eta_n$, $\mu_n^1(h) = \int_0^1 \eta_n(r)\, dr$ $(n > 0)$. For sufficiently large n

$$\frac{1}{4n} e^{-M(2n)} \leq \mu_n^1(h) \leq \mu_n^\infty(h) \leq e^{-M(n)} . \tag{8}$$

Without pausing to give the proofs we note that for many specific majorants $\mathcal{M}$ (compare with Sect. 3.1B)) it is easy to verify these assertions directly.

B)

Theorem. *If a continuous minus-function defined on the circle $\mathbb{T}$ satisfies the condition*

$$\sum n^2 |\hat{f}_-(n)| e^{M(2|n|)} < +\infty , \tag{9}$$

then there exists a function F_- that is continuous in the closed disk $\mathbb{D} \cup \mathbb{T}$ and such that

$$\text{(a) } F_-|\mathbb{D} \in C^1; \text{ (b) } |\bar{\partial} F_-(\zeta)| \leq \operatorname{const} h(1-|\zeta|) \ (\zeta \in \mathbb{D}); \text{ (c) } F_-\big|\mathbb{T} = f_- .$$

If $\mathcal{M}$ satisfies condition (1), *then by* A) *the function F_- provides an almost-analytic continuation of the function f_- from the circle $\mathbb{T}$ to the disk $\mathbb{D}$.*

Proof. We set

$$k(z) := |z|^2 \Omega(z|z|^{-1}) h(1-|z|) \ (z \in \mathbb{D}, z \neq 0) , \ \ k(0) = 0 ,$$

and

$$F_-(\zeta) := (2\pi)^{-1} \int_{\mathbb{D}} k(z)(\zeta - z)^{-1}\, dm_2(z) \ \ (\zeta \in \mathbb{C})$$

(m_2 denotes area). Here Ω is a function of class $C^1(\mathbb{T})$, whose choice is to be made more precise. In any case

$$F_- \in C^1(\mathbb{D}), \ \ F_- \in C(\mathbb{C}) , \ \ |\bar{\partial} F(\zeta)| = 1/2|k(\zeta)| \leq \operatorname{const} h(-|\zeta|) \ \ (\zeta \in \mathbb{D}) .$$

It remains to achieve the equality $g := F_-\big|\mathbb{T} = f_-$. We set

$$(F_-)_r(\zeta) := F_-(r\zeta) \ \ (r > 1) ,$$

so that

$$(F_-)_r(\zeta) = \sum_{n=0}^{\infty} \mu_{n+3}^1(h) \widehat{\Omega}(-n)(r\zeta)^{-(n+1)} \ \ (r > 1, \zeta \in \mathbb{T}) .$$

Further $g = \lim_{r\downarrow 1}(F_-)_r|\mathbb{T}$ (uniformly on $\mathbb{T}$). Therefore g is a minus-function and $\hat{g}(n-1)(\mu^1_{|n|+3})^{-1} = \widehat{\Omega}(n)$ $(n \le 0)$. It remains to define Ω as the sum of the trigonometric series $\sum \widehat{\Omega}(n)z^n$, where $\widehat{\Omega}(n) := \hat{f}_-(n-1)(\mu^1_{|n|+3}(h))^{-1}$. Using (9) and (8), it is easy to show that $\sum |n||\widehat{\Omega}(n)| < +\infty$, so that this series converges uniformly to a function of the class $C^1(\mathbb{T})$.

C) Let us verify that Vol'berg's first theorem follows from his second theorem.

Let the nonzero function $f \in L^1(\mathbb{T})$ satisfy condition (2). It is easy to see that

$$f_- := \sum_{n<0} \hat{f}(n)z^n \in C^\infty(\mathbb{T}) .$$

Therefore $f - f_- \in H^1(\mathbb{T})$ and hence $f - f_- = F_+^*$, where F_+ is a certain function of class $H^1(\mathbb{D})$ (cf. §2). We apply Dyn'kin's theorem (cf. B)) and construct an almost-analytic continuation F_- of the function f_-. This is possible since

$$\sum_{n<0} n^2|\hat{f}_-(n)| \exp(M(|n|)/2) \le c \sum_{n<0} n^2 \exp(-M(|n|)/2) < +\infty .$$

The function $F := F_+ + F_-$ satisfies conditions (4b) and (6), where $h := \exp(-m_1)$ and m_1 is the Legendre transform of the function $\mathcal{M}_1 : x \mapsto \frac{1}{2}\mathcal{M}(\frac{x}{2})$ (we remark that $F^* = f$ m-almost everywhere). The majorant h_1 satisfies the regularity condition R_2 (cf. A)), since $\mathcal{M}$ (and hence also $\mathcal{M}_1$) is subject to condition R_1. By Vol'berg's second theorem $\mathcal{L}(f) = \mathcal{L}(F^*) > -\infty$.

4.4. On the Proof of Vol'berg's Second Theorem

A) Let O be a bounded plane domain with boundary Γ, and let $C_r(\Gamma)$ be the set of real-valued functions that are continuous on Γ and $\mathcal{H}(O)$ the set of real-valued functions that are continuous on $O \cup \Gamma$ and harmonic in O. We need the following fact from potential theory (the *Wiener-Keldysh* theorem): there exists a unique positive operator $h = h_O$ mapping $C_r(\Gamma)$ into $\mathcal{H}(O)$ and such that if $f \in \mathcal{H}(O)$, then $h(f|\Gamma)$ coincides with f on Γ. In other words, the function $u := h(\varphi)$, where $\varphi \in C_r(\Gamma)$, is a solution of the classical *Dirichlet problem* in O ($\Delta u = 0$ in O and $u|\Gamma = \varphi$) if a solution exists.

This theorem leads us to the concept of a *harmonic measure* with respect to the domain O computed at the point $\zeta \in O$: this is the name given to a Borel probability measure ω_ζ^O on Γ that generates, via to the Riesz representation theorem, a positive linear functional

$$\varphi \mapsto h(\varphi)(\zeta) \ \ (\varphi \in C_r(\Gamma))$$

on the space $C_r(\Gamma)$. Thus

$$h(\varphi)(\zeta) = \int\limits_{\Gamma} \varphi \, d\omega_\zeta^O \,, \quad (\zeta \in O, \varphi \in C_r(\Gamma)) \,.$$

B) A subdomain O of the disk $\mathbb{D}$ is called a *boundary layer* if (a) for each point $\zeta \in \mathbb{T}$ there exist positive numbers r arbitrarily close to one and such that $r\zeta \in O$ (in particular $\operatorname{Fr} O \supset \mathbb{T}$); (b) for each point $\zeta \in O$ there exists a number $\alpha > 0$ such that $\omega_\zeta^O(E) \geq \alpha m(E)$ for any Borel subset E of the circle $\mathbb{T}$.

A boundary layer is, rougly speaking, a domain O in $\mathbb{D}$ whose complement (in $\mathbb{D}$) thins out extraordinarily rapidly near $\mathbb{T}$. (We remark that the "full" disk $\mathbb{D}$ is obviously a boundary layer; one can take $(1+|\zeta|)/(1-|\zeta|)$ as $\alpha = \alpha(\zeta)$).

C) Assume for simplicity that a function F that satisfies condition (4b) is continuous in the closed disk $\mathbb{D} \cup \mathbb{T}$. We shall try to choose a multiplier of the form $\exp g$, where $g \in C(\mathbb{D} \cup \mathbb{T})$ in such a way that the function $\Phi := (\exp g) \cdot F$ becomes analytic in some boundary layer O. It follows from (6) that $\Phi \big| \mathbb{T} \not\equiv 0$. Therefore $\Phi(\zeta) \neq 0$ for some point $\zeta \in O$ (since $\operatorname{Fr} O \supset \mathbb{T}$). The following version of Jensen's inequality applies to the domain O and the point ζ:

$$\log |\Phi(\xi)| \leq \int\limits_{\operatorname{Fr} O} \log |\Phi| \, d\omega_\zeta^O$$

(compare with Sect. 2.2). But then $\int\limits_{\mathbb{T}} \log |\Phi| \, d\omega_\zeta^O > -\infty$, and by part (b) we have $\int\limits_{\mathbb{T}} \log |\Phi| \, dm > -\infty$ by definition of a boundary layer. But $\log |\Phi| = \log |F| + \operatorname{Re} g$, and $g \big| \mathbb{T} \in C$. Hence $\mathcal{L}(F^*) > -\infty$.

D) To carry out this plan we set

$$G := \{\zeta \in \mathbb{D} : |F(\zeta)| > h(1 - |\zeta|)\}, \;\; b(\zeta) := -\frac{(\overline{\partial} F)(\zeta)}{F(\zeta)} \;\; (\zeta \in G) \,,$$

$$b(\zeta) := 0 \;\; (\zeta \in \mathbb{D} \setminus G) \,, \;\; g(\zeta) := \frac{1}{\pi} \int\limits_{\mathbb{D}} \frac{b(\xi) \, dm_2(\xi)}{\zeta - \xi} \;\; (\zeta \in \mathbb{C}) \,.$$

The function Φ defined in Sect. C) is analytic in $G : \overline{\partial} \Phi = \overline{\partial} F \cdot e^g + F e^g \cdot \overline{\partial} g = 0$, since $\overline{\partial} g = b$ in G. The key point (and a very delicate one) in the proof is the verification that under conditions (4), (6), and R_2 the set G contains a boundary layer.

It is interesting that the regularity condition R_2 cannot be omitted (or even significantly weakened).

4.5. The Borichev Approach

A) Up to now we have been dealing with almost analytic functions "of bounded type" (cf. (5)). In this section, following the method of Borichev, we shall discuss a uniqueness theorem for almost-analytic functions that increase significantly faster. This theorem in turn, leads to a certain form of the uncertainty principle for the hyperfunctions Φ_H (cf. Sect. 3.5B), formula (24)), where the two-sided sequence of complex numbers $(h_n)_{n\in\mathbb{Z}}$ decays rapidly as $n \to -\infty$ and may increase almost as rapidly when $n \to +\infty$. We recall that Vol'berg's first theorem refers to hyperfunctions (24) with $h_n \equiv \hat{f}(n)$, where $f \in L^1(\mathbb{T})$. The latter inclusion prevents any growth in the amplitudes of h_n as $n \to +\infty$ (since $\sup|\hat{f}| < +\infty$). In this respect Borichev's theorem, which we shall now state (cf. B)), is stronger than Vol'berg's theorem, although it does not contain it completely because of the difference in "regularity conditions" imposed on the majorant $\mathcal{M}$. In this section we shall not state these conditions explicitly (the classical series of analytic majorants from §3 satisfies them; cf. formula (8) in Sect. 3.1C)).

B) Consider a function f defined on the disc $\mathbb{D}$. We shall call a number L the *nontangential limit* of the function f at the point ζ (and write $L = \lim \operatorname{ang}_\zeta f$) if $\lim\limits_{t\to\zeta, t\in\Delta_\zeta\cap\mathbb{D}} f(t) = L$, for any angle Δ_ζ less than π with vertex ζ and bisector directed from ζ to the origin. The nontangential limit of a function defined in $\mathbb{D}_-$ is defined analogously.

Theorem (Borichev's First Theorem). *Let $\mathcal{M} : [1, +\infty) \to (0, +\infty)$ be a sufficiently regular function satisfying condition* (1). *We set $H_+ := H|\mathbb{D}_+$, $H_- := H|\mathbb{D}_-$, where H is the analytic function defined in Sect. 3.5. Let*

$$\text{(a)}\ \lim_{n\to-\infty} \frac{\log|h_n|}{\mathcal{M}(|n|)} = -\infty\,,\ \ \text{(b)}\ \overline{\lim_{n\to+\infty}} \frac{\log|h_n|}{\mathcal{M}(n)|} < +\infty\,, \tag{10}$$

and let $E \subset \mathbb{T}$ be a set of positive length. If

$$\lim \operatorname{ang}_\zeta H_+ = \lim \operatorname{ang}_\zeta H_-\,,\ \ (\zeta \in E)\,, \tag{11}$$

then $H \equiv 0$.

It follows from (10a) that the function H_- can be assumed infinitely differentiable in the closed set $\mathbb{D}_- \cup \mathbb{T}$. As for the function H_+, condition (10b) by itself does not guarantee that it has nontangential limits at points of the circle $\mathbb{T}$. The existence of these limits is an additional assumption making up part of the hypothesis (11), which can be interpreted as the existence of a gap of positive length in the support of the hyperfunction Φ_H (compare with Sect. 3.5, where a "real" gap, containing a whole arc, was discussed). Condition (10a) characterizes the rate of decay of the amplitudes $|h_n|$ as $n \to -\infty$. Examples show that the conclusion of the theorem fails if one weakens condition (10b), which restricts the rate of growth of the amplitudes $|h_n|$ as $n \to +\infty$.

C) Borichev's first theorem can be deduced from his second theorem on almost-analytic functions (using Dyn'kin's theorem). To formulate this theorem we fix the function $w : (0,1) \to (0,+\infty)$, and we assume that $\lim_{r\uparrow 1} w(r) = 0$. It must be subjected to certain regularity conditions, which we shall not formulate explicitly, limiting ourselves to indicating a typical and rather substantial series of examples:

$$w(r) := \exp\Big[-\exp\Big(\log\frac{1}{r}\cdot\log\log\frac{1}{r}\cdot\ldots\cdot\underbrace{\log\log\cdots\log\frac{1}{r}}_{n}\Big)^{-1}\Big] .$$

All of these functions satisfy the condition

$$\int^{1} \log\log\frac{1}{w(r)}\,dr = +\infty . \tag{12}$$

The ordered hierarchy of infinitesimals (as 1 is approached) here will be the family of functions $w_c : \zeta \mapsto w(|\zeta|^c)$ $(c > 0)$, and functions that increase as $\mathbb{T}$ is approached will be compared with the infinitely large w_c^{-1}.

We shall say that the function F defined in $\mathbb{D}$ is *rapidly decreasing* (as $\mathbb{T}$ is approached), if $|F(\zeta)| = O(w_c(|\zeta|))$ as $|\zeta| \uparrow 1$ for any $c > 0$; we shall say that F is *slowly increasing* if $|F(\zeta)| = O(w_c^{-1}(|\zeta|))$ for some $c > 0$. We denote by Q the class of slowly increasing functions $f \in C^1(\mathbb{D})$ such that $\overline{\partial} f$ is rapidly decreasing and by J the class of all rapidly decreasing functions of class Q. It can be shown that

$$J = \{f \in Q : \lim_{r\uparrow 1} \max_{r\mathbb{T}} |f| = 0\} ,$$

Theorem (Borichev's Second Theorem). *If the function w is sufficiently regular and satisfies condition* (12), *then for $f \in Q$*

$$m(\{\zeta \in \mathbb{T} : \lim \operatorname{ang}_\zeta f = 0\}) > 0 \Rightarrow f \in J . \tag{13}$$

This result is the simplest of a series of theorems of the same type contained in (Borichev 1988) and (Borichev, Vol'berg 1989), where analogues of this result relating to the logarithmic integral $\mathcal{L}(f)$ can be found for functions $f \in Q$.

D) Let us discuss the proof of theorem C). Consider the "zone of smallness" E_w of the function f:

$$E_w(f) := \{\zeta \in \mathbb{D} : |f(\zeta)| \le w(|\zeta|)\} .$$

If $f \in J$, then $E_w(f)$ contains a certain annulus $\mathbb{D} \setminus r_0\mathbb{D}$, $0 < r_0 < 1$.

We call the circle $r\mathbb{T}$ a *w-circle* if $r\mathbb{T} \cap \mathbb{E}_w$ contains an arc γ whose angular measure is larger than $\log(1/r)$:

$$\exists\, \gamma := \{re^{i\theta} : \alpha_1 \le \theta \le \alpha_2\},\ \alpha_1, \alpha_2 \in [0, 2\pi],\ \alpha_2 - \alpha_1 > \log\frac{1}{r},\ \ \gamma \subset E_w .$$

If $r < 1$, then $\log(1/r) > 1 - r$. Thus a w-circle intersects the zone of smallness in an arc whose length exceeds the distance between $r\mathbb{T}$ and $\mathbb{T}$. The main points in the proof of the theorem of C) are as follows. It follows from the fact that the nontangential limits of the function f are zero on a set of positive length that there exist w-circles arbitrarily close to $\mathbb{T}$. The implication (13) is deduced from this fact using the so-called "lemma on extension of an estimate," which asserts that the zone of smallness tends to enlarge (as one moves into the interior of the disk $\mathbb{D}$): for each w-circle $r\mathbb{T}$ there exists a certain circle $\rho(r)\mathbb{T}$ of slightly smaller radius $\rho(r)$ entirely contained in E_w. (The function ρ is defined in an interval of the form $(r^*, 1)$, $0 < r^* < 1$, and $\lim_1 \rho = 1$). To complete the proof of Borichev's theorem it remains to construct a sequence of w-circles $(r_j\mathbb{T})$, where $\lim r_j = 1$, and then apply the "extension lemma." We then obtain: $\max_{\rho(r_j)\mathbb{T}} |f| \leq w(\rho(r_j))$, from which the inclusion $f \in J$ now follows. Condition (12) can be used in the proof of the "extension lemma." For the proof (together with a series of more precise assertions of this type) see (Borichev, 1988) and (Borichev, Vol'berg 1989).

The results of this section and the one preceding (the theorems of Levinson-Cartwright, Beurling, Vol'berg, and Borichev) are closely related to one another. They all characterize the degree of smallness of a nonzero (hyper)function that is consistent with rapid one-sided decay in its amplitudes. The logarithmic integral occurs (more or less explicitly) in all of them, and each strengthens some significant part of the preceding result. But each has a unique proof, and none of the theorems contains the others. It would be interesting to find a point of view that would make it possible to get rid of the feeling of "embarrassment of riches" and give all of these facts a unified interpretation.

§5. Some Forms of the Uncertainty Principle for Charges with a Spectral Gap

Assume that there exists a nondegenerate interval I that does not meet the spectrum of the charge $\mu \in M(\mathbb{R})$:

$$I \cap \operatorname{Spec} \mu = \varnothing .$$

In this case we shall say that the charge μ has a *spectral gap*. It follows from Beurling's theorem (cf. Sect. 3.4A)) that a rapidly decreasing nonzero charge (i.e., one subject to condition (17) of §3) cannot have a spectral gap. In this section we shall discuss another approach to this result and other forms of the uncertainty principle for charges with a spectral gap.

5.1. Spectral Gaps and Orthogonality to Certain Entire Functions. The Pollard Function

A) Let X be a certain set of functions bounded and continuous on the line $\mathbb{R}$. We set

$$X^{\perp} := \left\{\mu \in M(\mathbb{R}) : \int_{\mathbb{R}} x\, d\mu = 0 \ \forall x \in X\right\}.$$

We shall also have need of sets of entire functions $\mathcal{E}_\sigma$ and B_σ: the set $\mathcal{E}_\sigma$ consists of all entire functions f of degree at most equal to the positive number σ (i.e., $f(z) = O(\exp \sigma|z|)$ as $|z| \to +\infty$), and $B_\sigma := \{f \in \mathcal{E}_\sigma : \sup_{\mathbb{R}} |f| < +\infty\}$. Finally we set

$$M_\sigma := \{\mu \in M(\mathbb{R}) : \operatorname{Spec} \mu \cap [-\sigma, \sigma] = \varnothing\}.$$

It is easy to show that

$$M_\sigma = B_\sigma^{\perp}. \tag{1}$$

B) Let $\mu \in M(\mathbb{R})$. We endow the set B_σ with the seminorm $\| \, \|_{\mu,1}$ given by $\|\varphi\|_{\mu,1} := \int |\varphi|\, d|\mu|$ for $\varphi \in B_\sigma$, and we set

$$B^1_{\sigma,\mu} := \{\varphi \in B_\sigma : \|\varphi\|_{\mu,1} \le 1\}, \ \ p_{\mu,\sigma}(\zeta) := \sup\{|\varphi(\zeta)| : \varphi \in B^1_{\sigma,\mu}\} \ \ (\zeta \in \mathbb{C}).$$

The mapping $\zeta \mapsto p_{\mu,\sigma}(\zeta)$ is called the *Pollard function* (of the charge μ). It characterizes in some (rather unexplicit) manner the massiveness of the charge. The "bigger" the charge μ, the more significant the influence of the estimate $\|\varphi\|_{\mu,1} \le 1$ on the possible values of the absolute value of the entire function $\varphi \in B_\sigma$ and the smaller the function $p_{\mu,\sigma}$.

Theorem (Pollard). *If $\mu \in M_\sigma$, $\mu \neq 0$, then*

$$\int_{-\infty}^{+\infty} \frac{\log^+ p_{\mu,\sigma}(x)}{1+x^2}\, dx < +\infty. \tag{2}$$

C) Before turning to the applications of Pollard's theorem, we state a simple assertion of a technical nature. Let $\mathcal{M}$ be a nonnegative left-continuous function defined and increasing on the ray $[0, +\infty)$. We use the symbol $\mathcal{F}_{\mathcal{M}}$ to denote the set of Lipschitz nonnegative minorants of the function $\mathcal{M}$ with derivative that is almost everywhere less than or equal to one in absolute value, i.e., for all $x', x'' \ge 0$:

$$f \in \mathcal{F}_{\mathcal{M}} \Leftrightarrow f \in C([0, +\infty)),\ 0 \le f \le \mathcal{M},\ \ |f(x') - f(x'')| \le |x' - x''|.$$

Consider the upper envelope $\mathcal{M}_*$ of the set $\mathcal{F}_{\mathcal{M}}$ ($\mathcal{M}_*(x) := \sup\{f(x) : f \in \mathcal{F}_{\mathcal{M}}\}$, $x \ge 0$). It is easy to see that $\mathcal{M}_*$ is increasing and belongs to $\mathcal{F}_{\mathcal{M}}$.

The integrals $\int^{+\infty} \frac{\mathcal{M}(x)}{x^2}\,dx$ and $\int^{+\infty} \frac{\mathcal{M}_*(x)}{x^2}\,dx$ either both converge or both diverge.

5.2. Applications to the Uncertainty Principle for Charges with a Spectral Gap

A) Let $\mu \in M_\sigma$, and let the function n be such that $\int^{+\infty} x^{-2}\log^+ n(x)\,dx = +\infty$. Assume that for any sufficiently large $x > 0$ we have constructed an entire function $\varphi_x \in B^1_{\sigma,\mu}$ satisfying the inequality $\varphi_x(x) \geq n(x)$. In that case $p_{\mu,\sigma}(x) \geq n(x)$, and by Pollard's theorem $\mu = 0$. Thus to deduce specific forms of the uncertainty principle using *Pollard's theorem* it is necessary to study the conditions for "smallness" of the charge μ that guarantee the existence of an entire function $\varphi_x \in B_\sigma$ that is small in the norm $\|\ \|_{\mu,1}$ but has a "peak" of prescribed height (not less than $n(x)$) at a prescribed point x.

B) Now assume that $\mu \in M_\sigma$ and $\operatorname{var}\mu \leq 1$, and the condition (17) of §3 holds. Then, as we already know, $\mu = 0$. We note the derivation of this result from the theorem of 1B). Let $\varphi \in B_\sigma$ and $x_0 \in \mathbb{R}$, $R > 0$, and $I(= I(x_0, R)) := [x_0 - R, x_0 + R]$. Assume that $|\varphi(x_0)| = \max_I |\varphi|$ and $\sup_{\mathbb{R}\setminus I} |\varphi| \leq 1$. Then

$$\begin{aligned}\|\varphi\|_{\mu,1} &\leq \Big(\int_I + \int_{\mathbb{R}\setminus I}\Big)|\varphi|\,d|\mu| \leq |\varphi(x_0)|\,|\mu|(I) + 1 \leq \\ &\leq |\varphi(x_0)|\rho_\mu(x_0 - R) + 1 = |\varphi(x_0)|\exp(-\mathcal{M}(x_0 - R)) + 1 \leq \\ &\leq |\varphi(x_0)|\exp(-\mathcal{M}_*(x_0) + R) + 1\,. \qquad (3)\end{aligned}$$

Here $\mathcal{M} := \log(1/\rho_\mu)$ and $\mathcal{M}_*$ is the Lipschitz regularization of the function $\mathcal{M}$ constructed in Sect. 5.1C).

Assume that for a given Lipschitz increasing function $\mathcal{M}_*$ and any $x_0 \in \mathbb{R}$ we have constructed a function $\varphi(= \varphi_{x_0}) \in B_\sigma$ and a number $R > 0$ such that

$$\left.\begin{array}{ll}\text{(a)} & |\varphi(x_0)| = \max_I |\varphi|, \sup_{\mathbb{R}\setminus I} |\varphi| \leq 1\,, \\ \text{(b)} & \frac{1}{2}e^{c'\mathcal{M}_*(x_0)} \leq |\varphi(x_0)| \leq e^{c\mathcal{M}_*(x_0)}\,, \\ \text{(c)} & (c-1)\mathcal{M}_*(x_0) + R \leq 0\,,\end{array}\right\} \qquad (4)$$

where the constants c and c' are independent of x_0. It will then follow from (3) that $\varphi/2 \in B^1_{\sigma,\mu}$ and $p_{\mu,\sigma}(x_0) \geq \frac{1}{4}\exp(c'\mathcal{M}_*(x_0))$. It follows from condition (17) of §3 and Sect. 5.1C) that condition (2) fails and $\mu = 0$.

The required function φ can be constructed explicitly by setting

$$\varphi := \cos\left[\sigma\sqrt{(z - x_0)^2 - R^2}\right]\,,\quad R := \mathcal{M}_*(x_0)(1 + \sigma^2)^{-1/2}\,.$$

This construction was used by Koosis (1988) to simplify the analogous construction of de Branges (1968). The function φ occurred much earlier in a work of S.N. Bernstein (see his *Collected Works*, Vol. II (1954)), who showed that it gives a unique solution of the problem of finding the quantity $\min\{\sup_{\mathbb{R}} |\psi| : \psi \in B_\sigma, |\psi(x)| \le 1 \text{ for } |x - x_0| \ge R\}$.

C) Using the same construction, it is easy to obtain another version of the uncertainty principle. Let W be a real-valued function defined on the whole line $\mathbb{R}$, and $\inf_{\mathbb{R}} W \ge 1$ and let the function $\log W$ be uniformly continuous. If $\mathcal{L}(W) = +\infty$, then a nonzero charge $\mu \in M(\mathbb{R})$ satisfying the condition $\int_{\mathbb{R}} W\, d|\mu| < +\infty$ cannot have a spectral gap. Using duality considerations, one can easily convert this result into an assertion about weighted approximation by trigonometric sums.

Let us set $C_W := \{f \in C(\mathbb{R}) : f(x) = o(W(x))\ (|x| \to +\infty)\}$ and $\|f\|_{C_W} := \max_{\mathbb{R}}(|f|W^{-1})$. We denote by E^σ the closure of the linear span of the set of functions of the form e^{iux} in the space C_W with respect to the norm $\| \, \|_{C_W}$, where $u \in [-\sigma, \sigma]$. The assertion formulated above means that $E^\sigma = C_W$ for any $\sigma > 0$.

5.3. Spectral Gaps and Sparseness of Support. Up to now the smallness of a charge with a spectral gap was expressed in terms of its variation. In this section we shall interpret the smallness of a charge as the presence of rather long gaps in its support. We shall be studying the following question: how sparse can a set be that is the support of a nonzero charge with a spectral gap?

A) We use $\delta(j)$ to denote the distance from the interval j to the origin: $\delta(j) := \inf\{|t| : t \in j\}$, $|j|$ to denote the length of the interval, and $\Delta(j)$ to denote its "relative length": $\Delta(j) := |j|/\delta(j)$. The letter J will denote a system of pairwise disjoint nondegenerate bounded open intervals bounded away from zero: $\inf_{j \in J} \delta(j) > 0$. We set

$$\mathcal{P}(J) := \sum_{j \in J} (\Delta(j))^2 .$$

From Pollard's theorem one can derive the following assertion (due to Beurling): *a finite charge with a spectral gap concentrated on the complement S of the system J is zero if $\mathcal{P}(J) = +\infty$.* This last condition can be written in the following equivalent form: $\int_{-\infty}^{+\infty} \frac{\operatorname{dist}(x, S)}{1 + x^2}\, dx = +\infty$.

It is shown in (Jöricke 1982) that an analogous theorem holds for charges with a slightly more general class of supports (not necessarily subject to the condition $\mathcal{P}(J) = +\infty$). On the other hand the presence of a spectral gap in the statement cannot be replaced by the requirement that the charge vanish on a set of positive length (Kargaev 1985).

B) Beurling obtained the theorem of A) as a corollary of another form of the uncertainty principle applied to certain charges with infinite total variation. Assume that μ is a locally finite charge on the line $\mathbb{R}$ and $\int_{\mathbb{R}} e^{-|t|}\, d|\mu|(t) < +\infty$.

We set

$$\sigma(x)(=\sigma_\mu(x)) := \log \int_{\mathbb{R}} e^{-|x-t|}\, d|\mu|(t)\,.$$

It is easy to see that the function σ is defined everywhere in $\mathbb{R}$ and satisfies a Lipschitz condition: $|\sigma(x)-\sigma(x')| \le |x-x'|$ $(x,x'\in\mathbb{R})$.

Theorem (Beurling). *Assume that*

$$\int_{-\infty}^{\infty} \sigma^+(x)(1+x^2)^{-1}\, dx < +\infty\,.$$

If μ is a nonzero charge with a spectral gap, then

$$\int_{-\infty}^{\infty} \sigma^-(x)(1+x^2)^{-1}\, dx < +\infty\,.$$

The presence of a spectral gap for an infinite charge μ should be understood in the sense that $\lim_{\varepsilon\downarrow 0} \hat{\mu}_\varepsilon(\xi) = 0$ for all points ξ of some nondegenerate interval, where $\mu_\varepsilon := e^{-\varepsilon|x|}\mu$ (it follows from the hypothesis of the theorem that $\mu_\varepsilon \in M(\mathbb{R})$).

Now assume that the charge $\mu \in M(\mathbb{R})$ satisfies the conditions enumerated in A). Let $\operatorname{var}\mu \le 1$. Then $\sigma \equiv -\sigma^-$. It follows from the definition of the function σ that $\sigma(x) \le -\operatorname{dist}(x,S)$ $(x\in\mathbb{R})$. Thus we again arrive at the result stated in A), since

$$\int_{\mathbb{R}} \sigma^-(x)\cdot(1+x^2)^{-1}\, dx = +\infty \text{ if } \mathcal{P}(J) = +\infty\,.$$

For yet another approach to the theorem of A) see (Benedicks 1984).

C) The problem of describing the closed sets $E\subset\mathbb{R}$ that can support a nonzero charge with a spectral gap is closely connected with certain problems of potential theory.

Let $E \subsetneqq \mathbb{R}$ be a closed set. Assume that the domain $\mathbb{C}\setminus E =: G$ is regular (in the sense that any function $f\in C(E)$ admits a continuous and bounded extension to the entire plane that is harmonic in G). We use the symbol $\mathcal{P}_E$ to denote the cone of positive functions that are harmonic in G with zero limiting values on E. It can be shown that $1 \le \dim \mathcal{P}_E \le 2$ (Levin 1989, Benedicks 1980); the number $\dim \mathcal{P}_E$ equals the number of infinitely distant points of the Martin boundary of the domain $\mathbb{C}\setminus E$ (Brélot 1961).

Define $M(E) := \{\mu \in M(\mathbb{R}) : \operatorname{supp}\mu \subset E\}$. A function f defined in $\mathbb{R}$ will be called *slowly increasing* if it is Lebesgue-measurable and $|f(x)| \le C(1+|x|)^C$ for some $C > 0$ and all $x \in \mathbb{R}$.

The role of the quantity $\dim \mathcal{P}_E$ in this problem is illustrated by the following assertions (Kargaev 1983; Levin 1989; Benedicks 1980). If $\dim \mathcal{P}_E = 2$, then $M_0 \cap M(E) \ne \{0\}$ for any $\sigma > 0$. If there exists a nonzero slowly increasing function with a spectral gap concentrated on E, then $\dim \mathcal{P}_E = 2$ (in this last assertion it is the Fourier transform of a slowly increasing generalized function that is meant).

An important group of papers is connected with these ideas, summarized in the survey article (Levin 1989). It contains interesting applications to weighted approximation by polynomials, trigonometric sums, and entire functions of finite degree. The essential results in this direction are contained in the articles (Koosis 1979, Benedicks 1980, 1984, Kargaev 1983, 1990).

5.4. Sapogov's Problem (on the Characteristic Function of a Set with a Spectral Gap). N.A. Sapogov has posed the following question: can the characteristic function χ_E of a set $E \subset \mathbb{R}$ of finite positive measure have a spectral gap? (We remark that if $m(E) < +\infty$, then $\chi_E \in L^1(\mathbb{R})$). We recall two obvious obstacles that must be taken into account when constructing such a set E. First, it cannot be semi-bounded; second, by Beurling's theorem of Sect. 3.4A) it must be rather massive near the points $+\infty$ and $-\infty$ (i.e., it is necessary that $\int^{+\infty} R_E(t)\,dt > -\infty$, and $\int_{-\infty} R_E(t)\,dt > -\infty$, where $R_E(t) := t^{-2}\log|\{s \in E : |s| > |t|\}|$).

It can be asserted, however, that roughly speaking there are no other obstacles. This can be seen from the theorem of Kargaev (1982) stated below.

Theorem. *Let k be an even function that is increasing on the ray $[0, +\infty)$ and such that $k(x) \ge 1$, $k(x+y) \le Ck(x)k(y)$ $(x, y \in \mathbb{R})$. Assume that*

$$\int^{+\infty} \frac{k(x)}{x^2}\,dx < +\infty\,, \quad \sum_{n\in\mathbb{Z}} (k(n))^{-1} < +\infty\,, \quad 0 < a < b < \pi\,.$$

Then there exists a sequence of pairwise disjoint intervals $([a_n, a_n + h_n])_{n\in\mathbb{Z}}$ such that

$$c_1 < h_n k(n) < c_2\,, \quad |n - a_n| < c_3(k(n))^{-1} \ (n \in \mathbb{Z})\,, \quad \hat{\chi}_E\big|(a,b) = 0\,,$$

where $E := \bigcup_{n\in\mathbb{Z}} [a_n, a_n + h_n]$.

The following result is a corollary of this theorem (Sapogov 1974, 1978): *for any $n = 1, 2, \ldots$ there exists a Borel set $\mathcal{E} \subset \mathbb{R}^n$ of finite positive measure with respect to the n-dimensional Lebesgue measure m_n and two different Borel m_n-absolutely continuous probability measures μ_1, μ_2 in $\mathbb{R}^n$ such that*

$\mu_1(g(\mathcal{E})) = \mu_2(g(\mathcal{E}))$ for any rigid motion g (i.e., composition of a translation and an orthogonal transformation) in $\mathbb{R}^n$. Recently Vol'berg and Kargaev (unpublished), answering a question posed by Benedicks (1984), constructed a set $E \subset \mathbb{R}$ and a nonzero function $f \in L^1(E)$ having the following properties: $0 < m(E) < +\infty$, and $\hat{f}$ vanishes on the union of a system of pairwise disjoint intervals of nonzero length.

5.5. de Branges' Theorem on Extreme Points. If $M_\sigma \cap M(E) \neq \{0\}$, the set E must be rather dense at infinity. A precise characterization of this density (depending on σ) is given by the following result of de Branges.

Theorem. *Let $E \subset \mathbb{R}$ be a closed set and σ a positive number. The following assertions are equivalent.*

1) $M_\sigma \cap M(E) \neq \{0\}$;

2) *there exists an entire function $S \in \mathcal{E}_\sigma$ having the following properties:* (a) $S(\mathbb{R}) \subset \mathbb{R}$; (b) *all the zeros of the function S are simple, real, and contained in E*; (c) $\sum_n |S'(x_n)|^{-1} < +\infty$, *where (x_n) is the sequence of zeros of the function S*; (d) $1/S(\zeta) = \sum_n (S'(x_n)(\zeta - x_n))^{-1}$ $(\zeta \in \mathbb{C})$; (e) $\int_{-\infty}^{+\infty} \frac{\log^+ |S(x)|}{1+x^2}\, dx < +\infty$; (f) $\lim_{|y|\to+\infty} |S(iy)|/|y| = \sigma$. *(We recall that the smaller σ is, the "sparser" the distribution of zeros of a function of class $\mathcal{E}_\sigma$.)*

The proof of this profound theorem is based on the *Krein-Mil'man* theorem on extreme points of a compact convex set in a topological vector space. This theorem is applied to the intersection of the unit ball of the space $M(\mathbb{R})$ with the set $M_\sigma \cap M(E)$. Entire functions having the properties (c) and (d) had formed the subject of a special study in the papers of M.G. Krein on operator theory long before de Branges' theorem (cf. (Levin 1956), which contains references to the original papers).

5.6. On Fabry's Theorem. This section is yet another variation on the basic theme of this chapter. As we have seen in Sect. 3.5, one can take analytic continuability of a function f that is analytic in $\mathbb{D}$ to a function analytic in $\mathbb{D} \cup \gamma \cup \mathbb{D}_- \cup \{\infty\}$, where γ is an arc of the circle $\mathbb{T}$, as the definition of a "gap in the support" of the corresponding hyperfunction (the gap coincides with γ). In this connection it is appropriate to mention the following classical result:

Theorem (Fabry's Theorem). *Let (n_k) be a strictly increasing sequence of indices. If $\lim_{k\to\infty} n_k/k = +\infty$, and the radius of convergence of the power series $\sum_{k=1}^{\infty} a_k z^{n_k}$ is one, then all the points of the circle $\mathbb{T}$ are singular for this series*(Bieberbach 1955; Gamelin 1969; Mandelbrojt 1969).

In application to hyperfunctions (on $\mathbb{T}$) we thus arrive at a conclusion of the following type: "gap in support + sparse spectrum $\Rightarrow$ vanishing of hyperfunction". The sparseness of the spectrum of a hyperfunction means here that $\operatorname{Card}\{\operatorname{Spec} f \cap [0, k[\} = o(k)$ $(k \to +\infty)$. It is interesting that this condition is sharp: if $\lim n_k/k < +\infty$, the conclusion of the theorem ceases to hold.

§6. Some Methods of Constructing Small Functions with Small Fourier Transform

In this section it will be shown that the condition of divergence of the logarithmic integral that occurs in many forms of the uncertainty principle, is inherent in the subject and cannot be weakened.

6.1. Outer Functions

A) We begin with the following problem: given a nonnegative function h defined on the set X (where $X = \mathbb{T}$ or $\mathbb{R}$) and a number $p \in [1, +\infty]$, construct a function $f \in H^p(X)$ such that

$$|f| = h \ m\text{-almost everywhere on } X\,. \tag{1}$$

For this problem to be solvable it is obviously necessary that

$$(a)\ h \in L^p(X) \text{ and } (b)\ \mathcal{L}(h) > -\infty\,. \tag{2}$$

where $\mathcal{L}(h) := \int\limits_{\mathbb{T}} \log h\, dm$ if $X = \mathbb{T}$ and $\mathcal{L}(h) = \int\limits_{\mathbb{R}} \log h\, d\Pi$ if $X = \mathbb{R}$. (Here $d\Pi := \frac{1}{\pi}\frac{dx}{1+x^2}$). We shall see that conditions (2) are also sufficient.

B) Suppose $\mu \in M(\mathbb{T})$. We set

$$S(\mu)(\zeta) := \frac{1}{2\pi}\int\limits_{\mathbb{T}} \frac{z+\zeta}{z-\zeta}\, d\mu(z)\ \ (\zeta \in \mathbb{D})\,.$$

The function $S(\mu)$ is called the *Schwartz transform* of the charge μ (or the function $f \in L^1(\mathbb{T})$ if $\mu = fm$) If μ is a real charge then $\operatorname{Re} S(\mu) = \mathcal{P}(\mu)|\mathbb{D}$ and $S(\mu)(0) = \mu(\mathbb{T}) \in \mathbb{R}$ (here $\mathcal{P}(\mu)$ denotes the Poisson transform of the charge μ (cf. Sect. 2.13)).

C) Suppose a nonnegative Lebesgue-measurable function h defined on $\mathbb{T}$ satisfies condition (2b). The function

$$\operatorname{Ext} h := \exp S(\log h)$$

that is analytic in $\mathbb{D}$ is called the *outer function* corresponding to h. It has the following properties: a) $\operatorname{Ext} h \in \mathbb{N}$; (b) $|(\operatorname{Ext} h)^*| = h$ m-almost everywhere;

(c) if $p \in [1, +\infty]$ and (2a) holds, then $\operatorname{Ext} h \in H^p(\mathbb{D})$. Thus under condition (2) the function $f := (\operatorname{Ext} h)^*|$ has property (1) and thereby solves the problem posed in A) (when $X = \mathbb{T}$).

D) We now discuss the solvability of Eq. (1) when $X = \mathbb{R}$. We set $h_1 = h \circ Z$, where $Z(w) := \frac{i(1+w)}{1-w}$ $(w \in \mathbb{T})$. It is easy to see that if $\mathcal{L}_{\mathbb{R}}(h) > -\infty$, then $\mathcal{L}_{\mathbb{T}}(h_1) > -\infty$. It is clear that the function $f := (\operatorname{Ext} h_1)^*(Z^{-1})$ has property (1) (where $X = \mathbb{R}$). It is not difficult to verify that $f \in H^p(\mathbb{R})$ under condition (2a).

The simple explicit solution of problem (1) described above makes it possible to establish easily the sharpness of a variety of specific forms of the uncertainty principle. Using outer functions it is not difficult to construct a continuous plus-function that vanishes on a given closed subset $E \subset \mathbb{T}$ of zero length and is nonzero everywhere on $\mathbb{T} \setminus E$. The same construction proves that the classes $\mathcal{U}(C^\infty)$ and Carl are identical (cf. Sect. 1.5E) of Chapt. 1). Another application of outer functions is given in the next section.

6.2. Determining Majorants and the Logarithmic Integral

A) We now return to the problem we discussed in Sect. 3.1. Let $H : \mathbb{R} \to [0, +\infty)$ be a bounded function. We shall show that if $\mathcal{L}(H) > -\infty$, then H cannot be a determining majorant.

Proof. We set $h(t) := H(t)\exp(-\sqrt{|t|})$ $(t \in \mathbb{R})$. Then $h(t) = O(|t|^{-N})$ $(|t| \to +\infty)$ for any $N > 0$. We construct a function $f \in H^1(\mathbb{R})$ that satisfies condition (1) (this is possible by Sect. 6.2). The function $g := \mathcal{F}^{-1}(f)$ vanishes on the half-line $[0, +\infty)$ and belongs to $C^\infty(\mathbb{R})$, so that $g^{(j)}(0) = 0$ for $j \in \mathbb{Z}_+$. At the same time $|g| \le H$ and $g \not\equiv 0$.

B) The function f constructed in A) makes it possible to carry out an important derivation relating to the power moment problem and the problem of weighted approximation by polynomials (cf. Sect. 2.3). In fact

$$\int_{\mathbb{R}} f(u)u^j \, du = g^{(j)}(0)i^{-j} = 0 \quad (j \in \mathbb{Z}_+)\,,$$

and

$$\int_{\mathbb{R}} |f(u)| \frac{du}{H(u)} = \int_{\mathbb{R}} e^{-\sqrt{|u|}} \, du < +\infty\,.$$

Thus if $\mathcal{L}(H) > -\infty$, there exists a nonzero charge $\mu \in M(\mathbb{R})$ all of whose moments are zero and $\int H^{-1}|d\mu| < +\infty$. We conclude from duality considerations that under the same condition the set of polynomials is not dense in the weight space $\mathbb{C}_H^0$.

6.3. One-sided Polynomial Decay of Amplitudes and Smallness of the Support. Hruščev's Theorem. In this section we shall give another example of the application of the machinery of outer functions.

A) How can one describe the class of closed sets $K \subset \mathbb{T}$ that can support a nonzero charge $\mu \in M(\mathbb{T})$ such that

$$|\hat{\mu}(n)| = O(\gamma_n) \ (n \to -\infty) , \tag{4}$$

where $\gamma := (\gamma_n)_{n<0}$ is a given infinitesimal sequence? The answer of course depends on γ. Thus, for example, if $\gamma_n \equiv q^n$, where $q > 1$, then the support of a nonzero charge subject to condition (4) must fill up the entire circle $\mathbb{T}$ (since the function $C(\mu)$ defined in Sect. 2.1J) coincides in $\mathbb{C} \setminus \mathbb{D}$ with some entire function under the conditions (4) and $\operatorname{supp} \mu \neq \mathbb{T}$. More substantive assertions of this type follow from Beurling's theorem (Sect. 3.4B)).

B) We now consider a sequence γ with polynomial order of decay. Let α be a positive number. We shall say that a closed set $K \subset \mathbb{T}$ is *α-spacious* if there exists a charge $\mu \in M(\mathbb{T})$ such that

$$\text{(a) } \operatorname{supp} \mu \subset K , \quad \text{(b) } \mu \neq 0, \quad \text{(c) } |\hat{\mu}(n)| = O(|n|^{-\alpha}) \ (n \to -\infty) . \tag{5}$$

If there exist a charge having properties (5) for any $\alpha > 0$, the set K will be called *spacious.* The class of α-spacious sets will be denoted by the symbol $\mathcal{M}_\alpha$; the symbol $\mathcal{M}_\infty$ will denote the class of spacious sets.

C) We now compare condition (5c) with its "two-sided" version

$$|\hat{\mu}(n)| = O(|n|^{-\alpha}) \ (|n| \to +\infty) . \tag{6}$$

If condition (6) holds and the number α is sufficiently large, then $\mu = fm$, where f is a continuous function (if $\alpha > 1$) or even a smooth function (if $\alpha > 2$). Hence $\operatorname{supp} \mu$ contains interior points (if $\mu \neq 0$). But then one can concentrate a nonzero function $\varphi \in C^\infty(\mathbb{T})$ on the set $\operatorname{supp} \mu$, so that the coefficients of φ decay faster than any power. Thus the characterization of the sets K that support a nontrivial charge μ subject to the "two-sided" condition (6) (for some $\alpha > 1$ or for all $\alpha > 1$) is very simple: K has this property if and only if it contains a nondegenerate arc. The "one-sided" condition (5) is less restrictive than (3) and one might expect that it is satisfied by certain nonzero charges with "thin" (nowhere dense) support. On the other hand the length of the support is positive if $\alpha > 1/2$ and conditions (5b) and (5c) hold.

Proof. If $\alpha > 1/2$, then (5b) implies that $\sum_{n<0} \hat{\mu}(n) z^n$ is the Fourier series of some function $f \in L^2(\mathbb{T})$ and $\sum_{n \geq 0} \hat{\mu}(n) z^n$ is the Fourier series of the plus-charge $\mu - fm$, which is necessarily absolutely continuous by the theorem of F. and M. Riesz (cf. Sect. 1.2 of Chapt. 1).

Thus

$$\alpha > 1/2 , \ K \in \mathcal{M}_\alpha \Rightarrow |K| > 0 .$$

However the implication arrow in this assertion cannot be reversed (cf. D) below). To clarify the role of the condition $\alpha > 1/2$ we note that when $\alpha = 1/2$

condition (6) can be satisfied by a nonzero charge μ with $|\operatorname{supp}\mu| = 0$ (cf. Sect. 2.4 of Chapt. 1).

D) We use the symbol $\mathcal{L}(K)$ to denote the set of arcs of the circle $\mathbb{T}$ complementary to K. We define the *entropy* of the set K to be the quantity

$$\operatorname{Entr} K := \sum_{l \in \mathcal{L}(K)} |l| \log |l| \,,$$

which we have already encountered in Sect. 1.5E) of Chapt. 1.

Theorem (Hruščev). *Let $\alpha > 1/2$, and let $K \subset \mathbb{T}$ be a closed set. The following assertions are equivalent:* 1) *K is α-spacious;* 2) *K is spacious;* 3) *K contains a set of positive length and finite entropy. If condition* 3) *holds, then there exists a nonzero function $\varphi \in L^\infty(K, m)$ such that $\hat{\varphi}(n) = O(|n|^{-\alpha})$ $(n \to -\infty)$ for any $\alpha > 0$.*

In general assertion 3) is not equivalent to the condition that the set K itself has finite entropy and positive length, although this is the case for certain regular sets K (the so-called "sets of Cantor type"). The basic difficulty of the proof is the construction of a sequence of outer functions (G_n), that are continuous up to and including $\mathbb{T}$ and such that

$$\text{(a)}\, G_N(0) = 1\,, \quad \text{(b)}\, |G_n(\zeta)| \le (1-|\zeta|)^{-\alpha} \quad (\zeta \in \mathbb{D})\,, \quad \text{(c)}\, \max_K |G_n| \le \exp(-n)\,.$$

It turns out that such a sequence exists precisely when condition 3) fails. In this case K is so "small" that estimates (c) and (b) are compatible with (a) (cf. (Hruščev 1978) and also (Makarov 1989), where a simplification of the original proof is proposed).

§7. Two Theorems of Beurling and Malliavin

7.1. Statement of the Problem. Whereas in the preceding section we were interested in functions with a spectral gap, here we shall study the opposite situation, when the spectrum contains only a bounded band of frequencies. More precisely, we shall study the structure of the absolute value of a function (defined in $\mathbb{R}$) with a bounded spectrum. In this situation it is hopeless to seek visualizable and exhaustive conditions for solvability of the problem

$$|f| = h\,, \quad \operatorname{Spec} f \subset [A, B]\,, \tag{1}$$

where h is a given function and A and B given numbers, f being the required function. We limit ourselves to seeking an answer to a simpler question: how small can the function $|f|$ be if $f \neq 0$ and $\operatorname{Spec} f$ is a bounded set? Put another way the same question can be posed as follows: how small can the absolute value of the Fourier transform of a nonzero function of compact support be? "Smallness" will be understood as the existence of a prescribed majorant on the whole line $\mathbb{R}$. Thus we simplify the difficult problem (1) and replace it by the following:

$$\text{(a) } f \neq 0; \text{ (b) } \operatorname{diam} \operatorname{supp} f < +\infty; \text{ (c) } |f| \leq h\,. \tag{2}$$

We now pass to precise definitions.

7.2. Admissible Majorants

A) Let a be a positive number. We shall agree to say that a function h is *a-admissible* if $0 \leq h(\xi) \leq 1$ for $\xi \in \mathbb{R}$ and there exists a nonzero charge $\mu \in M(\mathbb{R})$ such that

$$\text{(a) } \sup_{\xi \in \mathbb{R}} \frac{|\hat{\mu}(\xi)|}{h(\xi)} < +\infty\,, \text{ (b) } \operatorname{diam} \operatorname{supp} \mu < 2a\,. \tag{3}$$

If there exists a charge $\mu \neq 0$ satisfying conditions (3b) and

$$\|\hat{\mu}/h\|_{L^p(\mathbb{R})} < +\infty\,. \tag{4}$$

we shall say that the function h is a-admissible in the mean of order p. If h is a-admissible for any positive a, we shall call it *admissible*. The concept of a function that is admissible in the mean of order p is defined analogously.

B) An obvious obstacle for a-admissibility (and for a-admissibility in mean) of the function h is the divergence of the logarithmic integral $\mathcal{L}(h)$. The inequality

$$\mathcal{L}(h) := \frac{1}{\pi} \int_{-\infty}^{\infty} \frac{\log h(t)}{1+t^2}\, dt > -\infty \tag{5}$$

is necessary for a-admissibility and a-admissibility in mean.

Proof. Under condition (3b) the set $\operatorname{supp} \mu$ is semi-bounded and if $\mu \neq 0$, then $\mathcal{L}(|\hat{\mu}|) > -\infty$. On the other hand $\mathcal{L}(|\hat{\mu}|) = \mathcal{L}\left(\frac{|\hat{\mu}|}{h}\right) + \mathcal{L}(h)$. It remains to take account of the fact that $\log \frac{|\hat{\mu}|}{h} \leq \frac{|\hat{\mu}|}{h}$, so that $\mathcal{L}(\hat{\mu}) > -\infty \Rightarrow \mathcal{L}(h) > -\infty$.

This simple reasoning uses only the semi-boundedness of the support. However passing from semi-bounded supports to bounded supports significantly complicates the problem: under the condition (3b) the function $\hat{\mu}$ is entire and of finite degree (and not simply the boundary value of a function that is analytic only in a half-plane, as is the case when $\sup \operatorname{supp} \mu < +\infty$). This generates obvious additional obstacles (in relation to condition (5)) to the admissibility of the majorant h. Thus a majorant h whose set of zeros has a point of density (in $\mathbb{R}$) or vanishes faster than polynomially at some point t_0 (i.e., so that $h(t) = O(|t-t_0|^j)$ as $t \to t_0$ for any $j > 0$) cannot be admissible. Both of these conditions are fully compatible with condition (5). A function h cannot be a-admissible if $h(\sqrt{k}) = 0$ $(k = 1, 2, \ldots)$, for a nonzero entire function of finite degree cannot vanish at all points of the form $\sqrt{k}$. Sharpening these considerations, one can easily show that h is not a-admissible if the sequence $(h(\sqrt{k}))$ tends to zero sufficiently rapidly.

C) Effects of this type can be excluded by subjecting the function h to additional regularity conditions besides condition (5). Thus under condition (5) the construction of a charge $\mu \neq 0$ satisfying conditions (3) becomes completely simple if h is an even function that decreases on the ray $[0, +\infty)$. It suffices to set $\Omega(\xi) := h(2\xi)(1+\xi^2)^{-1}$, construct a function $Q \in H^1(\mathbb{C}_+)$ such that $|Q|/\mathbb{R} = \Omega$ and then set $Q_a := e^{-iax}Q$, $f_a := Q_a * \overline{Q}_a$, $\mu := \hat{f}_a m$ (without loss of generality one can assume that $\widehat{Q}(a) \neq 0$, so that $\hat{f}_a(0) \neq 0$). For another explicit way of solving the problem (3) (with the same regularity properties as h) see (Mandelbrojt 1969).

To study an oscillating function h for admissibility is much more difficult. It can be said that the main interest in this problem is precisely the regularity conditions on the majorant h that conrol its oscillation and, in conjunction with (5), guarantee its admissibility.

D) We shall assume that $h \equiv 1$ near the origin and that the function is even. We set

$$W := 1/h, \quad \sigma(\xi) := \log(W(\xi)/\xi) \quad (\xi \in \mathbb{R})$$

(so that $W \geq 1$ and the function σ is odd and vanishes near the origin.) In the first Beurling-Malliavin theorem (the first B-M theorem, for short) the main role is played by the function σ. It turns out that the majorant h is admissible if σ is not too large at infinity and does not oscillate too strongly. The "smallness" of σ is expressed by the condition

$$\int_0^{+\infty} \frac{\log \sigma(\xi)}{\xi}\, d\xi < +\infty\,, \tag{6}$$

which is equivalent to condition (5) and the oscillations of the function σ are restricted by the condition

$$\int_{-\infty}^{+\infty}\int_{-\infty}^{+\infty} \left(\frac{\sigma(x)-\sigma(y)}{x-y}\right)^2 dx\, dy < +\infty\,. \tag{7}$$

The condition (7) is well-known in potential theory and means that the function σ can be extended to the upper half-plane $\mathbb{C}_+$ as a harmonic function u with finite Dirichlet integral:

$$\iint_{\mathbb{C}_+} |\operatorname{grad} u|^2\, dx\, dy < +\infty$$

(in the sense that $\sigma(x) = \lim_{y \downarrow 0} u(x+iy)$ for almost every $x \in \mathbb{R}$).

Theorem (First Beurling-Malliavin Theorem (Beurling and Malliavin 1962; Koosis 1964; Malliavin 1979)). *Under conditions* (6) *and* (7) *the majorant h is* *admissible in the mean of any positive order. As for admissibility, conditions* (6) *and* (7) *guarantee it for some smoothing* h_ε *of the function* h.

We shall not describe the smoothing procedure here explicitly, limiting ourselves to two corollaries of theorem that indicate sufficient conditions for admissibility of the majorant h itself.

E) The first corollary is the following. If h satisfies condition (5) and the function $\log h$ is uniformly continuous, then h is admissible and admissible in mean of any positive order.

To state the second corollary we require a definition. An entire function f of finite degree is said to belong to the *Cartwright class* if

$$\int_{-\infty}^{+\infty} \frac{|\log|f(x)||}{1+x^2}\,dx < +\infty\,.$$

The set of such functions will be denoted Cart.

The second corollary of the first B-M theorem is the following: if $F \in \text{Cart}$ and $\inf_{\mathbb{R}} |F| > 0$, then the function $h := (1/|F|)\big|\mathbb{R}$ is admissible and admissible in mean of any positive order.

7.3. On the Second Beurling-Malliavin Theorem. The admissible majorants h described by the first B-M theorem are strictly positive. Thus the results of Sect. 7.2 do not contain any information on the structure of the set of frequencies that may be absent in the spectrum of a nonzero charge of compact support. The present section is devoted to this question.

A) Let $\Lambda \subset \mathbb{R}$ and $X \subset \mathcal{D}'(\mathbb{R})$ (= the space of distributions on $\mathbb{R}$). We set

$$R_X(\Lambda) := \sup\{a \geq 0 : (\varphi \in X,\ \operatorname{diam}\operatorname{supp}\varphi < 2\pi a,\ \hat{\varphi}\big|\Lambda = 0) \Rightarrow \varphi = 0\}\,.$$

The quantity $R_X(\Lambda)$ gives an optimal upper bound for $\operatorname{diam}\operatorname{supp}\varphi$ that is compatible with the absence of frequencies belonging to Λ in $\operatorname{Spec}\varphi$, where $\varphi \in X$ (i.e., compatible with the equality $\hat{\varphi}\big|\Lambda = 0$). We shall call it the Λ-diamter (relative to the class X). If $\operatorname{diam} P > 2\pi R_X(\Lambda)$, then the interval P contains the support of some distribution $\varphi \in X$ with $\hat{\varphi}\big|\Lambda = 0$ and if $\operatorname{diam} P < 2\pi R_X(\Lambda)$, then P is too small for this. It is not difficult to show that $R_X(\Lambda)$ is independent of X, when X varies in a rather wide range:

$$C^\infty(\mathbb{R}) \subset X \subset \mathcal{D}'(\mathbb{R})\,.$$

Having in mind such classes, we whall omit the index X from now on in the notation for the Λ-diameter $R(\Lambda)$.

We use the symbol Cart_σ to denote the set of entire functions of the Cartwright class whose degree does not exceed σ (cf. Sect. 7.2). The following equalities hold:

$$\begin{aligned} R(\Lambda) &= \sup\{a \geq 0 : (f \in \text{Cart}_{\pi a},\ f\big|\Lambda = 0) \Rightarrow f = 0\} = \\ &= \sup\{a \geq 0 : (f \in \text{Cart}_{\pi a} \cap L^\infty(\mathbb{R}),\ f\big|\Lambda = 0) \Rightarrow f = 0\}\,. \end{aligned}$$

B) The Λ-diameter is of interest as a characterization of the approximation possibilities of the family of exponentials $(e^{i\lambda x})$ $(\lambda \in \Lambda)$. Using the Hahn-Banach theorem, one can easily establish that

$$R(\Lambda) = \sup\{a \geq 0 : \text{the family } (e^{i\lambda x})\ (\lambda \in \Lambda) \text{ is complete in } X_{\pi a}\},$$

where $X_{\pi a}$ denotes any of the spaces $C^{(m)}([-\pi a, \pi a])$ $(m = 0, 1, 2, \ldots, \infty)$ or $L^p([-\pi a, \pi a])$.

C) It is useful to compare this approximation characteristic of the Λ-diameter with the following elementary example. Let Λ be an arithmetic progression with step $1/b$: $\Lambda = \Lambda_b := (k/b)$ $(k \in \mathbb{Z})$, where $b > 0$. The family $(e^{i\lambda x})$ $(\lambda \in \Lambda_b)$ is complete in $L^2(-\pi b, \pi b)$ and not complete in $L^2(-\pi a, \pi a)$ if $a > b$. Hence $R(\Lambda_b) = b$.

Let Λ be a set of real numbers, and let $n_\Lambda(t) = \operatorname{Card}(\Lambda \cap [0, t))$ when $t > 0$ and $n_\Lambda(t) = -\operatorname{Card}(\Lambda \cap [t, 0))$ when $t < 0$. When $\Lambda = \Lambda_b$, the following equality holds:

$$R(\Lambda) = \lim_{|t| \to +\infty} \frac{n_\Lambda(t)}{t}.$$

It also holds for many Λ that are not arithmetic progressions, but resemble them in the sense that the quotient $n_\Lambda(t)/t$ tends to some limit rather rapidly as $|t| \uparrow +\infty$. In the general case the second B-M theorem expresses the Λ-diameter in terms of a certain quantity $D(\Lambda)$ that characterizes the density of the set Λ at infinity. A precise definition of this quantity is rather complicated and we shall not reproduce it here. We emphasize only that, in contrast to the purely spectral quantity $R(\Lambda)$, $D(\Lambda)$ is defined in purely geometric terms and, despite a certain cumbersomeness, is amenable to computation or estimation in many specific interesting situations (Redheffer 1977). The classical estimates (Paley, Wiener 1934)

$$\overline{\lim_{|t| \to +\infty}} \frac{n_\Lambda(t)}{t} \leq R(\Lambda) \leq \inf\{|\lambda' - \lambda''| : \lambda', \lambda'' \in \Lambda,\ \lambda' \neq \lambda''\}$$

can be obtained from the B-M formula as very simple corollaries.

Commentary on the References

The Heisenberg uncertainty principle is stated and discussed in any textbook on quantum mechanics (cf., for example, (Faddeev, Yakubovskij 1980; Dirac 1958; von Neumann 1932). For manifestations of the uncertainty principle in other areas of physics and engineering see (Gorelik 1959; Kharkevich 1957; Khurgin, Yakovlev 1971).

The number of specific analytic results that illustrate the uncertainty principle is truly unimaginable. This enormous mass of mathematical facts is hardly amenable to a comprehensive systematization and classification. As far as we know, there is no monograph specially and wholly devoted to the uncertainty principle in the interpretation to which we adhere in this article. A great deal can be found in the treatises (Bari 1961; Kahane 1962; Kahane, Salem 1963; Zygmund 1959) on Fourier analysis (especially in the sections devoted to lacunary series). A significant part of complex analysis – most of all the uniqueness theorems – is also closely connected with the uncertainty principle, which is discussed (more or less explicitly) in the following works: (Vladimirov 1964; Levin 1956; Privalov 1950; Bieberbach 1955; Boas 1954; Bremermann 1965; de Branges 1968; Duren 1970; Ehrenpreis 1970; Gamelin 1969; Garnett 1981; Koosis 1980, 1988; Levin 1956; Paley, Wiener 1934). In the books (Vladimirov 1964) and (Bremermann 1965) one can find, in particular, information on functions whose spectrum is concentrated in a multidimensional cone. The classical theorems on the distribution of the singularities of analytic functions can frequently be interpreted as some form of the uncertainty principle; in this connection see the monographs (Bieberbach 1955; Carleman 1944). Particularly close to our theme are the books (Levin 1956; Koosis 1988; Levinson 1940; Paley, Wiener 1934), and also (Mandelbrojt 1937, 1952, 1962, 1969). In many respects our article follows the book of the same name by the authors, which is in press.

We now give bibliographical references and commentaries for the individual sections of the article.

Chapter 1. §1. For the different proofs of the theorem of F. and M. Riesz see (Bari 1961; Nikol'skii 1980; Privalov 1950; Gamelin 1969; Hoffman 1962; Koosis 1980; Oberlin 1980; Zygmund 1959). Some multidimensional generalizations and analogues in the spirit of "abstract function theory" can be found in (Aleksandrov 1981; Barbey, König 1977; Bochner 1944; Brummelhuis 1988; Gamelin 1969). In §1 we used the proof found in (Doss 1981). The Rudin-Carleson theorem appeared in (Carleson 1957) and (Rudin 1956). To the literature already pointed out in the text we add here the articles (Vinogradov 1970; Dyn'kin 1979; Kotochigov 1972; Oberlin 1980), and the survey (Vinogradov, Khavin 1974). Determining sets (in the sense of Sect. 1.5) for various classes of plus-functions were studied in (Korenblyum 1971; Hewitt 1976; Carleson 1952; Hoffman 1962). Condition (7) first appeared in the article (Beurling 1939). The left-hand side of (7) can be given a form that resembles entropy (in the information-theoretic sense; see assertion (b) in Sect. 1.5E)). This observation was developed in the article (Khrushchev 1977b). The Aleksandrov-Shapiro approach was used for a series of generalizations (including group-theoretic generalizations) in (Brummelhuis 1988). Examples of $\mathcal{R}$-sets and various information about them can be found in (Jöricke 1980; Dressler, Pigno 1974ab; Meyer 1968; Rudin 1960).

§2. The survey article (Hewitt 1976), which contains interesting historical information and the article (Brown, Hewitt 1980) are devoted to this theme. On the works of Riesz see (Katznelson 1976; Peyrière 1975; Zygmund 1959). For the Ivashev-Musatov theorem see (Bari 1937; Ivashev-Musatov 1957; Brown, Hewitt 1980; Körner 1977; de Leeuw, Katznelson 1970; Zygmund 1959).

§3. The annihilation of a pair of sets of finite volume was established by Benedicks (1985), and the strong annihilation by Amrein, Berthier (1977). Essential progress in the problem of describing the spectra that annihilate with the complement of any set of positive length (on the circle) was made by I. M. Mikheev (1975). In connection with Sect. 3.2 see the book (Khurgin, Yakovlev 1971), where historical and bibliographical information can be obtained. The Logvinenko-Sereda theorem has quite a long history. The story of its origins is told in the article (Paneyakh 1966), where, in addition, the connections with the theory of differential operators are explained. Far-reaching generalizations of the Logvninenko-Sereda theorem were given by Vol'berg (1981); see also (Gorin 1985; D'yakonov 1988; Havin, Jöricke 1981). The papers (Katsnel'son 1973; Lin 1965; Logvinenko 1977) also relate to these problems.

Chapter 2. **§2.** Here we are following the classical theory of Hardy spaces in its "purely complex" form as in (Privalov 1950; Gamelin 1969; Garnett 1981; Koosis 1980), in contrast to the real-abstract approaches of (Barbey, König 1977; Gamelin 1969; Hoffman 1962). For the Hardy classes in a half-plane see (Krylov 1939; Nikol'skii 1980; Hoffman 1962).

§3. This section is closely connected with the theory of quasi-analytic function classes, with the Bernstein weight problem, and with the power moment problem. The monographs (Akhiezer 1961; Mandelbrojt 1937, 1962; Levinson 1940; Mandelbrojt 1952, 1969) and the papers (Akhiezer 1956; Mergelyan 1956; Beurling 1961; Geetha 1969; Koosis 1964; Levinson, McKean 1964) are devoted to (the many parts of) this topic. Beurling's theorem in Sect. 3.4 is proved in (Beurling 1961), which also contains an L^2-analogue of the theorem (more difficult than in Sect. 3.4B)) (cf. also (Jöricke, Khavin 1979; Koosis 1988)). The Levinson-Cartwright theorem was proved in (Levinson 1940).

§4. In Sect. 4.4 we sketch the outline of the proof of Vol'berg's first theorem, following (Vol'berg, Jöricke 1986); the first proof (Vol'berg 1982) differed in technical aspects. The regularity condition R_2 was proposed by Brennan ("Functions with rapidly decreasing negative coefficients," Preprint, University of Kentucky Mathematics Department, 1986). In implicit form it is essentially contained in (Vol'berg, Jöricke 1986). The technique of "extending estimates" described in Sect. 4.5G) was significantly improved in (Borichev, Vol'berg 1989). The articles (Borichev 1988; Borichev, Vol'berg 1989) contain results on the finiteness of the logarithmic integral of an increasing almost-analytic function.

§5. The works (Kargaev 1983, 1985; Benedicks 1980, 1984; Levin 1989) are devoted to charges with a spectral gap and related topics (the last-named work in a certain sense summarizes a large cycle of preceding publications of the author, partly in collaboration with co-author Akhiezer). In many aspects we have followed the book (Koosis 1988). Supplements to the theorem of de Branges on extreme points have been recently obtained by Koosis (1990a, 1990b). Ehrenpreis (1981) proposed an interesting approach to the description of functions (of several variables) with a spectral gap. Different proofs of Fabry's theorem can be found in (Bieberbach 1955), and (Mandelbrojt 1969). A proof based on an elementary estimate of the coefficients of a trigonometric sum in terms of the absolute value of its values on a closed interval proposed in (Turan 1953), was given by Gaier (1970). In connection with Fabry's theorem we mention the classical works (Polya 1918, 1929).

An important generalization of Turan's estimate and applications to the themes of §§1.3 and 2.1 are obtained in (Nazarov 1993).

§6. The term "outer function" was introduced by Beurling (1949). The machinery of outer functions had been widely applied in complex analysis and its applications long before this (cf., for example, the book (Smirnov 1988), which contains information of historical nature).

§7. The original proof of the first Beurling-Malliavin theorem was simplified in (Malliavin 1979),[5] where a (nearly adequate) reduction of the original spectral problem to a certain extremal problem of pure potential theory was carried out. Another approach was developed in (Koosis 1979, 1983), and also (de Branges 1968).

For the first results preceding the second Beurling-Malliavin theorem see (Levin 1956, 1972), (Levinson 1963), and (Paley, Wiener 1934). The original proof (Beurling, Malliavin 1967) was simplified in (Kahane 1962); for subsequent simplification

[5] P. Koosis has recently found a new proof of this theorem, to appear in *Annales de l'Institut Fourier.*

see (Krasichkov 1986). The topic of §7, in which the main subject is criteria for completeness of a system of harmonics with prescribed frequencies on a given interval, is related to a large circle of basis problems for such systems. For this see (Hruščev, Nikol'skii, Pavlov 1981), where the history of the problem can also be found.

The uncertainty principle frequently manifests itself as the assertion that certain linear operators are not local (mainly convolution operators) and occurs in the form of the following interdiction: the image of a nonzero function f under the action of the operator cannot vanish in the same place as f. For this see the book (de Branges 1968), where precise conditions are given for an operator to be local in terms of its symbol, and also the articles (Jöricke, Khavin 1979; Kargaev 1985; Makarov 1985; Khavin 1983; Havin, Jöricke 1981; Havin, Jöricke, Makarov 1984).

Important contributions to the theme of §7 are due to B.N. Khabibulin (see his article "A theorem on the least majorant and its applications," I, II. Izv. Ross. Akad. Nauk, Ser. Mat., *57* (1), 129–146; *57* (3), 70–91 (1993), where further references can be found).

References*

Arnol'd, V.I. (1974): Mathematical Methods of Classical Mechanics. Nauka, Moscow. English transl.: Springer-Verlag, Berlin-Heidelberg-New York 1978, Zbl. 647.70001

Akhiezer, N.I. (1956): On weighted approximation of continuous functions by polynomials on the entire number line. Usp. Mat. Nauk *11*(4), 3–43

Akhiezer, N.I. (1961): The Classical Moment Problem. Fizmatgiz, Moscow. English transl.: Oliver and Boyd, Edinburgh-London 1965, Zbl. 124,62

Akhiezer, N.I. (1965): Lectures on Approximation Theory. 2nd ed. Nauka, Moscow, Zbl. 31,157

Aleksandrov, A.B. (1981): Essays on non locally convex Hardy classes. Complex Analysis and Spectral Theory. Seminar, Leningrad 1978/80. Lect. Notes Math. *864*, 1–89, Zbl. 482.46035

Amrein, W.O., Berthier, A.M. (1977): On support properties of L^p-functions and their Fourier transforms. J. Funct. Anal. *24*(3), 258–267, Zbl. 355.42015

Barbey, K., König, H. (1977): Abstract analytic function theory and Hardy algebras. Lect. Notes Math. *593*, Zbl. 373.46062

Bari, N.K. (1937): Sur le rôle des lois diophantiques dans le problème d'unicité du développement trigonométrique. Mat. Sb. *2(44)*, 699–722, Zbl. 18,18

Bari, N.K. (1961): Trigonometric Series. Fizmatgiz, Moscow. English transl.: Pergamon Press, New York 1964, Zbl. 129,280

Belov, A.S. (1976): Quasianalyticity of the sum of a lacunary series. Mat. Sb., Nov. Ser. *99*(3), 433–467. English transl.: Math. USSR, Sb. *28*, 389–419 (1978), Zbl. 367.42014

Benedicks, M. (1980): Positive harmonic functions vanishing on the boundary of certain domains in $\mathbb{R}^n$. Ark. Mat. *18*(1), 53–72, Zbl. 455.31009

Benedicks, M. (1984): The support of functions and distributions with a spectral gap. Math. Scand. *55*(2), 285–309, Zbl. 577.42008

* For the convenience of the reader, references to reviews in *Zentralblatt für Mathematik* (Zbl.), compiled using the MATH database, and *Jahrbuch über die Fortschritte der Mathematik* (Jbuch) have, as far as possible, been included in this bibliography.

Benedicks, M. (1985): On Fourier transforms of functions supported on sets of finite Lebesgue measure. J. Math. Anal. Appl. *106*(1), 180–183, Zbl. 576.42016

Bernstein, S.N. (1924): Le problème de l'approximation des fonctions continues sur tout l'axe réel et l'une de ses applications. Bull. Soc. Math. Fr. *59*, 399–410, Jbuch. 50,195

Bernstein, S.N. (1954): Collected Works. Vol. II. Academy of Sciences Press, Moscow, Zbl. 56,60

Beurling, A. (1939): Ensembles exceptionnels. Acta Math. *72*, 1–13, Zbl. 23,142

Beurling, A. (1949): On two problems concerning linear transformations in Hilbert space. Acta Math. *81*, 239–255, Zbl. 33,377

Beurling, A. (1961): Quasianalyticity and general distributions. Stanford Univ. Lect. Notes *30*

Beurling, A., Malliavin, P. (1962): On Fourier transforms of measures with compact support. Acta Math. *107*(3–4), 291–309, Zbl. 127,326

Beurling, A., Malliavin, P. (1967): On the closure of characters and the zeros of entire functions. Acta Math. *118*(1–2), 79–93, Zbl. 171, 119

Bieberbach, L. (1955): Analytische Fortsetzung. Springer-Verlag, Berlin-Heidelberg-New York, Zbl. 64,69

Boas, R. P. (1954): Entire Functions. Academic Press, New York, Zbl. 58,302

Bochner, S. (1944): Boundary values of analytic functions in several variables and of almost periodic functions. Ann. Math., II. Ser. *45*, 708–722, Zbl. 60,243

Borichev, A.A. (1988): Boundary uniqueness theorems for almost analytic functions and asymmetric algebras of sequences. Mat. Sb., Nov. Ser. *136*(3), 324–340. English transl.: Math. USSR, Sb. *64*, 323–338 (1989), Zbl. 663.30002

Borichev, A.A., Vol'berg, A.L. (1989): Uniqueness theorems for almost analytic functions. Algebra Anal. *1*(1), 146–177. English transl.: Leningr. Math. J. *1*, No. 1, 157–191 (1990), Zbl. 725.30038

de Branges, L. (1959): The Bernstein problem. Proc. Am. Math. Soc. *10*(5), 825–832, Zbl. 92,69

de Branges, L. (1968): Hilbert Spaces of Entire Functions. Prentice-Hall, Englewood Cliffs, New Jersey, Zbl. 157,433

Brélot, M. (1961): Eléments de la théorie classique du potentiel. Centre Docum. Univ., Paris, Zbl. 84,309

Bremermann, H. (1965): Distributions, complex variables, and Fourier transforms. Addison-Wesley, Reading, Massachusetts, Zbl. 151,181

Brown, G., Hewitt, E. (1980): Continuous singular measures with small Fourier-Stieltjes transforms. Adv. Math. *37*(1), 28–60, Zbl. 445.42004

Brummelhuis, R. G. M. (1988): Variations on a theme of Frederic and Marcel Riesz. Univ. van Amsterdam

Carleman, T. (1944): L'intégrale de Fourier et questions qui s'y rattachent. Almqvist und Wiksels boktr., Uppsala, Zbl. 60,255

Carleson, L. (1952): Sets of uniqueness for functions regular in the unit disc. Acta Math. *87*, 325–345, Zbl. 46,300

Carleson, L. (1957): Representations of continuous functions. Math. Z. *66*(5), 444–451, Zbl. 86,277

Cassels, J.W.S. (1957): An Introduction to Diophantine Approximation. Cambridge University Press, Zbl. 77,48

Dirac, P.A.M. (1958): The Principles of Quantum Mechanics. Fourth ed. Clarendon Press, Oxford, Zbl. 80,220

Doss, R. (1981): Elementary proof of the Rudin-Carleson theorem and the F. and M. Riesz theorems. Proc. Am. Math. Soc. *82*(4), 599–602, Zbl. 478.30028

Dressler, R.E., Pigno, L. (1974a): On strong Riesz sets. Colloq. Math. *29*(1), 157–158, Zbl. 271.43004

Dressler, R.E., Pigno, L. (1974b): Sets of uniform convergence and strong Riesz sets. Math. Ann. *211*(3), 227–231, Zbl. 308.42006

Duren, P.L. (1970): Theory of H^p-spaces. Academic Press, New York London, Zbl. 215,202

D'yakonov, K.M. (1988): On an interpolation problem and equivalent norms in the spaces K_θ^p. Vestn. Leningr. Univ., Ser. I, 1988, No. 4, 104–105. English transl.: Vestn. Leningr. Univ., Math. *21*, No. 4, 59–62 (1988), Zbl. 663.31002

Dyn'kin, E.M. (1972): Functions with a given estimate of df/dz and the theorem of N. Levinson. Mat. Sb., Nov. Ser. *89*(2), 182–190. English transl.: Math. USSR, Sb. *18*, 181–189 (1973), Zbl. 251.30033

Dyn'kin, E.M. (1979): Sets of free interpolation for the Hölder classes. Mat. Sb., Nov. Ser. *109*(1), 107–128. English transl.: Math. USSR, Sb. *37*, 97–117 (1980), Zbl. 407.30024

Dzhrbashyan, M.M. (1958): Uniqueness theorems for Fourier transforms and for infinitely differentiable functions. Izv. Akad. Nauk Arm. SSR, Ser. Fiz.-Mat. *10*(6), 7–24, Zbl. 188,429

Ehrenpreis, L. (1970): Fourier Analysis in Several Complex Variables. Wiley-Interscience, New York, Zbl. 195,104

Ehrenpreis, L. (1981): Spectral gaps and lacunas. Bull. Sci. Math., II. Ser. *105*(1), 17–28, Zbl. 461.42005

Faddeev, L.D., Yakubovskij, O.A. (1980): Lectures on Quantum Mechanics for Student Mathematicians. Leningrad University Press

Fefferman, C.L. (1983): The uncertainty principle. Bull. Am. Math. Soc., New Ser. *9*(2), 129–206, Zbl. 526.35080

Gaier, D. (1970): Bemerkungen zum Turanschen Lemma. Abh. Math. Semin. Univ. Hamburg *35*(1–2), 1–7, Zbl. 205,378

Gamelin, T. (1969): Uniform Algebras. Prentice-Hall, Englewood Cliffs, New Jersey, Zbl. 213,404

Garnett, J. (1981): Bounded Analytic Functions. Academic Press, New York, Zbl. 469.30024

Geetha, P.K. (1969): On the Bernstein approximation problem. J. Math. Anal. Appl. *25*(2), 450–469, Zbl. 169,73

Gorelik, G.S. (1959): Vibrations and Waves. Fizmatgiz, Moscow

Gorin, E.A. (1985): Some remarks in connection with a problem of B. P. Paneyakh on equivalent norms in spaces of analytic functions. Teor. Funkts., Funkts. Anal. Prilozh. *44*, 23–32. English transl.: J. Sov. Math. *48*, No. 3, 259–266 (1990), Zbl. 595.32020

Graham, C.C., McGehee, O.C. (1979): Essays in Commutative Harmonic Analysis. Springer-Verlag, Berlin-Heidelberg-New York, Zbl. 439.43001

Havin, V.P. (=Khavin, V.P.), Jöricke, B. (1981): On a class of uniqueness theorems for convolutions. Complex Analysis and Spectral Theory. Seminar, Leningrad 1979/80. Lect. Notes Math. *864*, 141–170, Zbl. 474.46030

Havin, V.P. (=Khavin, V.P.), Jöricke, B., Makarov, N.G. (1984): Analytic functions stationary on a set, the uncertainty principle for convolutions, and algebras of Jordan operators. Lect. Notes Math. *1043*, (Eds. V.P. Havin, S.V. Hruščev, N.K. Nikolskii), 536–540

Hewitt, E. (1976): Remarks on singular measures. Usp. Mat. Nauk *31*(5), 167–176. English transl.: Russ. Math. Surv. *31*, No. 5, 86–95 (1976), Zbl. 361.43001

Hoffman, K. (1962): Banach Spaces of Analytic Functions. Prentice-Hall, Englewood Cliffs, New Jersey, Zbl. 117,340

Hruščev, S.V. (=Khrushchev, S.V.) (1977a): Sets of uniqueness for the Gevrey classes. Ark. Mat. *15*(2), 253–304, Zbl. 387.30021

Hruščev, S.V. (=Khrushchev, S.V.), Nikol'skii, N.K., Pavlov, B.S. (1981): Unconditional bases of exponentials and of reproducing kernels. Lect. Notes Math. *864*, 214–335, Zbl. 466.46018

Ivashev-Musatov, O.S. (1957): On the coefficients of trigonometric null-series. Izv. Akad. Nauk SSSR, Ser. Mat. *21*(4), 559–578. English transl.: Transl., II. Ser., Am. Math. Soc. *14*, 289–310 (1960), Zbl. 82,280

Jöricke, B. (1981): Uniqueness theorems for functions with sparse spectrum. I. Teor. Funkts., Funkts. Anal. Prilozh. *35*, 26–34, Zbl. 517.43003

Jöricke, B. (1982): Uniqueness theorems for functions with sparse spectrum. II. Teor. Funkts., Funkts. Anal. Prilozh. *37*, 26–31, Zbl. 599.44007

Jöricke, B., Khavin, V.P. (=Havin, V.P.) (1979): The uncertainty principle for operators that commute with translation. I. Zap. Nauchn. Semin. Leningr. Otd. Mat. Inst. Steklova *92*, 134–170, Zbl. 431.46031

Jöricke, B., Khavin, V.P. (=Havin, V.P.) (1981): The uncertainty principle for operators that commute with translation. II. Zap. Nauchn. Semin. Leningr. Otd. Mat. Inst. Steklova *113*, 97–134. English transl.: J. Sov. Math. *22*, 1758–1783 (1983), Zbl. 471.46055

Jöricke, B., Khavin, V.P. (=Havin, V.P.) (1985a): The uncertainty principle for operators that commute with translation. III. Geometric Questions of the Theory of Functions and Sets. Collect. Sci. Works, Kalinin, 62–80, Zbl. 623.47024

Jöricke, B., Havin, V.P. (=Khavin, V.P.) (1985b): Traces of harmonic functions and comparison of L^p-norms of analytic functions. Math. Nachr. *123*, 225–254

Kahane, J.P. (1962): Travaux de Beurling et Malliavin. Sémin. Bourbaki *14*, No. 225, Zbl. 156,147

Kahane, J.P. (1970): Séries de Fourier Absolument Convergentes. Springer-Verlag, Berlin-Heidelberg-New York, Zbl. 195,76

Kahane, J.P., Salem, R. (1963): Ensembles Parfaits et Séries Trigonométriques. Hermann, Paris, Zbl. 112,293

Kantorovich, L.V., Akilov, G.P. (1977). Functional Analysis. Nauka, Moscow. English transl.: Pergamon Press, Oxford 1982, Zbl. 127,61; Zbl. 484.46003

Kargaev, P.P. (1982): The Fourier transform of the characteristic function of a set which vanishes on an interval. Mat. Sb., Nov. Ser. *117*(3), 397–411. English transl.: Math. USSR, Sb. *45*, 397–410 (1983), Zbl. 523.42005

Kargaev, P.P. (1983): The existence of a Phragmén-Lindelöf function and some conditions for quasianalyticity. Zap. Nauchn. Semin. Leningr. Otd. Mat. Inst. Steklova *126*, 97–108. English transl.: J. Sov. Math. *27*, 2486–2495 (1984), Zbl. 512.31010

Kargaev, P.P. (1985): Nonlocal almost differential operators and interpolation by functions with sparse spectrum. Mat. Sb., Nov. Ser. *128*(1), 133–142. English transl.: Math. USSR, Sb. *56*, 131–140 (1987), Zbl. 622.42007

Kargaev, P.P. (1990): On the Martin boundary of a plane domain whose complement is contained in a line. Mat. Zametki *47* (2), 20–27. English transl.: Math. Notes *47*, No. 2, 121–128 (1990), Zbl. 701.31003

Katsnel'son, V.Eh. (1973): Equivalent norms in spaces of entire functions. Mat. Sb., Nov. Ser. *92*(1), 34–54. English transl.: Math. USSR, Sb. *21*, 33–55 (1974), Zbl. 288.46024

Katznelson, Y. (1976). An Introduction to Harmonic Analysis, Second corrected edition. Dover, New York, Zbl. 352.43001

Kharkevich, A.A. (1957): Spectra and Analysis. Gostekhizdat, Moscow

Khavin, V.P. (=Havin, V.P.) (1983): The uncertainty principle for the one-dimensional potentials of M. Riesz. Dokl. Akad. Nauk SSSR *264*(3), 559–563. English transl.: Sov. Math., Dokl. *25*, 694–698 (1982), Zbl. 509.31009

Khavin, V.P. (=Havin, V.P.) (1987): Methods and structure of commutative harmonic analysis. Itogi Nauki Tekh., Ser. Sovrem. Probl. Mat., Fundam. Napravleniya *15*, 6–133. English transl.: Encycl. Math. Sci. *15*, 1–111, Springer-Verlag, Berlin-Heidelberg-New York 1991, Zbl. 656.43001

Khrushchev, S.V. (=Hruščev, S.V.) (1977b): The entropy meaning of the integrability of the logarithm. Zap. Nauchn. Semin. Leningr. Otd. Mat. Inst. Steklova *73*, 152–187. English transl.: J. Sov. Math. *34*, 2112–2133 (1986), Zbl. 409.60035

Khrushchev, S.V. (=Hruščev, S.V.) (1978): The problem of simultaneous approximation and removal of singularities of integrals of Cauchy type. Tr. Mat. Inst. Steklova *130*, 124–195. English transl.: Proc. Steklov Inst. math. *130*, 133–203 (1979), Zbl. 444.30031

Khurgin, Ya.I., Yakovlev, V.P. (1971): Functions of Compact Support in Physics and Engineering. Nauka, Moscow, Zbl. 245,93001

Koosis, P. (1964): Sur l'approximation pondérée par des polynômes et par des sommes d'exponentielles imaginaires. Ann. Sci. Ec. Norm. Supér., III. Ser. *81*(4), 387–408, Zbl. 128,66

Koosis, P. (1979): Harmonic estimates in certain slit regions and a theorem of Beurling and Malliavin. Acta Math. *142*(3–4), 225–304, Zbl. 406.31001

Koosis, P. (1980): Introduction to H_p-spaces. Lond. Math. Soc. Lect. Notes Series *40*, Cambridge, Zbl. 435.30001

Koosis, P. (1983): La plus petite majorante surharmonic. Ann. Inst. Fourier *33*(1), 67–107, Zbl. 507.30021

Koosis, P. (1988): The logarithmic integral. I. Cambridge University Press, Zbl. 665.30038

Koosis, P. (1990a): Mesures orthogonales extrémales pour l'approximation pondérée par des polynômes. C.R. Acad. Sci. Paris, I. Sér. *311*, 503–506, Zbl. 717.41010

Koosis, P. (1990b): Mesures orthogonales extrémales pour l'approximation pondérée par des sommes d'exponentielles imaginaires. C. R. Acad. Sci. Paris, I. Sér. *311*, 161–164, Zbl. 704.41014

Korenblyum, B.I. (1971): On functions holomorphic in the disk and smooth up to the boundary. Dokl. Akad. Nauk SSSR *200*(1), 24–27. English transl.: Sov. Math., Dokl. *12*, 1312–1315 (1972), Zbl. 235.30036

Körner, T.W. (1977): On the theorem of Ivasev-Musatov. I. Ann. Inst. Fourier *27*(3), 97–116, Zbl. 361.28002

Körner, T.W. (1978): On the theorem of Ivasev-Musatov. II. Ann. Inst. Fourier *28*(3), 123–142, Zbl. 384.28001

Körner, T.W. (1987): Uniqueness for trigonometric series. Ann. Math., II. Ser. *126*(1), 1–34, Zbl. 658.42013

Kotochigov, A.M. (1972): Interpolation by analytic function that are smooth up to the boundary. Zap. Nauchn. Semin. Leningr. Otd. Inst. Mat. Steklova *30*, 167–169. English transl.: J. Sov. Math. *4*, 448–449 (1976), Zbl. 329.41002

Krasichkov-Ternovskij, I.F. (1968): Systems of functions with the dual orthogonality property. Mat. Zametki *4*(5), 551–556. English transl.: Math. Notes *4*, 821–824 (1969), Zbl. 181,137

Krasichkov-Ternovskij, I.F. (1986): On absolute completeness of systems of exponentials on an interval. Mat. Sb., Nov. Ser. *131*(3), 309–322. English transl.: Math. USSR, Sb. *59*, 303–315 (1988), Zbl. 652.42015

Krylov, V.I. (1939): On functions that are analytic in a half-plane. Mat. Sb. *6(48)*, 95–138, Zbl. 22,363

de Leeuw, R., Katznelson, Y. (1970): The two sides of a Fourier-Stieltjes transform and almost idempotent measures. Isr. J. Math. *8*(3), 213–229, Zbl. 198,479

Levin, B.Ya. (1956): Distribution of Zeros of Entire Functions. Gostekhizdat, Moscow. English transl.: Transl. Math. Monogr. American Mathematical Society, Providence (1972), Zbl. 111,74

Levin, B.Ya. (1989): Completeness of systems of functions, quasianalyticity and subharmonic majorants. Zap. Nauchn. Semin. Leningr. Otd. Mat. Inst. Steklova *170*, 102–156. English transl.: J. Sov. Math. *63*, 171–201 (1993), Zbl. 701.30006

Levin, B.Ya., Livshits, M.S. (1941): Quasianalytic classes of functions represented by Fourier series. Mat. Sb. *9(51)*(3), 693–710

Levinson, N. (1963): Gap and Density Theorems. Am. Math. Soc. Colloq. Publ. *26*, New York, Zbl. 145,80

Levinson, N., McKean, H.P., Jr. (1964): Weighted trigonometric approximation on $\mathbb{R}^1$ with application to the germ field of a stationary Gaussian noise. Acta Math. *112*(1–2), 99–143, Zbl. 126,139

Lin, V.Ya. (1965): On equivalent norms in the space of square-integrable functions of exponential type. Mat. Sb., Nov. Ser. *67*(5), 586–608, Zbl. 143,355

Logvinenko, V.N. (1977): On measures that generate an equivalent norm in certain Banach spaces of entire functions. Teor. Funkts., Funkts. Anal. Prilozh. *28*, 51–55, Zbl. 439.46019

Logvinenko, V.N., Sereda, Yu.F. (1974): Equivalent norms in the space of entire functions of exponential type. Teor. Funkts., Funkts. Anal. Prilozh. *20*, 102–111, Zbl. 312.46039

Makarov, N.G. (1985): On stationary functions. Vestn. Leningr. Univ., Math. *22*, 7–14. English transl.: Vestn. Leningr. Univ., Math. *18*, 7–14 (1985), Zbl. 614.47025

Makarov, N.G. (1989): On a class of exceptional sets in the theory of conformal mappings. Mat. Sb. *9*, 1171–1182. English transl.: Math. USSR, Sb. *68*, No. 1, 19–30 (1990), Zbl. 688.30006

Malliavin, P. (1979): On the multiplier theorem for Fourier transforms of measures with compact support. Ark. Mat. *17*(1), 69–81, Zbl. 423.42009

Mandelbrojt, S. (1937): Quasianalytic Function Classes. Gostekhizdat, Moscow-Leningrad (Russian)

Mandelbrojt, S. (1952): Séries Adhérentes, Régularisation des Suites, Applications. Gauthier-Villars, Paris, Zbl. 48,52

Mandelbrojt, S. (1962): Closure Theorems and Composition Theorems. Inostrannaya Literatura, Moscow, Zbl. 129,47

Mandelbrojt, S. (1969): Séries de Dirichlet. Principes et Méthodes. Gauthier-Villars, Paris, Zbl. 207,72

Mergelyan, S.N. (1956): Weighted approximation by polynomials. Usp. Mat. Nauk *11*(5), 107–152. English transl.: Transl., II. Ser., Am. Math. Soc. *10*, 59–106 (1958), Zbl. 74,49

Meyer, Y. (1968): Spectres des mesures et mesures absolument continues. Stud. Math. *30*(1), 87–99, Zbl. 159,425

Mikheev, I.M. (1975): On series with gaps. Mat. Sb., Nov. Ser. *98*(4), 538–563, Zbl. 314.42012

Morgan, G.W. (1934): A note on Fourier transforms. J. Lond. Math. Soc. *9*(3), 187–192, Zbl. 9,248

Natanson, I.P. (1974): Theory of Functions of a Real Variable (4th ed.). Nauka, Moscow. English transl.: Ungar, New York 1955, Zbl. 39,282

Nazarov, F.L. (1993): Local estiamtes of exponential polynomials and their applications to the uncertainty principle. Algebra Anal. *5* (4), 3–66

von Neumann, J. (1932): Mathematische Grundlagen der Quantenmechanik. Springer-Verlag, Berlin-Heidelberg-New York, Zbl. 5,91

Nikol'skij, N.K. (1980): Treatise on the Shift Operator. Nauka, Moscow, Zbl. 508.47001. English transl.: Springer-Verlag, Berlin-Heidelberg-New York 1986, Zbl. 587.47036

Oberlin, D.M. (1980): A Rudin-Carleson theorem for uniformly convergent Taylor series. Mich. Math. J. *27*(3), 309–313, Zbl. 438.30002

Paley, R.E.A.C., Wiener, N. (1934): Fourier Transforms in the Complex Domain. Am. Math. Soc. Colloq. Publ. *19*, New York, Zbl. 11,16

Paneyakh, B.P. (1966): Some inequalities for functions of exponential type and a priori estimates for general differential operators. Usp. Mat. Nauk *21*(3), 75–114. English transl.: Russ. Math. Surv. *21*, 75–114 (1967), Zbl. 173,159

Pełczynski, A (1962):. On the universality of certain Banach spaces. Vestn. Leningr. Univ., Math. *17*, No. 13(3), 22–29, Zbl. 192,226

Peyrière, J. (1975): Étude de quelques propriétés de produits de Riesz. Ann. Inst. Fourier *25*(2), 127–169, Zbl. 307.43003

Pigno, L, Smith, B. (1982): Quantitative behaviour of the norms of an analytic measure. Proc. Am. Math. Soc. *86*(4), 581–585, Zbl. 513.43003

Polya, G. (1918): Über die Potenzreihen, deren Konvergenzkreis natürliche Grenze ist. Acta Math. *41*, 99–118, Jbuch. 46,483

Polya, G. (1929): Untersuchungen über Lücken und Singularitäten der Potenzreihen. I. Math. Z. *29*(4), 549–640, Jbuch. 55,186

Polya, G. (1933): Untersuchungen über Lücken und Singularitäten der Potenzreihen. II. Math. Ann. *34*(4), 731–777, Zbl. 8,62

Privalov, I.I. (1950): Boundary properties of analytic functions. Gostekhizdat, Moscow-Leningrad, Zbl. 41,397

Redheffer, R.M. (1977): Completeness of sets of complex exponentials. Adv. Math. *24*, 1–62, Zbl. 358.42007

Rudin, W. (1956): Boundary values of continuous analytic functions. Proc. Am. Math. Soc. *7*, 808–811, Zbl. 73,.297

Rudin, W. (1960): Trigonometric series with gaps. J. Math. Mech. *9*(2), 203–227, Zbl. 91,58

Salem, R. (1963): Algebraic Numbers and Fourier Analysis. Heath, Boston, Zbl. 126,78

Sapogov, N.A. (1974): On a uniqueness problem for finite measures in Euclidean spaces. Zap. Nauchn. Semin. Leningr. Otd. Mat. Inst. Steklova *41*, 3–13. English transl.: J. Sov. Math. *3*, 1–8 (1978), Zbl. 351.28006

Sapogov, N.A. (1978): On the Fourier transform of the indicator of a set of finite Lebesgue measure in $\mathbb{R}^n$. Zap. Nauchn. Semin. Leningr. Otd. Mat. Inst. Steklova *81*, p. 73

Shapiro, J.H. (1978): Subspaces of $L^p(G)$ spanned by characters: $0 < p < 1$. Isr. J. Math. *29*(2–3), 248–264, Zbl. 382.46015

Shirokov, N.A. (1978): Ideals and factorization in algebras of analytic functions that are smooth up to the boundary. Tr. Mat. Inst. Steklova *130*, 196–222. English transl.: Proc. Steklov Inst. Math. *130*, 205–233 (1979), Zbl. 447.46047

Shirokov, N.A. (1982): Zero sets of functions in Λ_ω. Zap. Nauchn. Semin. Leningr. Otd. Mat. Inst. Steklova *107*, 178–188. English transl.: J. Sov. Math. *36*, 408–414 (1987), Zbl. 512.30006

Shirokov, N.A. (1988): Analytic functions smooth up to the boundary. Lect. Notes Math. *1312*, Zbl. 656.30029

Smirnov, V.I. (1988): Selected Works. Complex Analysis. Mathematical Theory of Diffraction. Leningrad University Press, Zbl. 699.01037

Stein, E.M. (1966): Classes H_p, multiplicateurs et fonctions de Littlewood-Paley. C. R. Acad. Sci. Paris, Sér. A *263*(20), 780–781, Zbl. 156,366

Tikhomirov, V.M. (1986): Convex Analysis. Itogi Nauki Tekh., Ser. Sovrem. Probl. Mat., Fundam. Napravleniya *14*, 5–101. English transl.: Encycl. Math. Sci. *14*, 1–92, Springer-Verlag, Berlin-Heidelberg-New York 1990, Zbl. 781.49001

Turan, P. (1953): Eine neue Methode in der Analysis und deren Anwendungen. Akadémiai Kiadó, Budapest, Zbl. 52,46

Vinogradov, S.A. (1970): The Banach-Rudin-Carleman interpolation theorems and norms of embedding operators for certain classes of analytic functions. Zap. Nauchn. Semin. Leningr. Otd. Mat. Inst. Steklova *19*, 6–54. English transl.: Semin. Math., Steklov Math. Inst. Leningr. *19*, 1–28 (1972), Zbl. 247.30027

Vinogradov, S.A., Khavin, V.P. (=Havin, V.P.) (1974): Free interpolation in H^∞ and in certain other function classes. I. Zap. Nauchn. Semin. Leningr. Otd. Mat. Inst. Steklova *47*, 15–54. English transl.: J. Sov. Math. *9*, 137–171 (1978), Zbl. 355.41005

Vinogradov, S.A., Khavin V.P. (=Havin, V.P.) (1976): Free interpolation in H^∞ and in certain other function classes. II. Zap. Nauchn. Semin. Leningr. Otd. Mat. Inst. Steklova *56*, 12–58. English transl.: J. Sov. Math. *14*, 1027–1065 (1980), Zbl. 355.41006

Vladimirov, V.S. (1964): Methods of the Theory of Functions of Several Complex Variables. Nauka, Moscow. English transl.: MIT Press, Cambridge, Mass. (1966), Zbl. 125,319

Vol'berg, A.L. (1981): Thin and thick families of rational fractions. Lect. Notes Math. *864*, 440–480, Zbl. 519.46029

Vol'berg, A.L. (1982): The logarithm of an almost analytic function is integrable. Dokl. Akad. Nauk SSSR *265*(6), 1297–1302. English transl.: Sov. Math. Dokl. *26*, 238–243 (1982), Zbl. 518.30034

Vol'berg, A.L., Jöricke, B. (1986): Integrability of the logarithm of an almost analytic function and a generalization of the Levinson-Cartwright theorem. Mat. Sb., Nov. Ser. *130*(2), 335–348. English transl.: Math. USSR, Sb. *58*, 337–349 (1987), Zbl. 644.30020

Zygmund, A. (1959): Trigonometric Series. Two vols. Cambridge University Press, Zbl. 85,56

Index

Abbe, E. 148, 149
Acousto-optical cell 167–168
Akhiezer, A.I. 80
Akilov, G.P. 184
Aleksandrov, A.B. 190, 191
α-pair 202
– strong 202
Amplitude 88
Amrein, W.O. 205
Analytic representation of generalized function 35
Antosik, P. 43, 60, 82
Aperture 133, 140, 147, 151
Arnol'd, V.I. 84, 99, 104, 230
Atiyah, M. 59, 62, 68, 101, 102
Autocorrelation 133

Babenko, K.I. 53
Babich, V.M. 83, 112, 120
Barbey, K. 188
Bari, N.K. 183
Belov, A.S. 212
Benedicks, M. 239–241
Berenstein, C.A. 73, 75
Berestetskij, V.B. 80
Bernstein, I.N. 62, 68, 69
Bernstein, S.N. 238
Berthier, A.M. 205
Beurling, A. 52, 190, 235, 238, 247
Bieberbach, L. 241
Björk, J.E. 69, 73
Bochner, S. 3, 40, 55
Bogoliubov, N.N. *see* Bogolyubov, N.N.
Bogolyubov, N.N. 78–80, 82, 83
Bony, J.M. 38
Borichev, A.A. 228, 233–235
Borovikov, V.A. 102, 112
Bott, R. 59, 101, 102
Boundary layer 232
Boundary value of holomorphic function 35
Bragg angle 156, 171, 172
de Branges, L. 237
Brélot, M. 239
Bros, J. 80
Browder, F.E. 80
Brychkov, Yu.A. 49, 59

Carleman, T. 52
Carleson, L. 184, 190
Carrier frequency 134, 163
Cartwright, M. 235
Cassells, J.W.S. 195
Cauchy, A. 3, 13, 14
Charge 181
– continuous 193
– discrete 192
– minus 186
– plus 182, 183
– spectral mass 185
– subordinate to sequence 198
Class
– Cartwright 227, 248
– Hardy 181, 182, 218
– Hörmander 90
– Nevanlinna 214
Colombeau, J.F. 83
Computerized tomography 58
Condition
– Blaschke 215
– Carleson 189
Connection 26
– canonical 26
Convolution 10, 42–45, 132, 144, 159, 164–167, 169, 171
– of generalized function with test density 10
Convolver 133, 148, 159, 164–166, 169
Correlator 133, 169
Courant, R. 80
Criterion
– Davydova-Borovikov 102
– Gårding 102
Cross-correlation 133, 144, 159, 170–172
Current 22
– even 22
– exterior differential 23
– odd 22
Cuspidal edge 105

Darboux coordinates 86
Davydova, A.M. 102
Delta-function 7, 11, 19, 27
– expansion in plane waves 57
– on hypersurface 24, 25
– on submanifold 25, 31
Denjoy, A. 52
Density bundle 15
Dirac, P. 3

Direct image
- of distribution 23, 28
Distribution 5, 6, 8, 12, 18
- connected with wave manifold 87–95
- direct image 23
- Lagrangian 83, 91
- of compact support 8
- semiregular 30
- slowly increasing 45
Divisor 215
- of function 215
Doss, R. 183
Dostal, M.A. 73
Duffieux, P.M. 150
Duistermaat, J. 37, 92, 120
Dyson, F. 78, 79
Dzhrbashyan, M.M. 211

Ehrenpreis, L. 21, 50, 54, 56, 70, 71, 73, 75
Ehskin, L.I. 56
Electro-acoustic transducer 160–161
Elias, P. 150
Entropy 245
Equation
- Mathieu 155, 156
- of Schrödinger type 137
- wave 136
Erdélyi, A. 45
Euler, L. 72
Exponent of convergence 211
Exponent of summability 213
Exponential representation 72–75, 77

F-Charge 190
Fedoryuk, M.V. 37, 41, 68, 92
Fefferman, C.L. 180
Field
- intensity 159
Filter 139–141
- amplitude 139, 151
-- complex 151–153
- double grating 152
- electronic 134, 163, 171
- grating 151–153
- holographic 145–146, 170, 171
- optical 168
- phase 139, 153
- Vander Lugt 153
Filtering
- spatial 148–151, 171, 172
Finite difference 152, 153
Focal length 131, 169
Foliation 31
Form
- even 21
-- smooth 22
- Léray 25
- odd 21
-- smooth 22
Formula
- Carleman 209
- Herglotz-Petrovskij 59
- John 57
- Kirchhoff 79
- lens 141
- Poisson summation 9, 47
- Poisson-Jensen 213, 215
- Radon 57
- Riemann-Mellin 213
- Wiener 192
Fourier processor 131, 133, 134, 139, 143–146, 148
- acoustic 166–167
- acousto-optical 135, 168–172
- optical 131–136
Fourier series
- of generalized function 9
Front
- wave manifold 84
Function
- *a*-admissible 246
- Airy 83
- almost-analytic 229
- Babich 83
- Bessel 153
- complementary in the sense of Young 208
- correlation 158
- entire 236
- fluent 198
- generating 86
- minus 181
- outer 242–243
- Pauli-Jordan 78
- Pearcey 83
- plus 181
- Pollard 236
- rapidly increasing
-- Fourier transform 56–57
- slowly increasing
-- Fourier transform 54–56

Gårding, L. 43, 45, 59, 83, 101, 102, 112
Gamelin, T. 184, 188, 241
Gel'fand, I.M. 14, 21, 25, 28, 42, 49, 52–54, 57–59, 64, 66, 90

– conjecture 68
Gel'fand, S.I. 62, 68
Generalized function 8
– addition of 6
– analytic representation 35
– causal 78–80
– convolution with test density 10
– definition 5
– differentiation 13
– expansion in Fourier series 9
– localization principle 7
– multiplication 81–83
– multiplication by infinitely differentiable function 6
– of compact support 8
– of finite order 7, 8
– periodic 9, 10
– restriction 6
– sharp 102
– slowly increasing 46–47
– – Fourier transform 47
– smooth across foliation 31
– smooth along foliation 31
– support 7
Gorin, E.A. 207
Graev, M.I. 42, 58, 59
Graham, C.C. 187
Grothendieck, A. 21, 118
Group
– cohomology 54
– M. Riesz 44–45
Guillemin, V. 36, 60, 83, 120
Gurevich, D.I. 75
Gurevich, S.B. 158
Gusejn-Zade, S.M. 84, 99, 104

Hadamard, J. 3, 12–14, 43, 83
Havin, V.P. *see* Khavin, V.P.
Heaviside, O. 7
Helgason, S. 42, 45, 58
Henkin, G.M. *see* Khenkin, G.M.
Hilbert, D. 80
Hironaka, M. 62
Hirschowitz, A. 102
Hoffman, K. 184, 188
Holographic measurement 145–146
Holography 151, 153
Hörmander, L. 34, 36–38, 41, 54, 67, 71, 73, 83, 86, 87, 90–92
Hruščev, S.V. *see* Khrushchev, S.V.
Huyghens, C. 45
Hyperfunction 54, 227, 233
– Fourier 54

Iagolnitzer, D. 80
Image
– formation 141–142, 148, 157, 159
– Radon 58
Integral
– Cauchy principal value 13, 14
– Cauchy type 215
– logarithmic 228, 235
– oscillatory 37
– Poisson 215, 216, 219
– Riemann-Liouville 43–44
– versal 114–120
Inverse image
– of generalized function 24

Jeanquartier, D. 62
Jensen's inequality 187
John, F. 57, 59
Jöricke, B. 207, 238
Jost, R. 78

Kaneko, A. 54
Kantorovich, L.V. 184
Kargaev, P.P. 238, 240, 241
Kashiwara, M. 38, 54
Katznelson, Y. 187, 195, 201
Kawai, T. 38, 54
Kernel 13
– Cauchy-Sokhotskij 14, 35
– Hadamard 12, 14, 83
Khavin, V.P. 207
Khenkin, G.M. 38, 59
Khrushchev, S.V. 245
Khurgin, Ya.I. 204
Kirchhoff, G. 3
Komatsu, H. 52, 53
König, H. 83, 188
Konstantinov, V.B. 158
Koosis, P. 237, 240, 247
Körner, T.W. 201
Kostyuchenko, A.G. 21
Krasichkov-Ternovskij, I.F. 204
Krein, M.G. 241
Kuchment, P.A. 77
Kushnirenko, A.G. 82

Lacuna 101
– weak 101
Laser 151, 158
Lasker, E. 74
Lax, P.D. 92
de Leeuw, K. 187
Lehmann, H. 78
Lens 131, 163, 169
– cylindrical 143, 169

– thin 139, 148
Leont'ev, A.F. 72
Léray, J. 25
Levin, B.Ya. 211, 212, 239–241
Levinson, N. 235
Lie derivative 18, 22, 26
Light beam 131, 136–138
Lions, J.L. 43, 52
Liouville number 72
Livshits, M.S. 211
Localization principle
– for generalized functions 7
Logvinenko, V.N. 207
Lojasiewicz, S. 61, 67
Lützen, J. 3

Magenes, E. 52
Majorant
– admissible 246–248
– determining 221, 222
– Legendre dual 54
Makarov, N.G. 245
Malgrange, B. 43, 62, 63, 67, 69–71, 73, 100, 117, 118
Malliavin, P. 247
Mandelbrojt, S. 52, 211, 241, 247
Manifold
– Lagrangian 91
– Legendre 84
Mapping
– characteristic 98
– transversal to wave manifold 96
Maréchal, A. 151
Martin boundary 239
Martineau, A. 71, 80
Maslov, V.P. 37, 41, 92, 113
McGehee, O.C. 187
McLachlan, N.W. 156
Measure
– Cantor 194, 195
– harmonic 231
– Lebesgue 181, 182, 188
– – normalized 181
– Poisson 221
Methée, P.D. 66
Method
– M. Riesz 67–69
Mikusinski, J. 43, 60, 82
Milnor number 104
Moment 222
Morgan, G.N. 211
Morse index 102, 112
Multiplication 166
Multiplier 166, 167

Natanson, I.P. 208
Natterer, F. 58
Nazarov, F.L. 206
von Neumann, J. 180
Noether, M. 74
Noetherian operator 74, 76, 77
– global 77

O'Neill, E.L. 150
Operator
– Cauchy-Riemann 229
– finite difference 152
– Laplace 153
– Maslov 83

Palamodov, V.P. 51, 54, 56, 59, 62–64, 70, 73, 74, 76, 77
Paley, R. 42, 49, 249
Parametrix 91
Parasiuk, O.S. 83
Pearcey, T. 120
Pelczynski, A. 184
Penrose, R. 38
Petrovskij, I.G. 59, 101
Petrovsky, I.G. *see* Petrovskij, I.G.
Peyrière, J. 197
Phase grating 153–157, 168
Phase-contrast object 149, 150
Photodetector 159, 170–172
Photoelastic effect 167
Picard-Lefschetz monodromy 63
Piezoelectric crystal 160, 167
Pigno, L. 184, 186
Piriou, A. 102
Pisot-Vijayaraghavan number 195
Plus-polynomial 183
Poisson, S. 9
Polar 72
Polyakov, P.L. 38
Porter, A.B. 149
Poston, T. 109
Privalov, I.I. 183, 216
Problem
– Bernstein 224
– Cauchy 137, 155
– – Green's function 137
– – resolvent operator 137
– Dirichlet 231
– division 67–71
– moment 223
– Sapogov 240
Product
– Blaschke 215
– of distributions 19

– Riesz 195–198
Profile 88
Prudnikov, A.P. 49, 59
Pullback
– of generalized function 24
– of wave manifold 84, 97–114

r-Charge 192
Radon, J. 38
Rajchman, A. 192
Rajkov, D.A. 21
Ramanujan, S. 47
Ray approximation 137
Rayleigh-wave transducer 160–167
Redheffer, R.M. 249
Refractive index 139, 153, 167, 168
Regularization of divergent integral 11–13
Relative density 26, 28
Representation
– Jost-Lehmann-Dyson 78–79
Retarded Green's function 82
de Rham, G. 3, 21, 22
Riemann, B. 3
Riesz, M. 12, 45, 68, 83
de Roever, J.W. 73
Roumieu, M.C. 52–54
Rudin, W. 184, 191

Salem, R. 195
Sapogov, N.A. 240
Sato, M. 38, 54
Schaefer, H.H. 21
Schapira, P. 35, 54
Schwartz, L. 3, 8, 11, 12, 18, 19, 21, 30, 42, 50, 55, 56, 67, 68, 72, 75, 81
Sebastião a Silva, J. 52, 54, 56
Semidensity 41
Sereda, Yu.F. 207
Set
– α-spacious 244
– Cantor 193, 194
– determining 189
– Hadamard lacunary 191
– relatively dense 207
Shannon, C. 150
Shapiro, J. 190, 191
Shapiro, Z.Ya. 28
Sheaf 53
– of holomorphic functions 54
– of hyperfunctions 54
Shilov, G.E. 14, 25, 28, 49, 52–54, 57, 64, 66, 90
Shirkov, D.V. 78, 79, 82
Shirokov, Yu.M. 83
Shubin, M.A. 37
Signal 131
– acoustic 131
– coherent 158
– complex optical 145
– electronic 131, 160, 161, 163, 164, 167–172
– incoherent 158–159
– optical 131, 151, 153, 157
– radio-electronic 131, 134, 160, 164, 166, 169
Signature of point 106
Sikorski, R. 43, 60, 82
Simple point 104
Singular support 34, 38
Smith, B. 184, 186
Sobolev, S.L. 3, 16, 83
Sokhotskij, Yu.V. 14
Solution
– Bloch 76, 77, 155, 156
– Floquet 75, 77
Space
– fundamental 15, 17
– of test functions 6
– Schwartz 46–47, 55
– Shilov 54
– Sobolev 76
– Stein 77
– test 52–54
– translation-invariant 71–77
Spectral gap 238–240
Spectrum
– essential 202
– lacunary 206
– real 63, 64, 69
Stein, E. 49, 186
Sternberg, S. 36, 60, 83, 120
Stewart, I. 109
Struppa, D.C. 75
Submersion 23
Support
– essential 202
– of density 5, 15, 16, 24
– of distribution 8, 28
– of generalized function 7-9, 11, 17
– of relative density 28

Taratorin, A.M. 158
Tarski-Seidenberg principle 103
Taylor, B.A. 75
Theorem
– Amrein-Berthier 204–206
– Bari 194

- Beurling 224, 226–228, 235, 239, 240, 244
- de Branges 241
- Denjoy-Carleman-Mandelbrojt 52
- Dyn'kin 229, 231, 234
- Dzhrbashyan 208, 210
- edge of the wedge 80
- Ehrenpreis 75
- F. and M. Riesz 182, 184
- Fabry 241
- first Beurling-Malliavin 247
- first Borichev 233
- first Vol'berg 228
- Hironaka 68
- Ivashev-Musatov 198, 201
- Khrushchev 245
- Krein-Mil'man 241
- Lasker-Noether 76
- de Leeuw-Katznelson 185, 187
- Levinson-Cartwright 225–228
- Logvinenko-Sereda 207
- Malgrange preparation 100
- Mandelbrojt 211–213
- Morgan 206
- Nazarov 206, 211
- Paley-Wiener 42
- Paley-Wiener-Schwartz 42, 52
- Phragmén-Lindelöf 53, 211
- Pigno-Smith 186
- Pollard 236–238
- Riemann-Lebesgue 185, 192
- Rudin-Carleson 184
- Sard 104
- Schwartz 19
- second Beurling-Malliavin 248–249
- second Borichev 234
- second Vol'berg 229
- Slepyan-Pollak 205
- Tougeron 117
- Wiener-Keldysh 231
- Zygmund 206

Tikhomirov, V.M. 180, 230
Tillmann, H.G. 35
Tougeron, J.C. 117
Transform
- Fourier 38–59, 131, 201
-- direct 50
-- inverse 50
- Hankel 59
- Hilbert 59
- Laplace 59
- Legendre 230
- Mellin 59, 64, 120
- Penrose 59
- Radon 57–59, 92
-- of generalized function 58
-- on symmetric spaces 58
- Schwartz 242
- Stieltjes 59

Transmittance 139–141, 143–146, 168–170
Transmittance coefficient 144
Transparent 139, 151, 153, 171
Trèves, F. 21, 37, 66
Tube domain 79
Tvorogov, V.B. 102
Twistor 59

Ultradistribution 52–54

Van der Corput's lemma 201
Varchenko, A.N. 69, 84, 99, 104
Vasil'ev, V.A. 102
Vasilenko, G.N. 158
Vilenkin, N.Ya. 21, 42, 58, 59
Vladimirov, V.S. 78, 80
Vol'berg, A.L. 234, 235, 241

W-distribution 90
- cosharp 110

Wave
- elastic 160, 167
- plane 131, 149, 154, 156, 157, 172
- Rayleigh 160, 161, 164
- spatial acoustic 168
- surface acoustic 164
- ultrasonic 160

Wave front 34, 38, 84
- analytic 38
- extended 34
- nonsharp 102

Wave manifold 84
- finite 98
- inverse image 98, 101
- versal 96–101
-- minimal 105

Wave submanifold 83
Weiss, G. 49
Whitney, H. 67
Wiener, N. 42, 49, 151, 249

Yakovlev, V.P. 204
Yano, T. 69

Zemanian, A.N. 59
Zernike, F. 149
Zygmund, A. 193, 201

Encyclopaedia of Mathematical Sciences
Editor-in-Chief: R. V. Gamkrelidze

Analysis

Volume 13: **R.V. Gamkrelidze** (Ed.)
Analysis I
Integral Representations and Asymptotic Methods
1989. VII, 238 pp. 3 figs. ISBN 3-540-17008-1

Volume 14: **R.V. Gamkrelidze** (Ed.)
Analysis II
Convex Analysis and Approximation Theory
1990. VII, 255 pp. 21 figs. ISBN 3-540-18179-2

Volume 26: **S.M. Nikol'skiĭ** (Ed.)
Analysis III
Spaces of Differentiable Functions
1991. VII, 221 pp. 22 figs. ISBN 3-540-51866-5

Volume 27: **V.G. Maz'ya, S.M. Nikol'skiĭ** (Eds.)
Analysis IV
Linear and Boundary Integral Equations
1991. VII, 233 pp. 4 figs. ISBN 3-540-51997-1

Volume 19: **N.K. Nikol'skij** (Ed.)
Functional Analysis I
Linear Functional Analysis
1992. V, 283 pp. ISBN 3-540-50584-9

Volume 20: **A.L. Onishchik** (Ed.)
Lie Groups and Lie Algebras I
Foundations of Lie Theory.
Lie Transformation Groups
1993. VII, 235 pp. 4 tabs. ISBN 3-540-18697-2

Volume 41: **A.L. Onishchik, E.B. Vinberg** (Eds.)
Lie Groups and Lie Algebras III
Structure of Lie Groups and Lie Algebras
1994. V, 248 pp. 1 fig., 7 tabs. ISBN 3-540-54683-9

Volume 22: **A.A. Kirillov** (Ed.)
Representation Theory and Non-commutative Harmonic Analysis I
Fundamental Concepts. Representations of Virasoro and Affine Algebras
1994. VII, 234 pp. 11 figs. ISBN 3-540-18698-0

Volume 15: **V. P. Khavin, N. K. Nikol'skij** (Eds.)
Commutative Harmonic Analysis I
General Survey, Classical Aspects
1991. IX, 268 pp. 1 fig. ISBN 3-540-18180-6

Volume 72: **V. P. Havin, N. K. Nikol'skij** (Eds.)
Commutative Harmonic Analysis III
Generalized Functions. Applications
1995. VII, 266 pp. 34 fig. ISBN 3-540-57034-9

Volume 42: **V. P. Khavin, N. K. Nikol'skij** (Eds.)
Commutative Harmonic Analysis IV
Harmonic Analysis in $\mathbb{R}^n$
1992. IX, 228 pp. 1 fig. ISBN 3-540-53379-6

Several Complex Variables

Volume 7: **A.G. Vitushkin** (Ed.)
Several Complex Variables I
Introduction to Complex Analysis
1990. VII, 248 pp. ISBN 3-540-17004-9

Volume 8: **G.M. Khenkin, A.G. Vitushkin** (Eds.)
Several Complex Variables II
Function Theory in Classical Domains. Complex Potential Theory
1994. VII, 260 pp. 19 figs. ISBN 3-540-18175-X

Volume 9: **G.M. Khenkin** (Ed.)
Several Complex Variables III
Geometric Function Theory
1989. VII, 261 pp. ISBN 3-540-17005-7

Volume 10: **S.G. Gindikin, G.M. Khenkin** (Eds.)
Several Complex Variables IV
Algebraic Aspects of Complex Analysis
1990. VII, 251 pp. ISBN 3-540-18174-1

Volume 54: **G.M. Khenkin** (Ed.)
Several Complex Variables V
Complex Analysis in Partial Differential Equations and Mathematical Physics
1993. VII, 286 pp. ISBN 3-540-54451-8

Volume 69: **W. Barth, R. Narasimhan** (Eds.)
Several Complex Variables VI
Complex Manifolds
1990. IX, 310 pp. 4 figs. ISBN 3-540-52788-5

Volume 74: **H. Grauert, T. Peternell, R. Remmert** (Eds.)
Several Complex Variables VII
Sheaf-Theoretical Methods in Complex Analysis
1994. VIII, 369 pp. ISBN 3-540-56259-1

Preisänderungen vorbehalten

Tm.BA95.03.17

Encyclopaedia of Mathematical Sciences
Editor-in-Chief: R. V. Gamkrelidze

Dynamical Systems

Volume 1: **D.V. Anosov, V.I. Arnol'd** (Eds.)
Dynamical Systems I
Ordinary Differential Equations and Smooth Dynamical Systems
2nd printing 1994. IX, 233 pp. 25 figs. ISBN 3-540-17000-6

Volume 2: **Ya.G. Sinai** (Ed.)
Dynamical Systems II
Ergodic Theory with Applications to Dynamical Systems and Statistical Mechanics
1989. IX, 281 pp. 25 figs. ISBN 3-540-17001-4

Volume 3: **V.I. Arnold, V.V. Kozlov, A.I. Neishtadt**
Dynamical Systems III
Mathematical Aspects of Classical and Celestial Mechanics
2nd ed. 1993. XIV, 291 pp. 81 figs. ISBN 3-540-57241-4

Volume 4: **V.I. Arnol'd, S.P. Novikov** (Eds.)
Dynamical Systems IV
Symplectic Geometry and its Applications
1990. VII, 283 pp. 62 figs. ISBN 3-540-17003-0

Volume 5: **V.I. Arnol'd** (Ed.)
Dynamical Systems V
Bifurcation Theory and Catastrophe Theory
1994. IX, 271 pp. 130 figs. ISBN 3-540-18173-3

Volume 6: **V.I. Arnol'd** (Ed.)
Dynamical Systems VI
Singularity Theory I
1993. V, 245 pp. 55 figs. ISBN 3-540-50583-0

Volume 16: **V.I. Arnol'd, S.P. Novikov** (Eds.)
Dynamical Systems VII
Nonholonomic Dynamical Systems. Integrable Hamiltonian Systems.
1994. VII, 341 pp. 9 figs. ISBN 3-540-18176-8

Vol. 39: **V.I. Arnol'd** (Ed.)
Dynamical Systems VIII
Singularity Theory II. Applications
1993. V, 235 pp. 134 figs. ISBN 3-540-53376-1

Vol. 66: **D.V. Anosov** (Ed.)
Dynamical Systems IX
Dynamical Systems with Hyperbolic Behavior
1995. VII, 235 pp. 39 figs. ISBN 3-540-57043-8

Partial Differential Equations

Volume 30: **Yu.V. Egorov, M.A. Shubin** (Eds.)
Partial Differential Equations I
Foundations of the Classical Theory
1991. V, 259 pp. 4 figs. ISBN 3-540-52002-3

Volume 31: **Yu.V. Egorov, M.A. Shubin** (Eds.)
Partial Differential Equations II
Elements of the Modern Theory. Equations with Constant Coefficients
1995. VII, 263 pp. 5 figs. ISBN 3-540-52001-5

Volume 32: **Yu.V. Egorov, M.A. Shubin** (Eds.)
Partial Differential Equations III
The Cauchy Problem. Qualitative Theory of Partial Differential Equations
1991. VII, 197 pp. ISBN 3-540-52003-1

Volume 33: **Yu.V. Egorov, M.A. Shubin** (Eds.)
Partial Differential Equations IV
Microlocal Analysis and Hyperbolic Equations
1993. VII, 241 pp. 6 figs. ISBN 3-540-53363-X

Volume 63: **Yu.V. Egorov, M.A. Shubin** (Eds.)
Partial Differential Equations VI
Elliptic and Parabolic Operators
1994. VII, 325 pp. 5 figs. ISBN 3-540-54678-2

Volume 64: **M.A. Shubin** (Ed.)
Partial Differential Equations VII
Spectral Theory of Differential Operators
1994. V, 272 pp. ISBN 3-540-54677-4

Preisänderungen vorbehalten

Tm.BA95.06.13